Random Evolutionary Systems

In memory of our teacher and mentor,

Vladimir Semenovich Korolyuk

Series Editor
Nikolaos Limnios

Random Evolutionary Systems

Asymptotic Properties and Large Deviations

Dmitri Koroliouk
Igor Samoilenko

iSTE

WILEY

First published 2021 in Great Britain and the United States by ISTE Ltd and John Wiley & Sons, Inc.

ISTE Ltd
27-37 St George's Road
London SW19 4EU
UK

www.iste.co.uk

John Wiley & Sons, Inc.
111 River Street
Hoboken, NJ 07030
USA

www.wiley.com

Library of Congress Control Number: 2021937583

British Library Cataloguing-in-Publication Data
A CIP record for this book is available from the British Library
ISBN 978-1-78630-752-1

MIX
Paper from responsible sources
FSC www.fsc.org FSC® C013604

Contents

Preface

This book examines random evolutionary systems and their asymptotic properties, as well as large deviations.

Our study of random evolutionary systems is based on the martingale characterization of their trajectories, the method of associated semigroups and generating operators, the method of inversion on the spectrum of reducible-invertible operators, as well as the singular perturbation problem and the phase merging method. A special role is played, not only by classical approximation schemes for random evolutionary systems such as, in particular, diffusion approximation, but also by new ones: the Poisson approximation and the Lévy approximation.

The diffusion and Poisson approximation of random evolutions plays a special role.

The study of random evolutionary systems in terms of operators constructed in the schemes of the diffusion and Poisson approximation allows us to obtain a number of limit theorems and asymptotic expansions of processes that model complex stochastic systems, both those that are autonomous and those dependent on an external random environment. In this case, various orders of scaling, both of processes and their time parameters, are used to obtain different limit results.

April 2021

Introduction

The theory of complex systems is a modern field of natural sciences, which considers systems that consist of a large number of interacting parts. There is not a large number of quality properties of a system that are observed in the study of its individual parts; they arise only as a result of their interactions.

The theory of complex systems is based on methods of systems analysis as an applied scientific methodology, consisting, in particular, of mathematical methods, algorithmic software and computing tools, which provide the formation of holistic knowledge about the object, as a set of interconnected processes of different natures for further decision-making on its development and behavior, taking into account conflicting criteria, the presence of risk factors and inaccurate information. Due to the application of systems analysis methods, the theory of complex systems has been actively developing in recent decades and consists, in particular, of the development of methods for structuring, modeling, analysis and synthesis of deterministic and stochastic systems in problems of mathematical (including statistical) physics, applied questions of probability theory and fractal analysis and mathematical problems of biology, sociology, ecology, economics and medicine. The main motivation that unites all these diverse mathematical and natural sciences in a single discipline is closeness, as well as the kinship laws and methods of studying the collective behavior of such systems.

Mathematical problems associated with the theory of complex systems relate to the development of mathematical methods of system simplification, which can be very complex, even for computer analysis. In this case, the simplified system should be such that, first, its local characteristics are determined by simple enough functionals of the local characteristics of the original system, and, second, that the simplified model could be qualitatively analyzed by mathematical methods and its global characteristics are an effective approximation of the corresponding characteristics of the original system.

Typical examples of complex systems are evolutionary systems, which, in particular, are modeled using random evolutionary systems and random processes. For example, impulsive processes describe various models in the queuing theory, namely, network structures and production systems; processes with independent increments are used in modeling conflict systems, such as birth–death processes in ecological and biological systems, as well as in models of statistical physics; and stochastic evolutionary systems model problems that arise in reliability theory, control theory, financial mathematics and so on.

We provide an analysis of the asymptotic properties of such models, which are considered and studied from different points of view, using several approximation schemes. That is, the question of the convergence of evolutionary systems in Poisson approximation schemes is analyzed, the problem of large deviations for evolutionary systems in different approximation schemes is analyzed and some applications are discussed: the probability of leaving an interval, the conditions of equilibrium, stationarity, the classification of equilibria and so on.

The development of the theory of random evolutions began in the late 1960s, probably with the work of R. Griego and R. Hersh. They introduced the concept of random evolution in Griego and Hersh (1971).

Applications of this model follow from the work of R.Z. Khasminskii (1966a, 1966b), which were stimulated by the problems of stability of stochastic systems, in particular, by the works of R.L. Stratonovich (1967) on the problems of the nonlinear theory of oscillations in the presence of noise. In the 1960s and 1970s, problems related to the theory of random evolutions were actively studied by American mathematicians R. Hersh, M. Pinsky, G. Papanikolaou, T. Kurtz, R. Griego, L. Gorostiza (Pinsky 1968; Griego and Hersh 1971; Papanicolaou and Kellek 1971; Gorostiza 1972, 1973a, 1973b; Hersh and Papanicolaou 1972; Hersh and Pinsky 1972; Kurtz 1973; Hersh 1974; Papanicolaou 1975) and others. In particular, G. Papanicolaou, D. Strook and S. Varadhan proposed a martingale approach for proving limit theorems (Papanicolaou *et al.* 1977) using methods similar to solving the singular perturbation problem.

An effective method for proving the limit theorems in the theory of random evolutions is the theory of phase merging of complex systems, developed by V.S. Korolyuk and A.F. Turbin (1975, 1993). A.V. Skorokhod (2008) investigated several important problems of the theory of dynamic systems, denoted by stochastic differential equations. The problem of the stability of dynamic systems in the case of random perturbation of their parameters was investigated by R.Z. Khasminskii (2012) and V.S. Korolyuk (1991, 1998). Problems of stability and problems of stochastic approximation of evolutionary systems with switching were investigated by V.S. Korolyuk and Y.M. Chabanyuk (2007, 2007). V.S. Korolyuk and A.V. Swishchuk also developed a theory of semi-Markov random evolutions, based on martingale theory (Korolyuk and Swishchuk 1995a, 1995b). Many applications of processes with switching to the analysis of network systems can be found in the works of V.V. Anisimov (2008).

Asymptotic methods are collected and described in detail in the book by V.S. Korolyuk and N. Limnios (2005), where these methods are mainly applied to models in the schemes of averaging and diffusion approximation. At increasing time intervals, the averaging scheme demonstrates the deterministic averaging behavior of the system, and the diffusion approximation scheme shows stochastic fluctuations around the deterministic averaged trajectory. These two schemes differ in the normalization of the switching process, namely, in the case of the averaging scheme, the acceleration of time by the parameter ε^{-1} is considered, while, in the case of diffusion, approximation by the parameter ε^{-2} or ε^{-3} is considered. An important element of the algorithm is the assumption of ergodicity of the switching (Markov or semi-Markov) process. In their book, V.S. Korolyuk and N. Limnios (2005) also considered several models for processes with independent increments and impulsive processes with switching in the schemes of Poisson and Lévy approximation. Here we generalize the models in the schemes of Poisson and Lévy approximation to the case of processes with locally independent increments, impulsive recurrent processes and so on.

We also present another important area of research, namely, the solution of the large deviations problem, using the methods of asymptotic analysis of nonlinear exponential generators, associated with the singular perturbation problem. The large deviations problem arose as a method of solving statistical problems, related to estimating the probabilities of rare events. The first work in this direction was, obviously, the article by H. Cramér (1938), but this idea was finally fully formed in the work of G. Chernoff (1952).

The main method of studying such problems is the technique of substituting the measure and using the Chebyshev inequality to obtain estimates of the probabilities of rare events, and to calculate the rate functional. In this case, a new measure is introduced, in relation to which the studied events have a high probability, while the probabilities of these events relative to the initial measure are determined in terms of the Radon–Nicodemus derivative, which connects the two mentioned measures. An important role in determining the rate functional is also played by the Legendre transform, which links the rate functional and the cumulant of the process. The most famous works in this technique, which relate to Markov processes, are the works of M. Donsker and S. Varadan (1975a, 1975b, 1976) and M.Y. Freidlin and O.D. Wentzell (1976, 1978, 1979, 1990) (see also Deuschel and Stroock 1989; Dupuis and Ellis 1997; Dembo and Zeitouni 2010). These works contain comprehensive bibliographic references on this topic.

Another approach to solving the problem is related to the convergence and compactness of probabilistic measures. The works in which such methods are used include articles by A.A. Puhalsky (1991), G. O'Brien and W. Vervaat (1995) and A. de Acosta (1997).

Here, we use a method that has emerged and developed recently and is closely related to the control problem. The idea of such an approach was presented in Hopf (1950) and Çinlar *et al.* (1980) and was finally developed by Fleming and Suganidis (1986). The most developed and generalized version of this approach can be found in the book by J. Feng and T. Kurtz (2006), which also contains a relevant bibliography and historical review. It should be noted that the classical normalization scheme in which the large deviations problem is studied is the small diffusion scheme (see Feng and Kurtz (2006) and Freidlin and Ventzell (2012) and the corresponding bibliographic sections of these works). One of the rare models where another possible normalization for the process with independent increments was investigated is the work of A.A. Mogulsky (1993) (compare this to other works by A.A. Borovkov and A.A. Mogulsky (1992, 2012)).

We investigate the large deviations problem for processes with independent increments and impulsive processes with switching in the schemes of the Poisson and Lévy approximations. Models with phase merging are studied for the first time. This formulation of the problem is completely new and allows us to obtain some significant generalizations. For example, the presence of a diffusion component in the small diffusion scheme is embedded in the definition of the limit process. Conversely, the Poisson approximation scheme considers processes with independent increments with switching without a diffusion component, which occurs only after the limit transition. The Poisson approximation scheme, which takes into account rare large jumps in the process, is certainly a natural object for study, especially in terms of the problem of large deviations.

1

Basic Tools for Asymptotic Analysis

1.1. Basic concepts of operator asymptotic analysis

This first chapter introduces the known auxiliary concepts, which are actively used in the following sections. In particular, the concepts of the Poisson approximation and Lévy approximation, impulsive recurrent process, random evolutionary system with locally independent increments, compensating operator, Nisio semigroup, nonlinear exponential operator, martingale control problem, rate functional, compactness, etc. are introduced. The following results are new:

1. The previously known results were used for a detailed substantiation of the relationship between the Nisio semigroup, the martingale control problem, the Brick formula, the nonlinear exponential operator and the corresponding rate functional. Some comparative examples are given for the classical formulation of the problem of large deviations in the scheme of small diffusion.

2. The problems related to the study of compactness and exponential compactness of processes, as well as the application of the comparison condition to the limit nonlinear exponential generator are described. For the processes that will be studied further, the corresponding compactness conditions are tested using the methods from the stability theory, in particular the Lyapunov function. For the corresponding nonlinear exponential generator, the comparison condition is checked.

3. For the first time, the problem of studying the problem of large deviations in the Poisson approximation scheme is set, and the corresponding nonlinear exponential generators for the process with independent increments are determined.

1.1.1. *Reducibly invertible and potential operators*

We denote by $\mathbf{B}$ the Banach space of real-valued measurable functions, defined on a state space E of a random process, with sup-norm

$$||\varphi|| = \sup_{x \in E} |\varphi(x)|, \varphi \in \mathbf{B}.$$

Let $Q : \mathbf{B} \to \mathbf{B}$ be a linear operator on $\mathbf{B}$. We introduce the following subspaces:

$$\mathcal{D}_Q := \{\varphi : \varphi \in \mathbf{B}, Q\varphi \in \mathbf{B}\} \text{ – domain of definition } Q,$$

$$\mathcal{R}_Q := \{\psi : \psi = Q\varphi, \varphi \in \mathbf{B}\} \text{ – subspace of values } Q,$$

$$\mathcal{N}_Q := \{\varphi : Q\varphi = 0, \varphi \in \mathbf{B}\} \text{ – subspace of zeros } Q.$$

The operator Q will be called bounded if there exists a constant $C > 0$ such that $||Q\varphi|| \leq C||\varphi||, \varphi \in \mathcal{D}_Q$.

DEFINITION 1.1.– *A bounded linear operator Q is called reducibly invertible if the Banach space* $\mathbf{B}$ *can be represented as a direct sum of two subspaces*

$$\mathbf{B} = \mathcal{N}_Q \oplus \mathcal{R}_Q,$$

where the null-subspace has a non-trivial dimension

$$\dim \mathcal{N}_Q \geq 1.$$

More detailed information on the properties of reducibly invertible operators can be found in the works of Nashed (1976), Korolyuk and Turbin (1993) and Korolyuk and Limnios (2005); for possible generalizations in the Hilbert space, see Boichuk and Samoilenko (2004).

The following representation defines a projector on a subspace $\mathcal{N}_Q$:

$$\Pi\varphi := \begin{cases} \varphi, \varphi \in \mathcal{N}_Q, \\ 0, \varphi \in \mathcal{R}_Q. \end{cases}$$

Instead, the operator $I - \Pi$ is a projector on a subspace $\mathcal{R}_Q$

$$(I - \Pi)\varphi := \begin{cases} 0, \varphi \in \mathcal{N}_Q, \\ \varphi, \varphi \in \mathcal{R}_Q. \end{cases}$$

We can also define a projector by means of the resolvent

$$R_\lambda := [\lambda I - Q]^{-1}.$$

DEFINITION 1.2.– *A projector on a subspace $\mathcal{N}_Q$ is determined by the following ratio:*

$$\Pi := \lim_{\lambda \to 0+} \lambda R_\lambda.$$

When applied to the study of Markov processes, another way of defining the corresponding projector is important. It becomes possible in the presence of a Feller semigroup $P_t, t \geq 0$, which corresponds to the reducibly invertible operator Q (see, for example, Ethier and Kurtz (1986, p. 473)).

DEFINITION 1.3.– *The projector on a subspace $\mathcal{N}_Q$ is determined by the following ratio:*

$$\Pi\varphi := \lim_{\lambda \to 0+} \lambda \int_0^\infty e^{-\lambda t} P_t \varphi dt.$$

DEFINITION 1.4.– *The potential operator of a reducibly invertible operator Q is called the operator*

$$R_0 := \Pi - (Q + \Pi)^{-1} = (\Pi - Q)^{-1} - \Pi,$$

or in terms of resolvent

$$R_0 := \lim_{\lambda \to 0+} [R_\lambda - \Pi/\lambda].$$

For details, see Nashed (1976), Korolyuk and Turbin (1993) and Korolyuk and Limnios (2005).

If for a known semigroup $P_t, t \geq 0$, which corresponds to a reducibly invertible operator Q, the ergodicity condition

$$\lim_{t \to \infty} P_t = \Pi \neq 0,$$

verifies, then the potential operator R_0 is bounded and can be defined as

$$R_0 := \int_0^\infty (P_t - \Pi) dt.$$

The potential operator has the following properties:

$$QR_0 = R_0 Q = \Pi - I, \qquad [1.1]$$

$$\Pi R_0 = R_0 \Pi = 0. \qquad [1.2]$$

1.1.2. *Singular perturbation problem*

The singular perturbation problem (see Kurtz 1973; Papanicolaou *et al.* 1977; Ethier and Kurtz 1986; Korolyuk and Limnios 2005) for a reducibly invertible operator Q, corresponding to the switching process in the series scheme, with a small series parameter $\varepsilon \to 0 (\varepsilon > 0)$ and the perturbing operator Q_1 is solved in the following steps.

You need to build a vector $\varphi^\varepsilon = \varphi + \varepsilon\varphi_1$ and a vector ψ, which satisfy the asymptotic representation

$$[\varepsilon^{-1}Q + Q_1]\varphi^\varepsilon = \psi + \varepsilon\theta^\varepsilon \qquad [1.3]$$

with a uniformly bounded vector θ^ε, such that

$$||\theta^\varepsilon|| \leq C, \varepsilon \to 0.$$

The left part of the obtained equation can be rewritten as

$$[\varepsilon^{-1}Q + Q_1](\varphi + \varepsilon\varphi_1) = \varepsilon^{-1}Q\varphi + [Q\varphi_1 + Q_1\varphi] + \varepsilon Q_1\varphi_1.$$

Equating to the right part of [1.3], we have:

$$\begin{cases} Q\varphi = 0, \\ Q\varphi_1 + Q_1\varphi = \psi, \\ Q_1\varphi_1 = \theta^\varepsilon. \end{cases} \qquad [1.4]$$

The last equation shows that the function $\varphi_1 \in \mathcal{D}_{Q_1}$. In addition, from the first equation, we have function φ as an arbitrary function from the null-subspace of operator Q.

Thus, the main question remains a solution of equation

$$Q\varphi_1 = \psi - Q_1\varphi.$$

The solvability condition for the reducibly invertible operator Q has the form:

$$\Pi Q\Pi\varphi_1 = 0 = \Pi\psi - \Pi Q_1\Pi\varphi,$$

where we finally get

$$\Pi\psi = \Pi Q_1\Pi\varphi.$$

Note that the operator $\Pi Q_1\Pi$ acts in subspace $\mathcal{N}_Q$; therefore, we can consider the reduced operator $\widehat{Q}_1$ on the reduced subspace $\widehat{\mathcal{N}}_Q$:

$$\Pi Q_1\Pi = \widehat{Q}_1\Pi.$$

We also use $\widehat{\psi} := \widehat{\Pi\psi} \in \widehat{\mathcal{N}}_Q$.

Then, the last equation takes the form

$$\widehat{\psi} = \widehat{Q}_1\widehat{\varphi}.$$

Since we have a relation in subspace $\widehat{\mathcal{N}}_Q$, we can solve the second equation of system [1.4] with respect to φ_1 (see [1.1]):

$$\varphi_1 = R_0(Q_1\varphi - \psi), \Pi\varphi_1 = 0.$$

So,

$$\varphi_1 = R_0\widetilde{Q}_1\varphi, \widetilde{Q}_1 := Q_1 - \widehat{Q}_1,$$

and finally we have

$$\theta^\varepsilon = Q_1\varphi_1 = Q_1R_0\widetilde{Q}_1\varphi.$$

Thus, the explicit expressions for functions $\psi, \varphi_1, \theta^\varepsilon$, are obtained, which give the solution of the singular perturbation problem.

1.1.3. *Markov process*

Let $x(t), t \geq 0$ be a Markov process on a standard state space $(E, \mathcal{E})$ (where E is a Polish space and $\mathcal{E}$ is its Borel σ-algebra), defined by means of the generator

$$Q\varphi(x) = q(x)\int_{\mathbf{E}}[\varphi(y) - \varphi(x)]P(x, dy), \; x \in E, \varphi(u) \in \mathcal{B}_E. \qquad [1.5]$$

The semi-Markov kernel

$$Q(x, B, t) = P(x, B)(1 - e^{-q(x)t}), x \in E, B \in \mathcal{E}, t \geq 0,$$

determines the associated Markov renewal process (x_k, τ_k), $k \geq 0$, where $x_k, k \geq 0$, is called the embedded Markov chain, given by the stochastic kernel

$$P(x, B) = P(x_{k+1} \in B | x_k = x),$$

and $\tau_k, k \geq 0$, is the point process of jump moments, which is determined by the time distribution function of sojourn time $\theta_{k+1} = \tau_{k+1} - \tau_k, k \geq 0$:

$$P(\theta_{k+1} \leq t | x_k = x) = 1 - e^{-q(x)t}.$$

The corresponding counting process of jumps is

$$\nu(t) := \max\{k \geq 0 : \tau_k \leq t\}.$$

The main assumption about the Markov process is the following condition:

CM: The Markov process $x(t), t \geq 0$, is uniformly ergodic with a stationary distribution $\pi(A),\ A \in \mathcal{E}$.

REMARK 1.1.– *Let* Π *be a projector at null-subspace of a reducibly invertible operator* Q *(defined in [1.5]):*

$$\Pi\varphi(x) = \int_E \pi(dx)\varphi(x).$$

The Markov process $x(t), t \geq 0$, is supposed to be *uniformly ergodic*, if for a semigroup P_t, determined by such a process, exists

$$\lim_{t\to\infty} P_t = \Pi \neq 0$$

in uniform operator topology. The main property of uniform ergodicity is the exponential speed of this convergence:

$$||P_t - \Pi|| \leq Me^{-\alpha t}, t > 0$$

for some $M > 1, \alpha > 0$. For details, see Doob (1990) and Korolyuk and Turbin (1993).

We should note that in this case the Markov process $x(t), t \geq 0$, has a stationary distribution $\pi(x)$, the embedded Markov chain $x_n, n \geq 1$ has a stationary distribution $\rho(x)$ and the following relationships are found:

$$\pi(dx)q(x) = q\rho(dx), q := \int_E \pi(dx)q(x).$$

1.1.4. *Semi-Markov process*

We call the Markov renewal process a two-component Markov chain $x_n, \tau_n, n \geq 0$ on $(E \times \mathbf{R}_+, \mathcal{E} \oplus \mathcal{B}_+)$, where $\tau_0 \leq ... \leq \tau_n \leq ...$ are renewal moments. Such a process is homogeneous by the second component, and its transient probabilities are determined by the semi-Markov kernel

$$Q(x, B, t) = P(x, B)F_x(t), x \in E, B \in \mathcal{E}, t \geq 0$$

through equality

$$Q(x, B, t) = P(x_{n+1} \in B, \theta_{n+1} \leq t | x_n = x)$$

$$= P(x_{n+1} \in B | x_n = x)P(\theta_{n+1} \leq t | x_n = x), \theta_{n+1} := \tau_{n+1} - \tau_n.$$

Introduce the counting process $\nu(t), t \geq 0$:

$$\nu(t) = \sup\{n \geq 0 : \tau_n \leq t\},$$

which counts the number of renewal moments of the Markov renewal process over time $(0, t]$.

We call the semi-Markov process $x(t), t \geq 0$, associated with the Markov renewal process $x_n, \tau_n, n \geq 0$ a random process

$$x(t) := x_{\nu(t)}, t \geq 0.$$

Denote by Q the generator of the associated Markov process:

$$Q = q(x)(P - I),$$

where the transition probability operator P is defined as

$$Pf(x) = \int_E P(x, dy)f(y), x \in E,$$

for all bounded measurable real-function functions f, defined on E, $q(x)$ is determined as

$$q(x) := 1/m_1(x), m_1(x) := E\theta_x = \int_0^\infty \overline{F}_x(t)dt,$$

or $m_k(x) = \int_0^\infty s^k F_x(ds)$.

The main assumption about the semi-Markov process is the following condition:

CSM: The semi-Markov process $x(t), t \geq 0$ is uniformly ergodic with a stationary distribution

$$\pi(dx)q(x) = q\rho(dx), q := 1/m, m := \int_E \rho(dx)m(x),$$

$$\rho(B) = \int_E \rho(dx)P(x, B), \rho(E) = 1.$$

Let us denote a projector Π :

$$\mathcal{N}_Q := \Pi\mathcal{B}(E), \mathcal{R}_Q := (I - \Pi)\mathcal{B}(E);$$

$$\Pi\varphi(x) := \widehat{\varphi}\mathbf{1}, \widehat{\varphi} := \int_E \varphi(x)\pi(dx).$$

1.1.5. *Phase merging*

A Markov process $x^\varepsilon(t), t \geq 0$ is determined on a standard phase space $(E, \mathcal{E})$ with splitting

$$E = \bigcup_{k=1}^{N} E_k, E_k \cap E_{k'} = \emptyset, k \neq k'$$

in the series scheme, with a small series parameter $\varepsilon \to 0, \varepsilon > 0$.

Its Markov kernel has the form

$$Q^\varepsilon(x, B, t) = P^\varepsilon(x, B)[1 - e^{-q(x)t}], \ x \in E, B \in \mathcal{E}, t \geq 0.$$

The following conditions should also be fulfilled:

ME1: A kernel describing the transient probabilities of the embedded Markov chain $x_n^\varepsilon, n \geq 0$, has the following representation:

$$P^\varepsilon(x, B) = P(x, B) + \varepsilon P_1(x, B).$$

The stochastic kernel $P(x, B)$ on the split phase space is defined as follows:

$$P(x, E_k) = \mathbf{1}_k(x) := \begin{cases} 1, x \in E_k, \\ 0, x \notin E_k. \end{cases}$$

The stochastic kernel $P(x, B)$ defines the accompanying Markov chain $x_n, n \geq 0$ on the classes $E_k, 1 \leq k \leq N$. The perturbing kernel $P_1(x, B)$ also satisfies the condition

$$P_1(x, E) = 0,$$

which is a direct consequence of the equality $P^\varepsilon(x, E) = P(x, E) = 1$.

ME2: The associated Markov process $x^0(t), t \geq 0$, set by the generator

$$Q\varphi(x) = q(x) \int_E P(x, dy)[\varphi(y) - \varphi(x)],$$

is uniformly ergodic inside each of the classes $E_k, 1 \leq k \leq N$, with stationary distributions $\pi_k(dx), 1 \leq k \leq N$, which satisfy the ratio:

$$\pi_k(dx)q(x) = q_k\rho_k(dx), \ q_k := \int_{E_k} \pi_k(dx)q(x).$$

ME3: The average exit probabilities

$$\widehat{p}_k := \int_{E_k} \rho_k(dx) P_1(x, E \backslash E_k) > 0, 1 \leq k \leq N.$$

Thus, the perturbing kernel $P_1(x, B)$ determines the transition probabilities between classes $E_k, 1 \leq k \leq N$. So, the equality $P^\varepsilon(x, B) = P(x, B) + \varepsilon P_1(x, B)$ means that the embedded Markov chain $x_n^\varepsilon, n \geq 0$ spends a long time in each of the classes E_k and jumps between the classes with small probabilities $\varepsilon P_1(x, E \backslash E_k)$.

Under conditions **ME1–ME3,** there is a weak convergence (see Korolyuk and Limnios (2005, Chapter 4))

$$v(x^\varepsilon(t)) \Rightarrow \widehat{x}(t), \varepsilon \to 0, \; v(x) = k \in \widehat{E} = \{1, ..., N\}, x \in E_k.$$

The limit Markov process $\widehat{x}(t), t \geq 0$ on a merging phase space $\widehat{E} = \{1, ..., N\}$ is determined by the generating matrix

$$\widehat{Q}_1 = (\widehat{q}_{kr}, 1 \leq k, r \leq N),$$

where:

$$\widehat{q}_{kr} = \widehat{q}_k \widehat{p}_{kr}, k \neq r, \widehat{q}_k = q_k \widehat{p}_k, 1 \leq k \leq N.$$

$$\widehat{p}_{kr} = p_{kr} / \widehat{p}_k, p_{kr} = \int_{E_k} \rho_k(dx) P_1(x, E_r), 1 \leq k, r \leq N, k \neq r,$$

$$\widehat{p}_k = - \int_{E_k} \rho_k(dx) P_1(x, E_k).$$

ME4: The merged Markov process $\widehat{x}(t), t \geq 0$ is ergodic, with the stationary distribution $\widehat{\pi} = (\pi_k, k \in \widehat{E})$.

Thus, the operator Q^ε can be represented as:

$$Q^\varepsilon = Q + \varepsilon Q_1, \; Q_1(x) = q(x) \int_E P_1(x, dy) \varphi(y).$$

REMARK 1.2.– *Generalizations of this approach can be found in Yin and Zhang (1998, 2005), where the operator* $Q^\varepsilon = Q + \varepsilon Q_1$,

$$Q = q(x) \int_E P(x, dy)[\varphi(y) - \varphi(x)], \; Q_1(x) = q_1(x) \int_E P_1(x, dy) \varphi(y).$$

Let Π be a projector on the null-subspace of a reducibly invertible operator Q. Its action on test functions φ is defined as follows:

$$\Pi\varphi(x) = \sum_{k=1}^{N} \widehat{\varphi}_k \mathbf{1}_k(x), \widehat{\varphi}_k := \int_{E_k} \pi_k(dx)\varphi(x).$$

Denote the reduced operator $\widehat{Q}_1$ using the ratio

$$\widehat{Q}_1\Pi = \Pi Q_1 \Pi.$$

Let $\widehat{\Pi}$ be a projector on the null-subspace of a reducibly invertible operator $\widehat{Q}_1$:

$$\widehat{\Pi}\widehat{\varphi} := \sum_{k\in\widehat{E}} \widehat{\pi}_k \widehat{\varphi}_k.$$

Let us denote the potential matrix $\widehat{R}_0 = [\widehat{R}^0_{kl}; 1 \leq k, l \leq N]$ by the following ratios:

$$\widehat{Q}_1\widehat{R}_0 = \widehat{R}_0\widehat{Q}_1 = \widehat{\Pi} - I.$$

1.1.6. *Processes with independent increments*

The Markov processes with independent increments in the Euclidean space $\mathbf{R}^d$ with the norm $|\cdot|$ will be denoted $\eta(t)$, $t \geq 0$. In the general case, such processes are determined by the corresponding generators (see Bertoin (1996, Chapter I.2) and Korolyuk and Limnios (2005, Chapter 1.2.4))

$$\widetilde{\mathbf{\Gamma}}\varphi(u) = b\varphi'(u) - \frac{\sigma^2}{2}\varphi''(u) + \int_{\mathbf{R}^d} [\varphi(u+v) - \varphi(u) - v\varphi'(u)\mathbf{1}_{(|v|\leq 1)}]\widetilde{\Gamma}(dv), \quad [1.6]$$

where $\varphi(u)$ is a real-valued, twice differentiable function in $\mathbf{R}^d$, which is equal to 0 at infinity, and has a sup-norm $||\varphi|| = \sup_{u\in\mathbf{R}^d} |\varphi(u)|, \varphi(u) \in C_0^2(\mathbf{R}^d)$,

$$b = \int_{\mathbf{R}^d} v\Gamma(dv),$$

where $\Gamma(dv)$ is the intensity kernel that satisfies $\Gamma(\{0\}) = 0$.

In the following, we will study the asymptotic behavior of type [1.6] processes, whose generators can be represented as

$$\widetilde{\mathbf{\Gamma}}\varphi(u) = \sum_{k=1}^{d} b_k\varphi_k'(u) + \int_{\mathbf{R}^d} [\varphi(u+v) - \varphi(u) - \sum_{k=1}^{d} v_k\varphi_k'(u)\mathbf{1}_{(|v|\leq 1)}]\widehat{\Gamma}(dv),$$

$$\varphi_k'(u) := \partial\varphi(u)/\partial u_k, 1 \leq k \leq d.$$

Note that despite the absence of a diffusion component at increasing time intervals in Poisson approximation schemes, the diffusion component in the limit process appears as a result of the averaging of small jumps in the initial process.

EXAMPLE 1.1.– *A compound Poisson process is an example of a Markov process with independent increments*

$$\eta(t) = \sum_{k=1}^{\nu(t)} \eta_k,$$

where $\nu(t), t \geq 0$, is a homogeneous Poisson process with intensity λ, and $\eta_k, k \geq 1$, are independent and identically distributed real-valued random variables that do not depend on $\nu(t), t \geq 0$, and have a distribution function $F(u)$. The corresponding generator has the form

$$\mathbf{\Gamma}\varphi(u) = \lambda \int_{\mathbf{R}} (\varphi(u+v) - \varphi(u))F(dv).$$

We will also denote by $\eta(t)$, $t \geq 0$, a Markov processes with locally independent increments (also known as PDMP, piecewise deterministic Markov process, Davis (1993)) in the Euclidean space $\mathbf{R}^d$. Such processes are determined by the generator (see Korolyuk and Limnios (2005, Chapter 1.2.5), and Korolyuk (1999))

$$\mathbf{\Gamma}\varphi(u) = b(u)\varphi'(u) + \int_{\mathbf{R}^d} [\varphi(u+v) - \varphi(u) - v\varphi'(u)\mathbf{1}_{(|v|\leq 1)}]\Gamma(u, dv). \quad [1.7]$$

When the intensity kernel is bounded, the form of the generator is simplified, namely:

$$\widetilde{\mathbf{\Gamma}}\varphi(u) = b\varphi'(u) + \int_{\mathbf{R}^d} [\varphi(u+v) - \varphi(u)]\widetilde{\Gamma}(dv) \quad [1.8]$$

for a process with independent increments, and

$$\mathbf{\Gamma}\varphi(u) = b(u)\varphi'(u) + \int_{\mathbf{R}^d} [\varphi(u+v) - \varphi(u)]\Gamma(u, dv) \quad [1.9]$$

for a processes with locally independent increments (see Korolyuk and Limnios (2005, Chapter 1.2.5)).

1.1.7. *Poisson approximation scheme for the processes with independent increments*

The term *Poisson approximation* should be understood in a context similar to the averaging scheme or diffusion approximation, which allows us to study the limit behavior of random processes at increasing time intervals. Therefore, it is not a classic problem of approximating a random process by Poisson processes.

The main idea of the Poisson approximation (see Korolyuk and Limnios (2005, Chapter 7)) is that the small series parameter normalizes the probabilities (or intensities) of jumps. Thus, the jumps are divided into two types: small jumps with probabilities close to 1 and large jumps that occur with a probability going to 0, together with a small series parameter $\varepsilon \to 0$.

EXAMPLE 1.2.– *Here is a simple example of a random variable with similar properties. For* α *let:*

$$P\{\alpha = b\} = \varepsilon^2 p,$$

$$P\{\alpha = \varepsilon a_1 + \varepsilon^2 b_1\} = 1 - \varepsilon^2 p.$$

Then, we have:

$$\mathbf{E}\alpha = \varepsilon a_1 + \varepsilon^2(bp + b_1) + o(\varepsilon^2),$$

$$\mathbf{E}\alpha^2 = \varepsilon^2(b^2 p + a_1^2) + o(\varepsilon^2).$$

Such moment conditions characterize the Lévy approximation.

When $a_1 = 0$, *we will have*

$$\mathbf{E}\alpha = \varepsilon^2(bp + b_1) + o(\varepsilon^2),$$

$$\mathbf{E}\alpha^2 = \varepsilon^2 b^2 p + o(\varepsilon^2),$$

and therefore, by putting $\widetilde{\varepsilon} = \varepsilon^2$, *we obtain the moment conditions that characterize the Poisson approximation:*

$$\mathbf{E}\alpha = \widetilde{\varepsilon}(bp + b_1) + o(\widetilde{\varepsilon}),$$

$$\mathbf{E}\alpha^2 = \widetilde{\varepsilon} b^2 p + o(\widetilde{\varepsilon}).$$

In the study of processes with independent increments, the moment conditions are imposed on the corresponding intensity kernel, normalized by a small series parameter.

Let $C_3(\mathbf{R}^d)$ be a class of functions that defines a measure and includes real-valued bounded functions, such as $g(u)/|u|^2 \to 0$, as $|u| \to 0$ if $g \in C_3(\mathbf{R}^d)$ (see Jacod and Shiryaev 2003; Korolyuk and Limnios 2005).

Consider a family of normalized Markov processes with trajectories in $D_{\mathbf{R}}[0,\infty)$ and independent increments in the series scheme, with a small series parameter $\varepsilon \to 0, \varepsilon > 0$:

$$\eta_\varepsilon(t) = \eta(t/\varepsilon), t \geq 0,$$

determined by generators (see [1.8])

$$\widetilde{\mathbf{\Gamma}}^\varepsilon \varphi(u) = \varepsilon^{-1} \int_{\mathbf{R}} [\varphi(u+v) - \varphi(u)] \widetilde{\Gamma}^\varepsilon(dv).$$

Let the conditions of the Poisson approximation be satisfied:

C1: *Poisson approximation.*

PA1 Approximation of means:

$$b_\varepsilon = \int_{\mathbf{R}} v \widetilde{\Gamma}^\varepsilon(dv) = \varepsilon[b + \theta_b^\varepsilon],$$

and

$$c_\varepsilon = \int_{\mathbf{R}} v^2 \widetilde{\Gamma}^\varepsilon(dv) = \varepsilon[c + \theta_c^\varepsilon],$$

where

$$b < +\infty, c < +\infty.$$

PA2 For the intensity kernel, there is an asymptotic representation:

$$\widetilde{\Gamma}_g^\varepsilon = \int_{\mathbf{R}} g(v) \widetilde{\Gamma}^\varepsilon(dv) = \varepsilon[\widetilde{\Gamma}_g + \theta_g^\varepsilon]$$

for all $g \in C_3(\mathbf{R})$, which is a class of functions that defines a measure (see Jacod and Shiryaev 2003).

The kernel $\widetilde{\Gamma}^0(dv)$ is set on the class of functions that defines the measure $C_3(\mathbf{R})$ by the ratio

$$\widetilde{\Gamma}_g = \int_{\mathbf{R}} g(v) \widetilde{\Gamma}^0(dv), \quad g \in C_3(\mathbf{R}).$$

The negligibly small terms $\theta_b^\varepsilon, \theta_c^\varepsilon, \theta_g^\varepsilon$ satisfy the condition

$$|\theta_\cdot^\varepsilon| \to 0, \quad \varepsilon \to 0.$$

PA3 There is a relationship:

$$c := \int_{\mathbf{R}} v^2 \widetilde{\Gamma}^0(dv),$$

which causes the absence of a diffusion component in the limit generator.

C2 *Uniform quadratic integrability*:

$$\lim_{c\to\infty}\int_{|v|>c} v^2\widetilde{\Gamma}^0(dv) = 0.$$

LEMMA 1.1.– *Generator of the process with independent increments*

$$\widetilde{\mathbf{\Gamma}}^{\varepsilon}\varphi(u) = \varepsilon^{-1}\int_{\mathbf{R}}[\varphi(u+v) - \varphi(u)]\widetilde{\Gamma}^{\varepsilon}(dv)$$

in the scheme of Poisson approximation has the following asymptotic representation:

$$\widetilde{\mathbf{\Gamma}}^{\varepsilon}\varphi(u) = b\varphi'(u) + \int_{\mathbf{R}}[\varphi(u+v) - \varphi(u) - v\varphi'(u)]\widetilde{\Gamma}^0(dv) + \theta^{\varepsilon}\varphi,$$

where $|\theta^{\varepsilon}\varphi| \to 0, \varepsilon \to 0, \varepsilon > 0.$

REMARK 1.3.– *In the space* $\mathbf{R}^d$, *the asymptotic representation has the form*

$$\widetilde{\mathbf{\Gamma}}^{\varepsilon}\varphi(u) = \sum_{k=1}^{d} b_k\varphi_k'(u) + \int_{\mathbf{R}^d}[\varphi(u+v) - \varphi(u) - \sum_{k=1}^{d} v_k\varphi_k'(u)]\widehat{\Gamma}^0(dv) + \theta^{\varepsilon}\varphi,$$

$$\varphi_k'(u) := \partial\varphi(u)/\partial u_k, 1 \le k \le d.$$

PROOF.– Rewrite the expression for the generator as follows:

$$\widetilde{\mathbf{\Gamma}}^{\varepsilon}\varphi(u) = \varepsilon^{-1}\int_{\mathbf{R}}[\varphi(u+v) - \varphi(u) - v\varphi'(u) - \frac{v^2}{2}\varphi''(u)]\widetilde{\Gamma}^{\varepsilon}(dv)$$

$$+\varepsilon^{-1}\int_{\mathbf{R}} v\varphi'(u)\widetilde{\Gamma}^{\varepsilon}(dv) + \frac{\varepsilon^{-1}}{2}\int_{\mathbf{R}} v^2\varphi'(u)\widetilde{\Gamma}^{\varepsilon}(dv).$$

It is easy to see that the function $\psi_u(v) = \varphi(u+v) - \varphi(u) - v\varphi'(u) - \frac{v^2}{2}\varphi''(u)$ belongs to the class $C_3(\mathbf{R})$. Really,

$$\psi_u(v)/v^2 \to 0, v \to 0$$

uniformly by u, by the condition of derivative boundedness of the function $\varphi(u)$ on a compact. In addition, this function is continuous and bounded for $\varphi(u) \in C_0^2(\mathbf{R})$ by condition **PA1**.

Thus, from conditions **PA1, PA2**, we have:

$$\widetilde{\mathbf{\Gamma}}^{\varepsilon}\varphi(u) = \int_{\mathbf{R}}[\varphi(u+v) - \varphi(u) - v\varphi'(u) - \frac{v^2}{2}\varphi''(u)]\widetilde{\Gamma}^0(dv)$$

$$+b\varphi'(u) + \frac{c}{2}\varphi''(u) + \theta_b^{\varepsilon}\varphi + \theta_c^{\varepsilon}\varphi + \theta_{\psi}^{\varepsilon}\varphi.$$

By applying condition **PA3**, we finally get an asymptotic representation:

$$\widetilde{\mathbf{\Gamma}}^{\varepsilon}\varphi(u) = b\varphi'(u) + \int_{\mathbf{R}} [\varphi(u+v) - \varphi(u) - v\varphi'(u)]\widetilde{\Gamma}^{0}(dv) + \theta^{\varepsilon}\varphi.$$

The lemma is proved.

Now consider another scaling of processes with independent increments in the series scheme, with a small series parameter $\varepsilon \to 0, \varepsilon > 0$:

$$\eta_{\varepsilon}(t) = \eta(t/\varepsilon^2), t \geq 0,$$

determined by generators

$$\widetilde{\mathbf{\Gamma}}^{\varepsilon}\varphi(u) = \varepsilon^{-2} \int_{\mathbf{R}} [\varphi(u+v) - \varphi(u)]\widetilde{\Gamma}^{\varepsilon}(dv).$$

Suppose that the conditions of the Lévy approximation hold:

C1': *Lévy approximation.*

LA1 Approximation of means:

$$b_{\varepsilon} = \int_{\mathbf{R}} v\widetilde{\Gamma}^{\varepsilon}(dv) = \varepsilon b_1 + \varepsilon^2[b + \theta_b^{\varepsilon}],$$

and

$$c_{\varepsilon} = \int_{\mathbf{R}} v^2\widetilde{\Gamma}^{\varepsilon}(dv) = \varepsilon^2[c + \theta_c^{\varepsilon}],$$

where

$$b_1 < +\infty, b < +\infty, c < +\infty.$$

LA2 For the intensity kernel, there is an asymptotic representation:

$$\widetilde{\Gamma}_g^{\varepsilon} = \int_{\mathbf{R}} g(v)\widetilde{\Gamma}^{\varepsilon}(dv) = \varepsilon^2[\widetilde{\Gamma}_g + \theta_g^{\varepsilon}]$$

for all $g \in C_3(\mathbf{R})$ – a class of functions that defines a measure (Jacod and Shiryaev 2003).

The kernel $\widetilde{\Gamma}^0(dv)$ is specified on the class of functions, which determines the measure $C_3(\mathbf{R})$ by means of the relation

$$\widetilde{\Gamma}_g = \int_{\mathbf{R}} g(v)\widetilde{\Gamma}^0(dv), \quad g \in C_3(\mathbf{R}).$$

The negligibly small terms $\theta_b^{\varepsilon}, \theta_c^{\varepsilon}, \theta_g^{\varepsilon}$ satisfy the condition

$$|\theta_{\cdot}^{\varepsilon}| \to 0, \quad \varepsilon \to 0.$$

C2 *Uniform quadratic integrability:*

$$\lim_{c\to\infty}\int_{|v|>c} v^2\widetilde{\Gamma}^0(dv) = 0.$$

LEMMA 1.2.– *The generator of the process with independent increments*

$$\widetilde{\mathbf{\Gamma}}^\varepsilon\varphi(u) = \varepsilon^{-2}\int_{\mathbf{R}}[\varphi(u+v)-\varphi(u)]\widetilde{\Gamma}^\varepsilon(dv)$$

in the Lévy approximation scheme, has the following asymptotic representation:

$$\widetilde{\mathbf{\Gamma}}^\varepsilon\varphi(u) = \varepsilon^{-1}b_1\varphi'(u) + (b-b_0)\varphi'(u) + \frac{c-c_0}{2}\varphi''(u)$$
$$+\int_{\mathbf{R}}[\varphi(u+v)-\varphi(u)]\widetilde{\Gamma}^0(dv) + \theta^\varepsilon\varphi,$$

where

$$b_0 := \int_{\mathbf{R}} v\widetilde{\Gamma}^0(dv), c_0 := \int_{\mathbf{R}} v^2\widetilde{\Gamma}^0(dv),$$

negligible term $|\theta^\varepsilon\varphi| \to 0, \varepsilon \to 0, \varepsilon > 0.$

REMARK 1.4.– *In the space* $\mathbf{R}^d$, *the asymptotic representation has the form*

$$\widetilde{\mathbf{\Gamma}}^\varepsilon\varphi(u) = \varepsilon^{-1}\sum_{k=1}^d b_1^k\varphi_k'(u) + \sum_{k=1}^d(b^k-b_0^k)\varphi_k'(u) + \frac{1}{2}\sum_{k,r=1}^d(c^{kr}-c_0^{kr})\varphi_{kr}''(u)$$
$$+\int_{\mathbf{R}^d}[\varphi(u+v)-\varphi(u)]\widetilde{\Gamma}^0(dv) + \theta^\varepsilon\varphi,$$

$$c_0^{kr} := \int_{\mathbf{R}} v_k v_r\widetilde{\Gamma}_0(dv),\ \varphi_k'(u) := \partial\varphi(u)/\partial u_k, 1\le k\le d,$$

$$\varphi_{kr}''(u) := \partial^2\varphi(u)/\partial u_k\partial u_r, 1\le k,r\le d.$$

PROOF.– Let us rewrite the expression for the generator as follows:

$$\widetilde{\mathbf{\Gamma}}^\varepsilon\varphi(u) = \varepsilon^{-2}\int_{\mathbf{R}}[\varphi(u+v)-\varphi(u)-v\varphi'(u)-\frac{v^2}{2}\varphi''(u)]\widetilde{\Gamma}^\varepsilon(dv)$$
$$+\varepsilon^{-2}\int_{\mathbf{R}} v\varphi'(u)\widetilde{\Gamma}^\varepsilon(dv) + \frac{\varepsilon^{-2}}{2}\int_{\mathbf{R}} v^2\varphi'(u)\widetilde{\Gamma}^\varepsilon(dv).$$

It is easy to see that the function $\psi_u(v) = \varphi(u+v) - \varphi(u) - v\varphi'(u) - \frac{v^2}{2}\varphi''(u)$ belongs to class $C_3(\mathbf{R})$. Indeed,

$$\psi_u(v)/v^2 \to 0, v \to 0.$$

In addition, this function is continuous and bounded for $\varphi(u) \in C_0^2(\mathbf{R})$ for condition **LA1**.

Thus, from conditions **LA1, LA2**, we have:

$$\widetilde{\mathbf{\Gamma}}^\varepsilon \varphi(u) = \int_{\mathbf{R}} [\varphi(u+v) - \varphi(u) - v\varphi'(u) - \frac{v^2}{2}\varphi''(u)]\widetilde{\Gamma}^0(dv)$$

$$+\varepsilon^{-1} b_1 \varphi'(u) + b\varphi'(u) + \frac{c}{2}\varphi''(u) + \theta_b^\varepsilon \varphi + \theta_c^\varepsilon \varphi + \theta_\psi^\varepsilon \varphi.$$

Therefore, we finally get an asymptotic representation:

$$\widetilde{\mathbf{\Gamma}}^\varepsilon \varphi(u) = \varepsilon^{-1} b_1 \varphi'(u) + (b - b_0)\varphi'(u) + \frac{c - c_0}{2}\varphi''(u)$$

$$+ \int_{\mathbf{R}} [\varphi(u+v) - \varphi(u)]\widetilde{\Gamma}^0(dv) + \theta^\varepsilon \varphi.$$

The lemma is proved.

1.1.8. *Random evolutionary systems with locally independent increments*

A random evolutionary system with locally independent increments is determined by the ratio (see Korolyuk and Limnios (2005, Chapter 1)):

$$\xi(t) = \xi_0 + \int_0^t \eta(ds; x(s)),\ t \geq 0, \qquad [1.10]$$

where a right-continuous Markov process $\eta(t, x),\ t \geq 0, x \in E$, which has a left limit, is a process with locally independent increments that is determined by the generator (see [1.9])

$$\mathbf{\Gamma}(x)\varphi(u) = b(u,x)\varphi'(u) + \int_{\mathbf{R}^d} [\varphi(u+v) - \varphi(u) - v\varphi'(u)]\Gamma(u, dv; x).$$

REMARK 1.5.– *Definition* [1.10] *can be rewritten as:*

$$\xi(t) = \xi_0 + \sum_{k=1}^{\nu(t)} [\eta(\tau_{k+1}, x_k) - \eta(\tau_k, x_k)] + \eta(t - \tau_{\nu(t)}, x(t)). \qquad [1.11]$$

It follows that the increment of the random evolutionary system at the interval between the jumps of the switching process is equal to:

$$\xi(\tau_k + t) - \xi(\tau_k) = \eta(t, x(t)), 0 \leq t < \tau_{k+1}.$$

We see that a random evolutionary system consists of parts of the trajectories of the Markov process with locally independent increments $\eta(t; x)$, which in turn depends on the switching process.

The type [1.11] *processes are used, for example, in the theory of queuing systems (see Anisimov 2008), and the Poisson approximation scheme, described below, simulates the appearance, with a low probability, of large packets of information.*

Some particular cases of such processes are the following stochastic systems:

- *Integral functional*

$$\alpha(t) = \int_0^t a(x(s))ds, t \geq 0$$

where a is a deterministic measurable function defined on $(E, \mathcal{E})$.

- *A dynamic system*

$$\dot{u}(t) = C(u(t), x(t)), t \geq 0,$$

where C is deterministic $\mathbf{R}^d$-function defined on $\mathbf{R}^d \times E$.

- *A compound Poisson process*

$$\zeta(t) = \sum_{k=1}^{\nu(t)} \alpha(x_k),$$

where x_k is an embedded Markov chain of the jump Markov process $x(t)$.

A random evolutionary system [1.10] in the case of Markov switching is characterized by a generator from a two-component Markov process $\xi(t), x(t), t \geq 0$ (see Korolyuk and Limnios (2005, Chapter 2))

$$\mathbf{L}\varphi(u, x) = Q\varphi(u, \cdot)(x) + \mathbf{\Gamma}(x)\varphi(\cdot, x)(u). \qquad [1.12]$$

When there is a switching semi-Markov process, a compensating operator is used.

DEFINITION 1.5.– *(see Sviridenko (1998)) The compensating operator* $\mathbb{L}$ *of a three-component Markov renewal process* $\xi_n, x_n, \tau_n, n \geq 0$ *is the next generator*

$$\mathbb{L}\varphi(\xi_0, x_0, \tau_0) = q(x_0)\mathbf{E}[\varphi(\xi_1, x_1, \tau_1) - \varphi(\xi_0, x_0, \tau_0)|\mathcal{F}_0], \qquad [1.13]$$

where

$$\mathcal{F}_t := \sigma(\xi(s), x(s), \tau(s); 0 \leq s \leq t),$$
$$q(x) := 1/m(x), m(x) := \mathbf{E}\theta_x = \int_0^\infty \overline{F}_x(t)dt, 1/q(x_0) = \mathbf{E}[\theta_1|\mathcal{F}_0].$$

An important role is also played by the representation formula for the compensating operator:

$$\begin{aligned}\mathbb{L}\varphi(x,t) &= q(x)\int_0^\infty \int_E Q(x,dy,ds)(\varphi(y,t+s) - \varphi(x,t)) \\ &= q(x)\left[\int_0^\infty F_x(ds)\int_E P(x,dy)\varphi(y,t+s) - \varphi(x,t)\right]. \end{aligned} \qquad [1.14]$$

In the phase merging scheme, the generator of the two-component Markov process $\xi(t), x^\varepsilon(t), t \geq 0$ has the form (see Korolyuk and Limnios (2005, Chapter 5))

$$\mathbf{L}\varphi(u,x) = Q^\varepsilon\varphi(u,\cdot)(x) + \mathbf{\Gamma}(x)\varphi(\cdot,x)(u). \qquad [1.15]$$

1.1.9. *Impulse recurrent processes*

Impulsive processes $\widetilde{\xi}(t), t \geq 0$, in the Euclidean space $\mathbf{R}^d$, with Markov or semi-Markov switching $x(t)$, are denoted by the sum on the embedded Markov chain (see, for example, Korolyuk and Limnios (2005, Chapter 2.3))

$$\widetilde{\xi}(t) = \widetilde{\xi}_0 + \sum_{k=1}^{\nu(t)} \alpha_k(x_{k-1}), \quad t \geq 0. \qquad [1.16]$$

For an arbitrary sequence of elements $x_k, k \geq 0$ from E, we will consider $\alpha_k(x_{k-1}), k \geq 1$ independent. Let us denote by $\widetilde{G}_x$ the distribution function of $\alpha_k(x)$, i.e.

$$\widetilde{G}_x(dv) := P(\alpha_k(x) \in dv), k \geq 0, \varepsilon > 0, x \in E.$$

In addition to impulsive processes, we will also consider impulsive recurrent processes, which are discrete analogues of the process with locally independent increments, and are denoted as

$$\widetilde{\xi}(t) = \widetilde{\xi}_0 + \sum_{k=1}^{\nu(t)} \alpha_k(\xi_k, x_{k-1}). \qquad [1.17]$$

Here, for an arbitrary sequence of elements $z_k, k \geq 0$ ç $\mathbf{R}^d \times E$, we will consider $\alpha_k(z_{k-1}), k \geq 1$ to be independent. We denote by $G_{u,x}$ the distribution functions of $\alpha_k(z)$, i.e.

$$G_{u,x}(dv) := P(\alpha_k(u,x) \in dv), k \geq 0, \varepsilon > 0, u \in \mathbf{R}^d, x \in E.$$

It is worth noting that the two-component process $\widetilde{\xi}(t), x(t), t \geq 0$, is an additive Markov process (see, for example, Korolyuk and Limnios (2005, Chapter 2.5)).

1.1.10. *Random evolutionary systems and impulsive processes in the Poisson approximation scheme*

In the following we will consider Euclidean spaces $\mathbf{R}^d$ with norm $|\cdot|$ $(d \geq 1)$, and $(E, \mathcal{E})$, – *standard phase space*, (i.e. E is a Polish space, with Borel σ-algebra $\mathcal{E}$). For an arbitrary vector $v \in \mathbf{R}^d$ and matrix $c \in \mathbf{R}^{d \times d}$, let us denote by v^* and c^* the corresponding transposed vector and matrix.

The random evolutionary system in the Poisson approximation scheme has the following normalization

$$\xi^\varepsilon(t) = \xi_0^\varepsilon + \int_0^t \eta_\varepsilon(ds; x(s/\varepsilon))$$

and satisfies the conditions:

C1: A family of processes with locally independent increments $\eta_\varepsilon(t;x), t \geq 0$, $x \in E$ satisfies the conditions of the Poisson approximation (see Korolyuk and Limnios (2005, Chapter 7.2.3)).

PA1 *Approximation of means*:

$$b_\varepsilon(u;x) = \int_{\mathbf{R}^d} v\Gamma^\varepsilon(u, dv; x) = \varepsilon[b(u;x) + \theta_b^\varepsilon(u;x)],$$

and

$$c_\varepsilon(u;x) = \int_{\mathbf{R}^d} vv^*\Gamma^\varepsilon(u, dv; x) = \varepsilon[c(u;x) + \theta_c^\varepsilon(u;x)].$$

PA2 *Condition for the intensity kernel*:

$$\Gamma_g^\varepsilon(u;x) = \int_{\mathbf{R}^d} g(v)\Gamma^\varepsilon(u, dv; x) = \varepsilon[\Gamma_g(u;x) + \theta_g^\varepsilon(u;x)]$$

for all $g \in C_3(\mathbf{R}^d)$, and the kernel $\Gamma_g(u;x)$ is bounded for all $g \in C_3(\mathbf{R}^d)$, so, $|\Gamma_g(u;x)| \leq \Gamma_g$ (a constant, dependent on g), where the kernel $\Gamma(u,dv;x)$ is defined in class $C_3(\mathbf{R}^d)$ by the relation

$$\Gamma_g(u;x) = \int_{\mathbf{R}^d} g(v)\Gamma(u,dv;x), \quad g \in C_3(\mathbf{R}^d).$$

The negligible terms $\theta_b^\varepsilon, \theta_c^\varepsilon, \theta_g^\varepsilon$ satisfy the condition

$$\sup_{x\in E} |\theta_\cdot^\varepsilon(u;x)| \to 0, \quad \varepsilon \to 0.$$

PA3: The $c(u,x) = \int_{\mathbf{R}^d} vv^*\Gamma(u,dv;x)$ relationship takes place, which causes the absence of a diffusion component in the limit process.

PA4 *The condition for initial values*:

$$\sup_{\varepsilon>0} E|\xi_0^\varepsilon| \leq C < \infty$$

and

$$\xi_0^\varepsilon \Rightarrow \xi_0, \varepsilon \to 0.$$

The random evolutionary system in the Lévy approximation scheme has the following normalization:

$$\xi^\varepsilon(t) = \xi_0^\varepsilon + \int_0^t \eta_\varepsilon(ds; x(s/\varepsilon^2))$$

and satisfies the conditions:

C1': A family of processes with locally independent increments $\eta_\varepsilon(t;x), t \geq 0$, $x \in E$ satisfies the Lévy approximation conditions (see Korolyuk and Limnios (2005, Chapter 9.2)).

L1 *Approximation of means*:

$$b_\varepsilon(u;x) = \int_{\mathbf{R}^d} v\Gamma^\varepsilon(u,dv;x) = \varepsilon b_1(u;x) + \varepsilon^2[b(u;x) + \theta_b^\varepsilon(u;x)],$$

and

$$c_\varepsilon(u;x) = \int_{\mathbf{R}^d} vv^*\Gamma^\varepsilon(u,dv;x) = \varepsilon^2[c(u;x) + \theta_c^\varepsilon(u;x)].$$

L2 *Condition for the intensity kernel*:

$$\Gamma_g^\varepsilon(u;x) = \int_{\mathbf{R}^d} g(v)\Gamma^\varepsilon(u,dv;x) = \varepsilon^2[\Gamma_g(u;x) + \theta_g^\varepsilon(u;x)]$$

for all $g \in C_3(\mathbf{R}^d)$, and the kernel $\Gamma_g(u;x)$ is bound for all $g \in C_3(\mathbf{R}^d)$, so $|\Gamma_g(u;x)| \leq \Gamma_g$ (a constant, dependent on g), where the kernel $\Gamma(u,dv;x)$ is defined on the class $C_3(\mathbf{R}^d)$ by the relation

$$\Gamma_g(u;x) = \int_{\mathbf{R}^d} g(v)\Gamma(u,dv;x), \quad g \in C_3(\mathbf{R}^d).$$

The negligible terms $\theta_b^\varepsilon, \theta_c^\varepsilon, \theta_g^\varepsilon$ satisfy the condition

$$\sup_{x\in E} |\theta_\cdot^\varepsilon(u;x)| \to 0, \quad \varepsilon \to 0.$$

L3 *Balance condition*: $\int_E \pi(dx) b_1(u;x) = 0$.

L4 *Condition for initial values*:

$$\sup_{\varepsilon>0} E|\xi_0^\varepsilon| \leq C < \infty$$

and

$$\xi_0^\varepsilon \Rightarrow \xi_0, \varepsilon \to 0.$$

In addition, the following conditions must be met in both schemes:

C2 *Uniform quadratic integrability*:

$$\lim_{c\to\infty} \sup_{x\in E} \int_{|v|>c} vv^* \Gamma(u,dv;x) = 0.$$

C3 *Growth condition*: there is a positive constant L, such that

$$|b(u;x)| \leq L(1+|u|), \quad \text{and} \quad |c(u;x)| \leq L(1+|u|^2),$$

for all real-valued non-negative functions $f(v), v \in \mathbf{R}^d$, such that $\int_{\mathbf{R}^d\setminus\{0\}} (1+f(v))|v|^2 dv < \infty$, we have

$$|\Lambda(u,v;x)| \leq Lf(v)(1+|u|),$$

where $\Lambda(u,v;x)$ is a Radon–Nicodemus derivative of the kernel $\Gamma(u,B;x)$ in relation to the Lebesgue measure dv in $\mathbf{R}^d$, i.e.

$$\Gamma(u,dv;x) = \Lambda(u,v;x)dv.$$

In the study of impulse recurrent processes, the conditions are imposed on the distribution function $G_{u,x}(dv)$, normalized by a small series parameter. That is, in the case of the Poisson approximation, we have the scaling

$$\widetilde{\xi}^\varepsilon(t) = \widetilde{\xi}_0^\varepsilon + \sum_{k=1}^{\nu(t/\varepsilon)} \alpha_k^\varepsilon(\widetilde{\xi}_{k-1}^\varepsilon, x_{k-1}^\varepsilon), \quad t \geq 0$$

and the conditions **C1, C2, C3**, with replacement $G^{\varepsilon}_{u,x}(dv)$ in place of $\Gamma^{\varepsilon}(u, dv; x)$, $G^{\varepsilon}_{g}(u, x)$ in place of $\Gamma^{\varepsilon}_{g}(u; x)$, $G_g(u, x)$ in place of $\Gamma_g(u; x)$, $G_{u,x}(dv)$ in place of $\Gamma(u, dv; x)$ and G_g in place of Γ_g.

Finally, in the case of the Lévy approximation, we will have the scaling

$$\widetilde{\xi}^{\varepsilon}(t) = \widetilde{\xi}^{\varepsilon}_0 + \sum_{k=1}^{\nu(t/\varepsilon^2)} \alpha^{\varepsilon}_k(\widetilde{\xi}^{\varepsilon}_{k-1}, x^{\varepsilon}_{k-1}), \quad t \geq 0.$$

and the conditions **C1', C2, C3**, with the corresponding replacements of the intensity kernels by the distribution functions.

1.2. Nonlinear exponential generator of large deviations, Nisio semigroup and control problem

The solution of the large deviations problem using a nonlinear exponential generator is usually implemented in four stages (see Feng and Kurtz (2006, Chapter 2)):

(1) *Calculation of the exponential limit (nonlinear) generator* $\mathbf{H}$, *which determines large deviations.*

(2) *Verification of the exponential compactness of a family of pre-limit processes.*

(3) *Verification of the comparison principle for the limit generator.*

(4) *Construction of a variational representation of the rate functional.*

At the third stage of this program, the uniqueness of the solution of the equation is proved

$$(I - \alpha \mathbf{H})f = h, \alpha > 0$$

for wide-enough class of functions h. The consequence of this fact is the convergence of nonlinear semigroups H^{ε}_t that determines a large deviations principle for the finite-dimensional distributions of pre-limit processes, and under the condition of exponential compactness of a pre-limit family of processes, and finally the large deviations principle in the space $D_E[0, \infty)$.

The stages (2)–(4) are implemented in Feng and Kurtz (2006) for an exponential generator corresponding to processes in our consideration. Namely, the issue of exponential compactness for processes with independent increments is considered in

Examples 1.5 and 4.23, and in paragraphs 10.1.2 and 10.3.2; the verification of the comparison principle for the limit exponential generator of the form $\widehat{\mathbf{H}}^0\varphi(u) = \mathbf{H}^0(\varphi'(u))$ is described in paragraph 10.1.3 using Lemma 9.15 and the variational representation of the limit exponential generator is described in paragraph 10.1.5, using Theorem 8.14.

Note that the interpretation of a nonlinear semigroup, such as a Nisio semigroup, and its connection with the control problem makes it possible to build the rate functional in the most convenient form (see Example 1.5).

1.2.1. *Martingale characterization and Brick's formula*

Consider a Markov process $x(t), t \geq 0$, on a standard state space $(E, \mathcal{E})$. Let the respective generator $\mathbf{L}$ have a dense domain $\mathcal{D}(\mathbf{L}) \subseteq \mathcal{B}_E$, comprising functions, continuous together with their derivative. Here, $\mathcal{B}_E$ is a Banach space of a real-valued bounded test function $\varphi(x) \in E$, with the norm: $\|\varphi\| := \sup_{x\in E} |\varphi(x)|$.

We also denote by $C_b(E)$ the space of continuous functions that belong to $\mathcal{B}_E$. The classical linear semigroup of the process $x(t), t \geq 0$ is given as

$$L_t\varphi(u) := E[\varphi(x(t))|x(0) = u],$$

and the nonlinear semigroup:

$$H_t\varphi(u) := \ln E[e^{\varphi(x(t))}|x(0) = u], \quad \text{or} \quad H_t\varphi(u) = \ln L_t e^{\varphi(u)}. \qquad [1.18]$$

This is a semigroup, because we have:

$$H_{t+s}\varphi(u) = \ln L_{t+s}e^{\varphi(u)} = \ln L_t L_s e^{\varphi(u)}$$

$$= \ln L_t e^{\ln L_s e^{\varphi(u)}} = \ln L_t e^{H_s\varphi(u)} = H_t H_s\varphi(u).$$

The following result is used, in particular, in Feng and Kurtz (2006) and Korolyuk (2010b, 2011c).

LEMMA 1.3.– *A nonlinear exponential generator that corresponds to semigroup* [1.18] *has the form:*

$$\mathbf{H}\varphi(u) = e^{-\varphi(u)}\mathbf{L}e^{\varphi(u)}, e^{\varphi(u)} \in \mathcal{D}(\mathbf{L}).$$

PROOF.– Using the relation

$$L_t - 1 = t\mathbf{L} + o(t), t \to 0,$$

we have:

$$\mathbf{H}\varphi(u) = \lim_{t\to 0}\frac{1}{t}(H_t - I)\varphi(u) = \lim_{t\to 0}\frac{1}{t}(H_t\varphi(u) - \varphi(u))$$

$$= \lim_{t\to 0}\frac{1}{t}(\ln L_t e^{\varphi(u)} - \ln e^{\varphi(u)}) = \lim_{t\to 0}\frac{1}{t}\ln e^{-\varphi(u)}L_t e^{\varphi(u)}$$

$$= \lim_{t\to 0}\frac{1}{t}\ln\{1+[e^{-\varphi(u)}L_t e^{\varphi(u)}-1]\} = \lim_{t\to 0}\frac{1}{t}\ln\{1+e^{-\varphi(u)}[L_t e^{\varphi(u)} - e^{\varphi(u)}]\}$$

$$=\lim_{t\to 0}\frac{1}{t}\ln\{1+e^{-\varphi(u)}[t\mathbf{L}e^{\varphi(u)}+o(t)]\}=\lim_{t\to 0}\frac{1}{t}e^{-\varphi(u)}t\mathbf{L}e^{\varphi(u)}=e^{-\varphi(u)}\mathbf{L}e^{\varphi(u)}.$$

The lemma is proved.

An important method used for studying the limit properties of Markov processes is a martingale characterization of such processes, while considering the following martingales (see, for example, Stroock and Varadhan (1979) and Ethier and Kurtz (1986)):

$$\mu_t = \varphi(x(t)) - \varphi(x(0)) - \int_0^t \mathbf{L}\varphi(x(s))ds. \qquad [1.19]$$

Exponential martingale characterization is used in the theory of large deviations (see Feng and Kurtz (2006, Chapter 1)):

$$\widetilde{\mu}_t = \exp\{\varphi(x(t)) - \varphi(x(0)) - \int_0^t \mathbf{H}\varphi(x(s))ds\} \qquad [1.20]$$

which is a martingale.

The equivalence of relations [1.19] and [1.20] follows from:

PROPOSITION 1.1.– *(see Ethier and Kurtz (1986, p. 66)) Let $x(t)$ and $y(t)$ be real-valued, right-continuous stochastic processes. Suppose that for any t, $\inf_{s\le t} x(s) > 0$. It is claimed that*

$$\mu(t) = x(t) - \int_0^t y(s)ds$$

is a local $\{\mathcal{F}_t\}$-martingale if and only if

$$\widetilde{\mu}(t) = x(t)exp\left\{-\int_0^t \frac{y(s)}{x(s)}ds\right\} \text{ is a local } \{\mathcal{F}_t\}\text{-martingale.}$$

PROOF.– According to Proposition 3.2 in Ethier and Kurtz (1986), if $M(t)$ is a local $\{\mathcal{F}_t\}$-martingale, and $v(t)$ a real-valued, continuous function with bounded variation, then $\int_0^t v(s)dM(s)$ is also a local $\{\mathcal{F}_t\}$-martingale.

First, suppose that if $\mu(t) = x(t) - \int_0^t y(s)ds$ is a local $\{\mathcal{F}_t\}$-martingale, then

$$\int_0^t \exp\left\{-\int_0^s \frac{y(u)}{x(u)}du\right\} d\mu(s) = \int_0^t \exp\left\{-\int_0^s \frac{y(u)}{x(u)}du\right\} dx(s)$$

$$-\int_0^t \exp\left\{-\int_0^s \frac{y(u)}{x(u)}du\right\} y(s)ds = x(t)\exp\left\{-\int_0^t \frac{y(u)}{x(u)}du\right\} - x(0)$$

$$-\int_0^t x(s)\left[\exp\left\{-\int_0^s \frac{y(u)}{x(u)}du\right\}\right]' ds - \int_0^t \exp\left\{-\int_0^s \frac{y(u)}{x(u)}du\right\} y(s)ds$$

$$= x(t)\exp\left\{-\int_0^t \frac{y(u)}{x(u)}du\right\} - x(0) - \int_0^t x(s)\exp\left\{-\int_0^s \frac{y(u)}{x(u)}du\right\}\left[-\frac{y(s)}{x(s)}\right] ds$$

$$-\int_0^t \exp\left\{-\int_0^s \frac{y(u)}{x(u)}du\right\} y(s)ds = x(t)\exp\left\{-\int_0^t \frac{y(u)}{x(u)}du\right\} - x(0)$$

is also a local $\{\mathcal{F}_t\}$-martingale.

In the opposite direction, if $\widetilde{\mu}(t) = x(t)exp\left\{-\int_0^t \frac{y(s)}{x(s)}ds\right\}$ is a local $\{\mathcal{F}_t\}$-martingale, then

$$\int_0^t \exp\left\{\int_0^s \frac{y(u)}{x(u)}du\right\} d\widetilde{\mu}(s) = \int_0^t \exp\left\{\int_0^s \frac{y(u)}{x(u)}du\right\}$$

$$\times \exp\left\{-\int_0^s \frac{y(u)}{x(u)}du\right\} dx(s) + \int_0^t \exp\left\{\int_0^s \frac{y(u)}{x(u)}du\right\} x(s)$$

$$\times \left[\exp\left\{-\int_0^s \frac{y(u)}{x(u)}du\right\}\right]' ds = x(t) - x(0) - \int_0^t y(s)ds$$

is also a local $\{\mathcal{F}_t\}$-martingale.

The proposition is proved.

The use of a nonlinear exponential generator in the study of the large deviations problem for Markov processes is motivated by the presence of a connection between the nonlinear semigroup corresponding to the generator and the rate functional. Here are the following considerations that demonstrate this connection.

The large deviations problem for the Markov process $x^\varepsilon(t), t \geq 0$, in the series scheme with a small parameter $\varepsilon \to 0 (\varepsilon > 0)$, is to find the rate functional $I(x)$. One possible definition $I(x)$ is as follows (see Feng and Kurtz (2006, Chapter 3)):

DEFINITION 1.6.–

$$-I(x) = \lim_{\delta\to 0} \liminf_{\varepsilon\to 0} \varepsilon \ln P\{x^\varepsilon(t) \in B_\delta(x)\}$$
$$= \lim_{\delta\to 0} \limsup_{\varepsilon\to 0} \varepsilon \ln P\{x^\varepsilon(t) \in \overline{B}_\delta(x)\},$$

where $B_\delta(x) := \{y \in E : r(x, y) < \delta\}$.

If the last ratio is true for all $x \in E$, then the *weak large deviations principle* holds true. Under the additional condition of exponential compactness (see section 1.3.1) of a Markov process $x^\varepsilon(t)$, the *large deviations principle* is verified.

PROPOSITION 1.2.– *(Bryc's formula, Bryc (1990)) (see Dembo and Zeitouni (2010, Theorems 4.3.1 and 4.4.2)). Let the Markov process* $x^\varepsilon(t), t \geq 0$, $\varepsilon \to 0$ $(\varepsilon > 0)$ *be exponentially compact in* $D_E[0, \infty)$ *and a limit*

$$\Lambda(\varphi) = \lim_{\varepsilon\to 0} \varepsilon \ln E[e^{\varphi(x^\varepsilon(t))/\varepsilon}] \tag{1.21}$$

for all $\varphi \in C_b(E)$.

Then, $x^\varepsilon(t)$ *satisfies the large deviations principle with rate functional*

$$I(x) = \sup_{\varphi\in C_b(E)} \{\varphi(x) - \Lambda(\varphi)\}.$$

PROOF.– (see Feng and Kurtz (2006, Proposition 3.8)): Let $x \in E$, and for every $\delta > 0$, we will require that $\theta_\delta > 0$ satisfies the ratio

$$\overline{B}_\delta(x) \subset \{y : \varphi(y) > \varphi(x) - \theta_\delta\}.$$

Using the continuity of φ, we can put $\lim_{\delta\to 0} \theta_\delta = 0$. Then, according to the expanded Chebyshev inequality

$$P\{x^\varepsilon(t) \in \overline{B}_\delta(x)\} \leq e^{(\theta_\delta - \varphi(x))/\varepsilon} E[e^{\varphi(x^\varepsilon(t))/\varepsilon}].$$

So, we have

$$\lim_{\delta\to 0} \liminf_{\varepsilon\to 0} \varepsilon \ln P\{x^\varepsilon(t) \in \overline{B}_\delta(x)\} \leq -\varphi(x) + \liminf_{\varepsilon\to 0} \varepsilon \ln E[e^{\varphi(x^\varepsilon(t))/\varepsilon}]$$

and

$$\lim_{\delta\to 0} \limsup_{\varepsilon\to 0} \varepsilon \ln P\{x^\varepsilon(t) \in \overline{B}_\delta(x)\} \leq -\varphi(x) + \limsup_{\varepsilon\to 0} \varepsilon \ln E[e^{\varphi(x^\varepsilon(t))/\varepsilon}].$$

Thus, for $x \in E$ and $\delta > 0$ we put $\varphi_{\delta,x} \in C_b(E)$, such that $\varphi_{\delta,x}(y) \leq \varphi_{\delta,x}(x) - r\mathbf{1}_{B_\delta^c(x)}(y)$ for all $y \in E$. Then

$$e^{-r/\varepsilon} + P\{x^\varepsilon(t) \in B_\delta(x)\} \geq e^{-\varphi_{\delta,x}(x)/\varepsilon} E[e^{\varphi_{\delta,x}(x^\varepsilon(t))/\varepsilon}],$$

namely

$$\limsup_{\varepsilon\to 0} \varepsilon \ln P\{x^\varepsilon(t) \in B_\delta(x)\} \geq -\varphi_{\delta,x}(x) + \limsup_{\varepsilon\to 0} \varepsilon \ln E[e^{\varphi_{\delta,x}(x^\varepsilon(t))/\varepsilon}]$$

and

$$\liminf_{\varepsilon\to 0} \varepsilon \ln P\{x^\varepsilon(t) \in B_\delta(x)\} \geq -\varphi_{\delta,x}(x) + \liminf_{\varepsilon\to 0} \varepsilon \ln E[e^{\varphi_{\delta,x}(x^\varepsilon(t))/\varepsilon}].$$

If [1.21] holds for all $\varphi \in C_b(E)$, then it follows from the last inequality

$$I(x) = \sup_{\varphi\in C_b(E)} \{\varphi(x) - \Lambda(\varphi)\},$$

which causes a weak large deviations principle. By applying the exponential compactness condition of the Markov process, we obtain the desired result.

The proposition is proved.

1.2.2. *Nonlinear exponential generator of large deviations*

Thus, the study of the large deviations problem is related to the boundary behavior of a nonlinear semigroup

$$H_t^\varepsilon \varphi(u) := \varepsilon \ln E e^{\varphi(x^\varepsilon(t))/\varepsilon}. \qquad [1.22]$$

The following result is also used in Feng and Kurtz (2006) and Korolyuk (2010b, 2011c).

LEMMA 1.4.– *The nonlinear exponential generator corresponding to the semigroup* [1.22] *has the form:*

$$\mathbf{H}^\varepsilon \varphi(u) = e^{-\varphi(u)/\varepsilon} \varepsilon \mathbf{L}^\varepsilon e^{\varphi(u)/\varepsilon}, e^{\varphi(u)/\varepsilon} \in \mathcal{D}(\mathbf{L}^\varepsilon).$$

PROOF.– Taking into account the relation

$$L_t^\varepsilon - 1 = t\mathbf{L}^\varepsilon + o(t), t \to 0,$$

we have:

$$\mathbf{H}^\varepsilon \varphi(u) = \lim_{t\to 0} \frac{1}{t}(H_t^\varepsilon - I)\varphi(u) = \lim_{t\to 0} \frac{1}{t}(H_t^\varepsilon \varphi(u) - \varphi(u)) = \lim_{t\to 0} \frac{1}{t}(\varepsilon \ln L_t^\varepsilon e^{\varphi(u)/\varepsilon}$$

$$-\varepsilon \ln e^{\varphi(u)/\varepsilon}) = \lim_{t\to 0} \frac{\varepsilon}{t} \ln e^{-\varphi(u)/\varepsilon} L_t^\varepsilon e^{\varphi(u)/\varepsilon} = \lim_{t\to 0} \frac{\varepsilon}{t} \ln\{1 + e^{-\varphi(u)/\varepsilon}[L_t^\varepsilon e^{\varphi(u)/\varepsilon}$$

$$-e^{\varphi(u)/\varepsilon}]\} = \lim_{t\to 0} \frac{\varepsilon}{t} e^{-\varphi(u)/\varepsilon} t\mathbf{L}^\varepsilon e^{\varphi(u)/\varepsilon} = e^{-\varphi(u)/\varepsilon} \varepsilon \mathbf{L}^\varepsilon e^{\varphi(u)/\varepsilon}.$$

The lemma is proved.

In the classical form, the solution of the large deviations problem is carried out using the cumulative process (see, for example, Freidlin and Ventzell (2012)). This approach can be used, in particular in the averaging scheme.

In the scheme of the Poisson approximation, the analysis is complicated and it is necessary to use exponential generators. The relationship between the cumulant and the exponential generator is as follows.

The Markov process generator can be represented as (e.g. Skorokhod (2008))

$$\mathbf{L}\varphi(x) = \int_{\mathbf{R}} e^{\lambda x} a(\lambda)\overline{\varphi}(\lambda)d\lambda,$$

where $a(\lambda)$ is the process cumulant, $\overline{\varphi}(\lambda) = \int_{\mathbf{R}} e^{\lambda x}\varphi(x)dx$.

The inverse transformation gives

$$\int_{\mathbf{R}} e^{-\lambda x}\mathbf{L}\varphi(x)dx = a(\lambda)\overline{\varphi}(\lambda).$$

Let us rewrite

$$\int_{\mathbf{R}} e^{-\lambda x}\mathbf{L}\varphi(x)dx = \int_{\mathbf{R}} e^{-\lambda x} a(\lambda)\varphi(x)dx,$$

and by replacement

$$e^{-\lambda x}\varphi(x) = \widetilde{\varphi}(x)$$

we obtain

$$\int_{\mathbf{R}} e^{-\lambda x}\mathbf{L}e^{\lambda x}\widetilde{\varphi}(x)dx = \int_{\mathbf{R}} a(\lambda)\widetilde{\varphi}(x)dx.$$

So,

$$e^{-\lambda x}\mathbf{L}e^{\lambda x} = a(\lambda),$$

or through an exponential generator:

$$\mathbf{H}\varphi_0(x) = a(\lambda), \quad \text{where} \quad \varphi_0(x) = \lambda x.$$

1.2.3. *Large deviations problem in the Poisson approximation scheme*

Consider the family of normalized (here $\varepsilon, \delta \to 0$, such that $\varepsilon^{-1}\delta \to 1$) Markov processes with trajectories in $D_{\mathbf{R}}[0, \infty)$ and independent increments

$$\eta_\varepsilon^\delta(t) = \varepsilon\eta^\delta(t/\varepsilon^2), t \geq 0, \qquad [1.23]$$

determined by generators

$$\widetilde{\mathbf{\Gamma}}^{\delta}_{\varepsilon}\varphi(u) = \varepsilon^{-2}\int_{\mathbf{R}}[\varphi(u+\varepsilon v) - \varphi(u)]\widetilde{\Gamma}^{\delta}(dv). \qquad [1.24]$$

Let the conditions of the Poisson approximation be satisfied:

C1: *Poisson approximation.*

PA1 Approximation of means:

$$b_\delta = \int_{\mathbf{R}} v\widetilde{\Gamma}^{\delta}(dv) = \delta[b + \theta^{\delta}_{b}],$$

and

$$c_\delta = \int_{\mathbf{R}} v^2\widetilde{\Gamma}^{\delta}(dv) = \delta[c + \theta^{\delta}_{c}],$$

where $b < +\infty, c < +\infty$.

PA2 For the intensity kernels, there is an asymptotic representation:

$$\widetilde{\Gamma}^{\delta}_{g} = \int_{\mathbf{R}} g(v)\widetilde{\Gamma}^{\delta}(dv) = \delta[\widetilde{\Gamma}_g + \theta^{\delta}_{g}]$$

for all $g \in C_3(\mathbf{R})$ – a class of functions that defines the measure (see Jacod and Shiryaev (2003)).

The kernel $\widetilde{\Gamma}^0(dv)$ is set on a class of functions, which determines the measure $C_3(\mathbf{R})$ by the relation

$$\widetilde{\Gamma}_g = \int_{\mathbf{R}} g(v)\widetilde{\Gamma}^0(dv), \quad g \in C_3(\mathbf{R}).$$

The negligibly small terms $\theta^{\delta}_{b}, \theta^{\delta}_{c}, \theta^{\delta}_{g}$ satisfy the condition

$$|\theta^{\delta}_{\cdot}| \to 0, \quad \delta \to 0.$$

PA3 A relation $c := \int_{\mathbf{R}} v^2\widetilde{\Gamma}^0(dv)$ takes place, which causes the absence of the diffusion component in the exponential generator, which determines the solution of the large deviations problem.

C2 *Uniform quadratic integrability*:

$$\lim_{c\to\infty}\int_{|v|>c} v^2\widetilde{\Gamma}^0(dv) = 0.$$

C3 *Exponential boundedness*:

$$\int_{\mathbf{R}} e^{p|v|}\widetilde{\Gamma}^{\delta}(dv) < \infty, \forall p \in \mathbf{R}.$$

The solution of the large deviations problem, for processes with independent increments in the Poisson approximation scheme, is determined by using a nonlinear exponential generator:

$$\mathbf{H}_\Gamma^{\varepsilon,\delta}\varphi^\varepsilon(u) = e^{-\varphi/\varepsilon}\varepsilon\mathbf{\Gamma}_\varepsilon^\delta e^{\varphi/\varepsilon}. \qquad [1.25]$$

LEMMA 1.5.– *The exponential generator* [1.25] *in the Poisson approximation scheme has the following asymptotic representation:*

$$\mathbf{H}_\Gamma^{\varepsilon,\delta}\varphi(u) = \mathbf{H}_\Gamma\varphi(u) + \theta_\Gamma^{\varepsilon,\delta}\varphi, \qquad [1.26]$$

where the functions $\varphi(u) \in C_c^3(\mathbf{R})$ *and* $|\theta_\Gamma^{\varepsilon,\delta}\varphi| \to 0, \varepsilon, \delta \to 0,$

$$\mathbf{H}_\Gamma\varphi(u) = b\varphi'(u) + \int_{\mathbf{R}} [e^{v\varphi'(u)} - 1 - v\varphi'(u)]\widetilde{\Gamma}^0(dv). \qquad [1.27]$$

REMARK 1.6.– *The generator can be expanded into a space of absolutely continuous functions (see Feng and Kurtz (2006))*

$$C_b^1(\mathbf{R}) = \{\varphi : \exists \lim_{|u|\to\infty} \varphi(u) = \varphi(\infty), \lim_{|u|\to\infty} \varphi'(u) = 0\}.$$

PROOF.– Considering the form of generator [1.24], for generator [1.26] we have:

$$\mathbf{H}_\Gamma^{\varepsilon,\delta}\varphi(u) = \varepsilon^{-1}\int_{\mathbf{R}} [e^{\Delta_\varepsilon\varphi(u)} - 1]\widetilde{\Gamma}^\delta(dv),$$

where

$$\Delta_\varepsilon\varphi(u) := \varepsilon^{-1}[\varphi(u+\varepsilon v) - \varphi(u)].$$

Rewrite the expression for the generator as follows:

$$\mathbf{H}_\Gamma^{\varepsilon,\delta}\varphi(u) = \varepsilon^{-1}\int_{\mathbf{R}} [e^{\Delta_\varepsilon\varphi(u)} - 1 - \Delta_\varepsilon\varphi(u) - \frac{1}{2}(\Delta_\varepsilon\varphi(u))^2]\widetilde{\Gamma}^\delta(dv)$$

$$+\varepsilon^{-1}\int_{\mathbf{R}} [\Delta_\varepsilon\varphi(u) + \frac{1}{2}(\Delta_\varepsilon\varphi(u))^2]\widetilde{\Gamma}^\delta(dv).$$

It is easy to see that the function $\psi_u^\varepsilon(v) = e^{\Delta_\varepsilon\varphi(u)} - 1 - \Delta_\varepsilon\varphi(u) - \frac{1}{2}(\Delta_\varepsilon\varphi(u))^2$ belongs to the class $C_3(\mathbf{R})$. Really,

$$\psi_u^\varepsilon(v)/v^2 \to 0, v \to 0.$$

In addition, this function is continuous and bounded for each ε by the condition of the boundedness of the function $\varphi(u)$. Moreover, the boundedness of the function $\psi_u^\varepsilon(v)$ is uniform over u under the conditions **C2, C3** and the boundedness of the derivative $\varphi'(u)$.

Thus, from conditions **PA1, PA2**, we have:

$$\mathbf{H}_\Gamma^{\varepsilon,\delta}\varphi(u) = \varepsilon^{-1}\delta\int_{\mathbf{R}}[e^{\Delta_\varepsilon\varphi(u)} - 1 - \Delta_\varepsilon\varphi(u) - \frac{1}{2}(\Delta_\varepsilon\varphi(u))^2]\widetilde{\Gamma}^0(dv)$$

$$+\varepsilon^{-1}\int_{\mathbf{R}}[\Delta_\varepsilon\varphi(u) - v\varphi'(u) - \varepsilon\frac{v^2}{2}\varphi''(u)]\widetilde{\Gamma}^\delta(dv) + \varepsilon^{-1}\delta b\varphi'(u) + \frac{1}{2}\delta c\varphi''(u)$$

$$+\varepsilon^{-1}\int_{\mathbf{R}}[\frac{1}{2}(\Delta_\varepsilon\varphi(u))^2 - \frac{v^2}{2}(\varphi'(u))^2]\widetilde{\Gamma}^\delta(dv) + \varepsilon^{-1}\delta\frac{1}{2}c(\varphi'(u))^2.$$

Applying Taylor's formula to test functions $\varphi(u) \in C_c^3(\mathbf{R})$, and condition **PA2**, we obtain:

$$\mathbf{H}_\Gamma^{\varepsilon,\delta}\varphi(u) = \varepsilon^{-1}\delta\int_{\mathbf{R}}[e^{v\varphi'(u)} - 1 - v\varphi'(u) - \frac{v^2}{2}(\varphi'(u))^2]\widetilde{\Gamma}^0(dv)$$

$$+\varepsilon^{-1}\delta\int_{\mathbf{R}}(e^{v\varphi'(u)}\varepsilon\frac{v^2}{2}\varphi''(\widetilde{u}) - \varepsilon\frac{v^2}{2}\varphi''(\widetilde{u}) - \varepsilon^2\frac{v^4}{8}(\varphi''(\widetilde{u}))^2)\widetilde{\Gamma}^0(dv)$$

$$+\varepsilon^{-1}\delta\int_{\mathbf{R}}\varepsilon^2\frac{v^3}{3!}\varphi'''(\widetilde{u})\widetilde{\Gamma}^0(dv) + \varepsilon^{-1}\delta b\varphi'(u) + \frac{1}{2}\delta c\varphi''(u)$$

$$+\varepsilon^{-1}\delta\int_{\mathbf{R}}\varepsilon^2\frac{v^4}{4}(\varphi''(\widetilde{u}))^2\widetilde{\Gamma}^0(dv) + \varepsilon^{-1}\delta\frac{1}{2}c(\varphi'(u))^2.$$

Finally, taking into account **PA3** and the limit condition $\varepsilon^{-1}\delta \to 1$, we have:

$$\mathbf{H}_\Gamma^{\varepsilon,\delta}(x)\varphi(u) = \mathbf{H}_\Gamma(x)\varphi(u) + \theta_\Gamma^{\varepsilon,\delta}\varphi,$$

where $|\theta_\Gamma^{\varepsilon,\delta}\varphi| \to 0, \varepsilon, \delta \to 0$.

The lemma is proved.

Thus, the solution of the large deviations problem for the processes defined by [1.23]–[1.24] is given by the exponential limit generator [1.27]:

$$\mathbf{H}_\Gamma\varphi(u) = b\varphi'(u) + \int_{\mathbf{R}}[e^{v\varphi'(u)} - 1 - v\varphi'(u)]\widetilde{\Gamma}^0(dv).$$

Let us now consider another normalization (here $\varepsilon, \delta \to 0$, so that $\varepsilon^{-1}\delta \to 1$) of a family of Markov processes with trajectories in $D_{\mathbf{R}}[0,\infty)$ and independent increments

$$\eta_\varepsilon^\delta(t) = \varepsilon\eta^\delta(t/\varepsilon^3), t \geq 0,$$

determined by generators

$$\mathbf{\Gamma}_\varepsilon^\delta\varphi(u) = \varepsilon^{-3}\int_{\mathbf{R}}[\varphi(u+\varepsilon v) - \varphi(u)]\widetilde{\Gamma}^\delta(dv).$$

Suppose that the conditions of the Lévy approximation hold:

C1': *Lévy approximation.*

LA1 Approximation of means:

$$b_\delta = \int_{\mathbf{R}} v\widetilde{\Gamma}^\delta(dv) = \delta b_1 + \delta^2[b + \theta_b^\delta],$$

and

$$c_\delta = \int_{\mathbf{R}} v^2\widetilde{\Gamma}^\delta(dv) = \delta^2[c + \theta_c^\delta],$$

where $b_1 < +\infty, b < +\infty, c < +\infty$.

LA2 For the intensity kernel, there is an asymptotic representation:

$$\widetilde{\Gamma}_g^\delta = \int_{\mathbf{R}} g(v)\widetilde{\Gamma}^\delta(dv) = \delta[\widetilde{\Gamma}_g + \theta_g^\delta]$$

for all $g \in C_3(\mathbf{R})$ – a class of functions that defines a measure (see Jacod and Shiryaev 2003).

The kernel $\widetilde{\Gamma}^0(dv)$ is set on the function class that defines the measure $C_3(\mathbf{R})$ by means of the relation

$$\widetilde{\Gamma}_g = \int_{\mathbf{R}} g(v)\widetilde{\Gamma}^0(dv), \quad g \in C_3(\mathbf{R}).$$

The negligibly small terms $\theta_b^\delta, \theta_c^\delta, \theta_g^\delta$ satisfy the condition

$$|\theta_\cdot^\delta| \to 0, \quad \delta \to 0.$$

The conditions **C2, C3** coincide with the case of the Poisson approximation.

LEMMA 1.6.– *The exponential generator*

$$\mathbf{H}_\Gamma^{\varepsilon,\delta}\varphi^\varepsilon(u) = e^{-\varphi/\varepsilon}\varepsilon\mathbf{\Gamma}_\varepsilon^\delta e^{\varphi/\varepsilon}$$

in the scheme of the Lévy approximation has the following asymptotic representation:

$$\mathbf{H}_\Gamma^{\varepsilon,\delta}\varphi_\varepsilon^\delta = \varepsilon^{-1}b_1\varphi'(u) + \mathbf{H}_\Gamma\varphi(u) + \theta_\Gamma^{\varepsilon,\delta}\varphi,$$

where the functions $\varphi(u) \in C_c^3(\mathbf{R})$ and

$$\mathbf{H}_\Gamma\varphi(u) = (b - b_0)\varphi'(u) + \frac{1}{2}(c - c_0)(\varphi'(u))^2 + \int_{\mathbf{R}} [e^{v\varphi'(u)} - 1]\widetilde{\Gamma}^0(dv),$$

$$b_0 := \int_{\mathbf{R}} v\widetilde{\Gamma}^0(dv), c_0 := \int_{\mathbf{R}} v^2\widetilde{\Gamma}^0(dv),$$

and the negligible term is $|\theta_\Gamma^{\varepsilon,\delta}\varphi| \to 0, \varepsilon, \delta \to 0$.

PROOF.– We have:

$$\mathbf{H}_\Gamma^{\varepsilon,\delta}\varphi(u) = \varepsilon^{-2}\int_{\mathbf{R}}[e^{\Delta_\varepsilon\varphi(u)} - 1]\widetilde{\Gamma}^\delta(dv),$$

where

$$\Delta_\varepsilon\varphi(u) := \varepsilon^{-1}[\varphi(u+\varepsilon v) - \varphi(u)].$$

Let us rewrite the expression for the generator as follows:

$$\mathbf{H}_\Gamma^{\varepsilon,\delta}\varphi(u) = \varepsilon^{-2}\int_{\mathbf{R}}[e^{\Delta_\varepsilon\varphi(u)} - 1 - \Delta_\varepsilon\varphi(u) - \frac{1}{2}(\Delta_\varepsilon\varphi(u))^2]\widetilde{\Gamma}^\delta(dv)$$

$$+\varepsilon^{-2}\int_{\mathbf{R}}[\Delta_\varepsilon\varphi(u) + \frac{1}{2}(\Delta_\varepsilon\varphi(u))^2]\widetilde{\Gamma}^\delta(dv).$$

It is easy to see that the function $\psi_u^\varepsilon(v) = e^{\Delta_\varepsilon\varphi(u)} - 1 - \Delta_\varepsilon\varphi(u) - \frac{1}{2}(\Delta_\varepsilon\varphi(u))^2$ belongs to the class $C_3(\mathbf{R})$. Indeed,

$$\psi_u^\varepsilon(v)/v^2 \to 0, v \to 0.$$

In addition, this function is continuous and limited for each ε under the condition of boundedness of the function $\varphi(u)$. Moreover, the boundedness of the function $\psi_u^\varepsilon(v)$ is uniform by u under conditions **C2'**, **C3'** and the boundedness of the derivative $\varphi'(u)$.

Thus, from the conditions **LA1, LA2**, we have:

$$\mathbf{H}_\Gamma^{\varepsilon,\delta}\varphi(u) = \varepsilon^{-2}\delta^2\int_{\mathbf{R}}[e^{\Delta_\varepsilon\varphi(u)} - 1 - \Delta_\varepsilon\varphi(u) - \frac{1}{2}(\Delta_\varepsilon\varphi(u))^2]\widetilde{\Gamma}^0(dv)$$

$$+\varepsilon^{-2}\int_{\mathbf{R}}[\Delta_\varepsilon\varphi(u) - v\varphi'(u) - \varepsilon\frac{v^2}{2}\varphi''(u)]\widetilde{\Gamma}^\delta(dv) + \varepsilon^{-2}\delta b_1\varphi'(u) + \varepsilon^{-2}\delta^2 b\varphi'(u)$$

$$+\varepsilon^{-1}\delta^2 c\varphi''(u) + \varepsilon^{-2}\int_{\mathbf{R}}[\frac{1}{2}(\Delta_\varepsilon\varphi(u))^2 - \frac{v^2}{2}(\varphi'(u))^2]\widetilde{\Gamma}^\delta(dv) + \varepsilon^{-2}\delta^2\frac{1}{2}c(\varphi'(u))^2.$$

Applying Taylor's formula to the test functions $\varphi(u) \in C_c^3(\mathbf{R})$, and the condition **LA2**, we obtain:

$$\mathbf{H}_\Gamma^{\varepsilon,\delta}\varphi(u) = \varepsilon^{-2}\delta^2\int_{\mathbf{R}}[e^{v\varphi'(u)} - 1 - v\varphi'(u) - \frac{v^2}{2}(\varphi'(u))^2]\widetilde{\Gamma}^0(dv)$$

$$+\varepsilon^{-2}\delta^2\int_{\mathbf{R}}(e^{v\varphi'(u)}\varepsilon\frac{v^2}{2}\varphi''(\widetilde{u}) - \varepsilon\frac{v^2}{2}\varphi''(\widetilde{u}) - \varepsilon^2\frac{v^4}{8}(\varphi''(\widetilde{u}))^2)\widetilde{\Gamma}^0(dv)$$

$$+\varepsilon^{-2}\delta^2 \int_{\mathbf{R}} \varepsilon^2 \frac{v^3}{3!}\varphi'''(\widetilde{u})\widetilde{\Gamma}^0(dv) + \varepsilon^{-2}\delta b_1\varphi'(u) + \varepsilon^{-2}\delta^2 b\varphi'(u) + \varepsilon^{-1}\delta^2 c\varphi''(u)$$

$$+\varepsilon^{-2}\delta^2 \int_{\mathbf{R}} \varepsilon^2 \frac{v^4}{4}(\varphi''(\widetilde{u}))^2\widetilde{\Gamma}^0(dv) + \varepsilon^{-2}\delta^2 \frac{1}{2}c(\varphi'(u))^2.$$

Finally, considering the limit condition $\varepsilon^{-1}\delta \to 1$, we have:

$$\mathbf{H}_\Gamma^{\varepsilon,\delta}(x)\varphi(u) = \varepsilon^{-1}b_1\varphi'(u) + \mathbf{H}_\Gamma(x)\varphi(u) + \theta_\Gamma^{\varepsilon,\delta}\varphi,$$

where $|\theta_\Gamma^{\varepsilon,\delta}\varphi| \to 0, \varepsilon, \delta \to 0$.

The Lemma is proved.

1.2.4. *Nisio semigroup and control problem*

The item 4) of the program for solving the large deviations problem (Feng and Kurtz 2006) consists of the construction of a variational representation of the rate functional. This construction follows from Bryc's formula (see Proposition 1.2) and is closely related to the martingale control problem.

We demonstrate how namely the nonlinear semigroup [1.18] is connected with the corresponding nonlinear operator and with the martingale control problem. To do this, we first establish a connection between the nonlinear semigroup [1.18] and the Nisio semigroup Nisio (1976a, 1976b).

In the following, for a probabilistic measure P, given on a space E, we denote

$$E^P f(x) := \int_E f(x)dP(x).$$

If for a probabilistic measure Q, absolutely continuous with respect to another probability measure P, the following relation takes place:

$$\frac{dQ}{dP} = e^{\varphi(u)}/C, C = E^P e^{\varphi(u)},$$

then we have

$$\ln \frac{dQ}{dP} = \varphi(u) - \ln E^P e^{\varphi(u)},$$

where we get the identity

$$\ln E^P e^{\varphi(u)} = \varphi(u) - \ln \frac{dQ}{dP}.$$

Consider a nonlinear semigroup [1.18] constructed by a Markov process $x(t)$:

$$H_t\varphi(u) = \ln E^P[e^{\varphi(x(t))}|x(0) = u].$$

We apply the following result.

PROPOSITION 1.3.– *(see Dupuis and Ellis (1997, Proposition 1.4.2)) Let E be a measurable space, $P(E)$ a family of probabilistic measures in the space E and $\varphi(x)$ a bounded measurable function $\varphi : E \to \mathbf{R}$.*

Then, a variational formula takes place

$$\ln E^P e^{\varphi(x(t))} = \sup_{Q \in P(E)} E^Q \left\{ \varphi(x(t)) - \ln \frac{dQ}{dP}(x(\cdot)) \right\}. \qquad [1.28]$$

The supremum is reached on a probabilistic measure Q, which is absolutely continuous with respect to measure P and satisfies the condition

$$\frac{dQ}{dP}(x) = e^{\varphi(x)} \frac{1}{\int_E e^{\varphi(x)} dP}.$$

The right-hand side of equation [1.28] can be interpreted as a nonlinear Nisio semigroup. This semigroup is used in the solution of the martingale control problems in Nisio (1976a, b). To construct the corresponding martingale with control, we will use the following lemma.

LEMMA 1.7.– *(see Feng and Kurtz (2006, Lemma 5.8)) Let $x(t) \in E$ be the solution of martingale problem for generator*

$$\mathbf{L}\varphi(x) = q(x) \int_E (\varphi(y) - \varphi(x)) P(x, dy)$$

distribution $P \in \mathcal{P}(D_E[0, \infty))$, and let $\mathbf{H}g(x) = e^{-g(x)} \mathbf{L} e^{g(x)}$.

For any $T > 0$, let us denote the measure $Q \in \mathcal{P}(D_E[0, \infty))$ as follows:

$$\frac{dQ}{dP}(x(\cdot)) = \exp \left\{ g(x(t)) - g(x(0)) - \int_0^T \mathbf{H}g(x(s)) ds \right\}. \qquad [1.29]$$

Then, the Markov process $\{x(t), 0 \leq t \leq T\}$, with respect to the measure Q, is a solution of the martingale problem for $\mathbf{L}^g$, where

$$\begin{aligned} \mathbf{L}^g f(x) &= e^{-g(x)} \mathbf{L}(\varphi(x) e^{g(x)}) - e^{-g(x)} \varphi(x) \mathbf{L} e^{g(x)} \\ &= q(x) \int_E e^{g(y) - g(x)} (\varphi(y) - \varphi(x)) P(x, dy). \qquad [1.30] \end{aligned}$$

The last assertion follows from a more general result (Girsanov transformation) after substitution $h(x) = e^{g(x)}, g(x) = \ln h(x)$:

LEMMA 1.8.– *(see Feng (1999, Lemma 3.6)) Let $x(t)$ be a solution of the martingale problem for generator $\mathbf{L}$ in the space $C_E[0;T]$, with distribution P (and filtration $\mathfrak{F}_t$).*

Let $h(x) \in B(E) \cap D(\mathbf{L}), h(x) \geq \varepsilon > 0$. Then

$$R(t) = \frac{h(x(t))}{h(x(0))} \exp\left\{-\int_0^t \frac{Lh}{h}(x(s))ds\right\}$$

and $\mathcal{F}_t$ is exponential martingale (with a mathematical expectation equal to 1) by the distribution P.

Let us define another measure $Q \in \mathcal{P}(C_E[0;T])$ by $dQ/dP = R(T)$. We will consider $\mathcal{D}(\mathbf{L})$ closed with respect to multiplication by $h(x)$. Then, with respect to the measure Q, $x(t)$ is the solution of the martingale problem for the generator $\mathbf{L}^{\ln(h)}$ in the space $C_E[0;T]$.

PROOF.– Let us denote a martingale

$$\mu_t^\varphi = \varphi(x(t)) - \varphi(x(0)) - \int_0^t \mathbf{L}\varphi(x(s))ds.$$

Since the measures P, Q are equivalent at any finite time interval, and $R(t)$ is bounded, then by using the Girsanov theorem (see Dellacherie and Meyer (1978, p. 238, Theorem 49), or Protter (1990, p. 134, Theorem 37)), we have:

$$\varphi(x(t))) - \varphi(x(0)) - \int_0^t \mathbf{L}\varphi(x(s))ds - \int_0^t \frac{1}{R(s)} d\langle R, \mu^\varphi\rangle(s) \qquad [1.31]$$

as a Q-martingale.

To calculate $\langle R, \mu^\varphi\rangle(s)$, let us integrate in parts

$$\int_0^t \frac{R(s-)}{h(x(s-))} dh(x(s)) = R(t) - 1 - \int_0^t h(x(s))d\frac{R(s-)}{h(x(s-))}$$

$$= R(t) - 1 - \int_0^t \frac{h(x(s))}{h(x(0))} \exp\left\{-\int_0^s \frac{Lh}{h}(x(\theta))d\theta\right\} \left(-\frac{Lh}{h}(x(s))\right) ds$$

$$= R(t) - 1 + \int_0^t R(s-)\frac{Lh}{h}(x(s))ds.$$

So, we see that: $R(t) = 1 + \int_0^t \frac{R(s-)}{h(x(s-))} dh(x(s)) - \int_0^t R(s-)\frac{Lh}{h}(x(s))ds = 1 + \int_0^t \frac{R(s-)}{h(x(s-))} d\mu^h(s)$.

It follows

$$\langle R, \mu^\varphi \rangle(t) = 0 + \int_0^t \frac{R(s-)}{h(x(s-))} d\langle \mu^\varphi(s), \mu^h(s) \rangle. \qquad [1.32]$$

We use the following theorem:

THEOREM 1.1.– *(see Fleming and Souganidis (1986, Theorem 1.1)) Let a stochastic process $x(t)$ in $C_E[0;T]$ be a solution of the martingale problem for the generator $\mathbf{L}$ and the operator's graph $\mathcal{G}(\mathbf{L}) \in B(E) \times B(E)$.*

Let us denote a martingale

$$\mu_t^\varphi = \varphi(x(t)) - \varphi(x(0)) - \int_0^t \mathbf{L}\varphi(x(s))ds,$$

for each $\varphi \in \mathcal{D}(\mathbf{L})$. Then, the predictable quadratic covariance μ_t^φ and μ_t^h for $\varphi, h, \varphi h \in \mathcal{D}(\mathbf{L})$ has the form

$$\langle \mu^\varphi, \mu^h \rangle = \int_0^t \langle \varphi, h \rangle_{\mathbf{L}}(x(s))ds,$$

where $\langle \varphi, h \rangle_{\mathbf{L}} := \mathbf{L}(\varphi h) - \varphi \mathbf{L}h - h\mathbf{L}\varphi$.

REMARK 1.7.– *In the case $\varphi = h$, we have a well-known fact (see, for example, Liptser and Shiryayev (1989); Skorokhod* et al. *(2002)) relative to the quadratic characteristic of the martingale μ_t^φ:*

$$\langle \mu^\varphi \rangle_t = \int_0^t [\mathbf{L}\varphi^2(x(s)) - 2\varphi(x(s))\mathbf{L}\varphi(x(s))]ds, t \geq 0.$$

Thus, we can rewrite [1.32]:

$$\langle R, \mu^\varphi \rangle(t) = \int_0^t \frac{R(s-)}{h(x(s-))} \langle \varphi, h \rangle_{\mathbf{L}}(x(s))ds.$$

And therefore, from [1.31], we obtain the Q-martingale:

$$\varphi(x(t))) - \varphi(x(0)) - \int_0^t \left\{ \mathbf{L}\varphi(x(s)) + \frac{\langle \varphi, h \rangle_{\mathbf{L}}}{h}(x(s)) \right\} ds$$

$$= \varphi(x(t))) - \varphi(x(0)) - \int_0^t \left\{ h^{-1}\mathbf{L}(fh) - h^{-1}f\mathbf{L}h \right\} ds$$

$$= \varphi(x(t))) - \varphi(x(0)) - \int_0^t \mathbf{L}^{\ln(h)}\varphi(x(s))ds.$$

Lemma 1.8 is proved.

From Lemma 1.8 follows Lemma 1.7; hence, by the measure Q, the Markov process $\{x(t), 0 \le t \le T\}$ is a solution of the martingale problem for $\mathbf{L}^g$, namely

$$\varphi(x(t)) - \varphi(x(0)) - \int_0^t \mathbf{L}^g \varphi(x(s))ds \qquad [1.33]$$

is a martingale with control $g(x)$ regarding the measure $Q \in \mathcal{P}(D_E[0,\infty))$.

Now, we show that the nonlinear Nisio semigroup corresponds to the nonlinear operator associated with the control problem.

According to Kurtz (1987, Theorem 3.1), there exists a probability distribution $Q_{t,x}$ on $\mathcal{P}(D_E[0,\infty))$, such that the Nisio semigroup [1.28] has the form

$$H_t\varphi(x) = E^{Q_{t,x}} \left\{ \varphi(x(t)) - \ln \frac{dQ_{t,x}}{dP}(x(\cdot)) \right\}.$$

Substituting as the measure Q in Lemma 1.7, the distribution $Q_{t,x}$ (given that [1.33] is martingale), we obtain:

$$H_t\varphi(x) = E^{Q_{t,x}} \left\{ \int_0^t \mathbf{L}^g \varphi(x(s))ds - \ln \frac{dQ_{t,x}}{dP}(x(\cdot)) \right\} + \varphi(x(0)). \qquad [1.34]$$

Let us choose a function $A^g(x)$, such that on the right-hand side of [1.29], we have

$$g(x(t)) - g(x(0)) - \int_0^t \mathbf{H}g(x(s))ds =: \int_0^t A^g(x(s))ds. \qquad [1.35]$$

Note that $\int_0^t A^g(x(s))ds$ is the functional for which the problem of minimization is solved in Kurtz (1987).

Let $R_{x,g}$ be a distribution of a solution of the martingale problem with $g(x) = g, x(0) = x$, then from the inequalities

$$E^{Q_{t,x}} \left\{ \int_0^t \inf_g (\mathbf{L}^g \varphi(x(s)) - A^g(x(s))ds) \right\} \le H(t)\varphi(x) - \varphi(x(0)) \le$$

$$\le \inf_g E^{R_{x,g}} \left\{ \int_0^t \mathbf{L}^g \varphi(x(s)) - A^g(x(s))ds \right\}$$

we can claim convergence

$$\lim_{t\to 0} t^{-1}(H_t - I)\varphi(x) = \mathbf{H}\varphi(x) = \inf_g (\mathbf{L}^g\varphi(x) - A^g(x)) \qquad [1.36]$$

uniformly on E under the conditions (i–vii) of Kurtz (1987), namely:

i) $D(\mathbf{L})$ is dense in $C_0(E)$ – a space of continuous functions, equal to 0 at infinity;

ii) for any $g \in G$ – a locally compact separable metric space and $\varphi \in D(\mathbf{L})$, we have $\mathbf{L}^g\varphi \in C_0(E)$;

iii) for any $\varphi \in D(\mathbf{L})$ and any compact $K \subset G$, we have in $E^\Delta = E \cup \{\Delta\}$ – a single-point compaction of locally compact space E:

$$\lim_{x \to \Delta} \sup_{g \in K} \mathbf{L}^g\varphi(x) = 0;$$

iv) for any $g \in G$ and $\varphi(x) = \sup_z \varphi(z) > 0$, we have $\mathbf{L}^g\varphi(x) \le 0$;

v) for any $\varphi \in D(\mathbf{L})$, there exist non-negative a, b and $\eta \in (0, 1)$, such that

$$|\mathbf{L}^g\varphi(x)| \le a + b(A^g(x))^\eta;$$

vi) if G is not compact, then there exists a strictly positive function $\Psi \in C_0(E)$, such that

$$1/\Psi(g) \le a + bA^g(x), x \in E, g \in G.$$

EXAMPLE 1.3.– *(see Kurtz (1987)) If* $E = G = \mathbf{R}^d$, $D(\mathbf{L}) = C_c^2(\mathbf{R}^d)$,

$$\mathbf{L}^g\varphi(x) = \sum_{i=1}^{d} \frac{\partial^2\varphi(x)}{\partial x_i^2} + g\partial\varphi(x),$$

and $A^g(x) = |x|^2 + |g|^2$, *then we can put* $\eta = 1/2$, $\Psi(u) = (1+|u|^2)^{-1}$, $A^g(\Delta) = \infty$;

vii) for any $g \in G$, $A^g(\cdot)$ *is limited and continuous.*

Note that conditions v) and vi) guarantee that the infimum is reached, and condition vii) guarantees the continuity of the limit generator.

Thus, conditions (i–vii) guarantee the convergence of nonlinear semigroups to the limiting nonlinear generator $\mathbf{H}$. *If we can find a generator with control* $\mathbf{L}^g\varphi(x)$ *and a function with control* $A^g(x)$, *which satisfy the conditions (i–vii), it is possible to construct the rate functional based on the following considerations (see Feng and Kurtz (2006, Theorems 5.15 and 8.14)).*

Therefore, let a limit exist for Nisio semigroups corresponding to pre-limit processes in the series scheme at $\varepsilon \to 0$

$$H_t^\varepsilon\varphi(u) := \varepsilon \ln Ee^{\varphi(x^\varepsilon(t))/\varepsilon}$$

and it is equal to H_t.

Let us construct a martingale control problem (see also [1.33]*)*

$$\int_{[0,t]} |\mathbf{L}^g\varphi(x(s))|ds < \infty, \forall g,$$

$$\varphi(x(t)) - \varphi(x(0)) - \int_{[0,t]} \mathbf{L}^g\varphi(x(s))ds = 0, \qquad [1.37]$$

which corresponds to the Nisio semigroup H_t, denoted in [1.34]. *According to* [1.28], *it can also be represented as*

$$H_t\varphi(u) = \ln E^P e^{\varphi(x(t))}$$

for the limit $H_t^\varepsilon\varphi(u)$.

PROPOSITION 1.4.– *(see Feng and Kurtz (2006, Theorem 8.14)) The rate functional for the pre-limit process $x^\varepsilon(t)$ has the form*

$$I(x) = I_0(x(0)) + \inf_{g:(x,g)\in\mathcal{T}} \int_{[0,\infty)} A^g(x(s))ds,$$

where $\mathcal{T}$ is the set of solutions of the martingale problem [1.37].

PROOF.– Following on from Proposition 1.2 and relations [1.28], [1.29] and [1.35].

When applying this method of limit transition, we do not deal with semigroups, but with pre-limit nonlinear exponential generators $\mathbf{H}^\varepsilon\varphi(u)$, which correspond to the Markov process $x^\varepsilon(t)$. Therefore, by having the limit generator $\mathbf{H}\varphi(x)$, we can use a generator with control $\mathbf{L}^g\varphi(x)$ and ratio [1.36] to find the function $A^g(x)$, and then use it to build the rate functional. It is clear that this approach raises questions about the convergence of semigroups to the semigroup of the limit generator and about the exponential compactness of the pre-limit process. These problems are solved on the above-mentioned stages 2) and 3), which are studied in Feng and Kurtz (2006). The following paragraph is devoted to the discussion of these issues.

We illustrate the approach described above with examples.

EXAMPLE 1.4.– *In Feng and Kurtz (2006, Example 1.4), the large deviations problem is solved for the Markov process in $\mathbf{R}^d$, which satisfies the Ito equation (see Freidlin and Ventzell (2012, Chapters 3 and 4))*

$$x^\varepsilon(t) = x + \int_0^t b(x^\varepsilon(s))ds + \sqrt{\varepsilon}\int_0^t \sigma(x^\varepsilon(s))dw(s).$$

Here is the same example with some additional explanations.

Let us put $a(x) = \sigma(x)\sigma^T(x)$, *then the linear generator has the form*

$$\mathbf{L}^{\varepsilon}\varphi(x) = \frac{\varepsilon}{2}\sum_{ij} a_{ij}(x)\partial_i\partial_j\varphi(x) + \sum_i b_i(x)\partial_i\varphi(x),$$

where we can take as $D(\mathbf{L}^{\varepsilon})$, *a set of functions* $c+\varphi, c \in \mathbf{R}, \varphi \in C_c^2(\mathbf{R}^d)$ – *the space of twice differentiating functions with compact support. It is easy to calculate*

$$\mathbf{H}^{\varepsilon}\varphi(x) = e^{-\varphi/\varepsilon}\varepsilon\mathbf{L}^{\varepsilon}e^{\varphi/\varepsilon}$$

$$= \frac{\varepsilon}{2}\sum_{ij} a_{ij}(x)\partial_i\partial_j\varphi(x) + \frac{1}{2}\sum_{ij} a_{ij}(x)\partial_i\varphi(x)\partial_j\varphi(x) + \sum_i b_i(x)\partial_i\varphi(x).$$

Accordingly,

$$\mathbf{H}\varphi(x) = \lim_{\varepsilon\to 0}\mathbf{H}^{\varepsilon}\varphi(x) = \frac{1}{2}(\nabla\varphi(x))^T a(x)\nabla\varphi(x) + b(x)\nabla\varphi(x). \qquad [1.38]$$

The classical approach to the search for the rate functional is carried out using a variational representation of the generator $\mathbf{H}$. *We introduce two functions on* $\mathbf{R}^d \times \mathbf{R}^d$:

$$H(x,p) := \frac{1}{2}p^T a(x)p + b(x)p = \frac{1}{2}|\sigma^T(x)p|^2 + b(x)p,$$

$$A(x,q) = \sup_{p\in\mathbf{R}^d}\{qp - H(x,p)\}.$$

The Legendre inverse transformation gives:

$$H(x,p) = \sup_{q\in\mathbf{R}^d}\{qp - A(x,q)\}. \qquad [1.39]$$

From the point of view of the generalized approach described above, this means that

$$\mathbf{H}\varphi(x) = H(x,\nabla\varphi(x)) = \sup_{g\in\mathbf{R}^d}\{\mathbf{L}^g\varphi(x) - A^g(x)\}, \qquad [1.40]$$

where

$$\mathbf{L}^g\varphi(x) := g\nabla\varphi(x),$$

and by denoting $g = q, \nabla\varphi(x) = p$, *we obtain* [1.39].

The inverse transformation has the form ($\mathbf{L}^g\varphi(x) = L^g(\nabla\varphi(x)) = L^q(p)$):

$$A^q(x) = \sup_{p\in\mathbf{R}^d}\{L^q(p) - H(x,p)\}.$$

Let us first calculate the explicit form of the function $A^q(x)$ *in the space* $\mathbf{R}$*:*

$$(qp - H(x,p))'_p = (qp - \frac{1}{2}(\sigma(x)p)^2 - b(x)p)'_p = q - \sigma^2(x)p - b(x) = 0,$$

which implies

$$p(q) = (q - b(x))/\sigma^2(x).$$

So,

$$A^q(x) = qp(q) - \frac{1}{2}\sigma^2(x)p^2(q) - b(x)p(q) = (q - b(x))^2/2\sigma^2(x).$$

In the case of space $\mathbf{R}^d$*, we have the ratio in the matrix form:*

$$((q,p) - H(x,p))'_p = q - a(x)p - b(x) = 0,$$

whence under the condition of non-degeneracy of the matrix $a(x)$

$$p(q) = a^{-1}(x)(q - b(x))^T,$$

therefore

$$A^q(x) = (q - b(x))p(q) - \frac{1}{2}(\sigma(x)p(q))^2 = \frac{1}{2}(q - b(x))a^{-1}(x)(q - b(x))^T.$$

The corresponding martingale control problem for generator $\mathbf{L}^q\varphi(x) = q\nabla\varphi(x)$ *has the following form:*

$$\varphi(x(t)) - \varphi(x(0)) = \int_0^t q\nabla\varphi(x(s))ds,$$

which implies

$$q = x'(t).$$

Thus, the classical variant of solving the large deviations problem leads to the following type of rate functional:

$$I(x) = I_0(x(0)) + \inf_{q\in T}\int_0^\infty A^q(x(s))ds$$

which in $\mathbf{R}$*, has the form*

$$I(x) = I_0(x(0)) + \int_0^\infty (x'(s) - b(x(s)))^2/2\sigma^2(x(s))ds,$$

instead, in $\mathbf{R}^d$*, we can write*

$$I(x) = I_0(x(0)) + \frac{1}{2}\int_0^\infty (x'(s) - b(x(s)))a^{-1}(x)(x'(s) - b(x(s))^T ds.$$

EXAMPLE 1.5.– *In the conditions of the previous example, we will consider an other generator with control*

$$\mathbf{L}^g\varphi(x) = L^g(\nabla\varphi(x)) = g(\sigma^T(x)\nabla\varphi(x)) + b(x)\nabla\varphi(x), \varphi \in C_0^2(\mathbf{R}^d).$$

Again, denote $g = q, \nabla\varphi(x) = p$, *then*

$$L^q(p) = q(\sigma^T(x)p) + b(x)p,$$

and to find the function $A^q(x)$ *in* $\mathbf{R}$, *it can be calculated:*

$$(L^q(p) - H(x,p))'_p = (q\sigma(x)p + b(x)p - \frac{1}{2}(\sigma(x)p)^2 - b(x)p)'_p$$
$$= q\sigma(x) - \sigma^2(x)p = 0,$$

from here we have

$$p(q) = q/\sigma(x).$$

Thus,

$$A^q(x) = q\sigma(x)p(q) + b(x)p(q) - \frac{1}{2}(\sigma(x)p(q))^2 - b(x)p(q) = q^2/2.$$

In the case of space $\mathbf{R}^d$, *the function* $A^q(x)$ *also has the explicit form*

$$A^q(x) = q^2/2,$$

that results from formula [1.40], *and after substituting* $A^g(x) = g^2/2$, *we obtain:*

$$(\mathbf{L}^g\varphi(x) - A^g)'_g = \sigma^T(x)\nabla\varphi(x) - g = 0,$$

whence it follows

$$g = \sigma^T(x)\nabla\varphi(x),$$

and therefore the operator has the form [1.38].

The corresponding martingale control problem for the operator $\mathbf{L}^q\varphi(x) = q(s)\sigma^T(x)\nabla\varphi(x) + b(x)\nabla\varphi(x)$ *has the form:*

$$\varphi(x(t)) - \varphi(x(0)) = \int_0^t [q(s)\sigma^T(x(s))\nabla\varphi(x(s)) + b(x(s))\nabla\varphi(x(s))]ds,$$

$$x'(t)\nabla\varphi(x(t)) = q(t)\sigma^T(x(t))\nabla\varphi(x(t)) + b(x(t))\nabla\varphi(x(t)),$$

thus we have:

$$x(t) = x(0) + \int_0^t b(x(s))ds + \int_0^t q(s)\sigma^T(x(s))ds, \qquad [1.41]$$

or in $\mathbf{R}$*:*

$$q(s) = (x'(s) - b(x(s)))/\sigma(x(s)).$$

The rate functional

$$I(x) = I_0(x(0)) + \inf_{q \in \mathcal{T}} \int_0^\infty A^q(x(s))ds$$

in $\mathbf{R}$*, again has the explicit form*

$$I(x) = I_0(x(0)) + \int_0^\infty (x'(s) - b(x(s)))^2/2\sigma^2(x(s))ds,$$

and in $\mathbf{R}^d$*, we have:*

$$I(x) = I_0(x(0)) + \frac{1}{2}\inf_q \int_0^\infty |q(s)|^2 ds,$$

where inf *is being taken by all possible* $q(s)$*, for which equation* [1.41] *has solutions.*

EXAMPLE 1.6.– *Asymptotically small diffusion* $\zeta^\varepsilon(t) = \sqrt{\varepsilon}\sigma w(t), t \geq 0, \varepsilon \to 0(\varepsilon > 0)$ *is determined by the generator*

$$\mathbf{L}^\varepsilon\varphi(u) = \varepsilon\frac{1}{2}\sigma^2\varphi''(u).$$

The exponential operator of asymptotically small diffusion is easily calculated:

$$\mathbf{H}^\varepsilon\varphi(u) = e^{-\varphi/\varepsilon}\varepsilon\mathbf{L}^\varepsilon e^{\varphi/\varepsilon} = \frac{1}{2}\sigma^2[\varphi'(u)]^2 + \varepsilon\sigma^2\varphi''(u).$$

Therefore, the limiting exponential operator of asymptotically small diffusion is determined by the equality

$$\mathbf{H}\varphi(u) = \frac{1}{2}\sigma^2[\varphi'(u)]^2.$$

Now, we calculate the generator with control using formula [1.30]*:*

$$\mathbf{L}_g^\varepsilon\varphi(u) = e^{-g/\varepsilon}\mathbf{L}^\varepsilon\varphi e^{g/\varepsilon} - e^{-g/\varepsilon}\varphi\mathbf{L}^\varepsilon e^{g/\varepsilon} = \frac{1}{2}\sigma^2\varepsilon e^{-g/\varepsilon}\Bigg(\varphi'' e^{g/\varepsilon}$$

$$+2\frac{\varphi' g'}{\varepsilon}e^{g/\varepsilon} + \frac{\varphi g''}{\varepsilon}e^{g/\varepsilon} + \varphi\left(\frac{g'}{\varepsilon}\right)^2 e^{g/\varepsilon} - \frac{\varphi g''}{\varepsilon}e^{g/\varepsilon} - \varphi\left(\frac{g'}{\varepsilon}\right)^2 e^{g/\varepsilon}\Bigg)$$

$$= \frac{1}{2}\sigma^2(\varepsilon\varphi'' + 2\varphi' g').$$

Therefore, the limit generator with control has the form:

$$\mathbf{L}^g\varphi(u) = \sigma^2\varphi' g'.$$

Let us denote $\varphi' = p, \sigma^2 g' = q$ *and calculate the function* $A(q)$ *by the formula* $(L(p,q) := \mathbf{L}_g\varphi(u))$*:*

$$A(q) = \sup_p(L(p,q) - H(p)) = \sup_p(qp - H(p)).$$

We have:

$$(qp - H(p))'_p = (qp - \frac{1}{2}\sigma^2 p^2)'_p = q - \sigma^2 p = 0,$$

i.e. $p = \sigma^{-2}q$*, and therefore, we finally obtain:*

$$A(q) = \sigma^{-2}(q^2 - \frac{1}{2}q^2) = \frac{1}{2}\sigma^{-2}q^2.$$

The rate functional has the form:

$$I(x) = I_0(x(0)) + \int_0^\infty A(x'(s))ds = I_0(x(0)) + \frac{1}{2}\sigma^{-2}\int_0^\infty (x'(s))^2 ds.$$

EXAMPLE 1.7.– *Let us return to the normalized Markov processes with independent increments*

$$\eta_\varepsilon^\delta(t) = \varepsilon\eta^\delta(t/\varepsilon^2), t \geq 0,$$

considered in section 1.2.3.

Recall that under the conditions of the Poisson approximation, the solution of the large deviations problem for such processes is determined using a nonlinear exponential generator [1.27]*:*

$$\mathbf{H}_\Gamma\varphi(u) = a\varphi'(u) + \int_{\mathbf{R}} [e^{v\varphi'(u)} - 1 - v\varphi'(u)]\widetilde{\Gamma}^0(dv).$$

Let us calculate a generator of the process with independent increments and control g using formula [1.30]*:*

$$\mathbf{\Gamma}_g^{\varepsilon,\delta}\varphi(u) = e^{-g/\varepsilon}\mathbf{\Gamma}_\varepsilon^\delta\varphi e^{g/\varepsilon} - e^{-g/\varepsilon}\varphi\mathbf{\Gamma}_\varepsilon^\delta e^{g/\varepsilon} = e^{-g/\varepsilon}\varepsilon^{-2}\int_{\mathbf{R}}[\varphi(u+\varepsilon v)e^{g(u+\varepsilon v)/\varepsilon}$$

$$-\varphi(u)e^{g(u)/\varepsilon} - \varphi(u)e^{g(u+\varepsilon v)/\varepsilon} + \varphi(u)e^{g(u)/\varepsilon}]\widetilde{\Gamma}^\delta(dv) = \varepsilon^{-2}\int_{\mathbf{R}} e^{\frac{g(u+\varepsilon v)-g(u)}{\varepsilon}}[\varphi(u$$

$$+\varepsilon v) - \varphi(u)]\widetilde{\Gamma}^\delta(dv) = \varepsilon^{-2}\int_{\mathbf{R}} e^{\frac{g(u+\varepsilon v)-g(u)}{\varepsilon}}[\varphi(u+\varepsilon v) - \varphi(u) - \varepsilon v\varphi'(u)$$

$$-\varepsilon^2\frac{v^2}{2}\varphi''(u)]\widetilde{\Gamma}^\delta(dv) + \varepsilon^{-2}\int_{\mathbf{R}} e^{\frac{g(u+\varepsilon v)-g(u)}{\varepsilon}}[\varepsilon v\varphi'(u) + \varepsilon^2\frac{v^2}{2}\varphi''(u)]\widetilde{\Gamma}^\delta(dv).$$

*Since the subintegral function in the first integral belongs to the class $C_3(\mathbf{R})$, we can apply the condition of the Poisson approximation **PA2**:*

$$\mathbf{\Gamma}_g^{\varepsilon,\delta}\varphi(u) = \varepsilon^{-2}\delta\int_{\mathbf{R}} e^{\frac{g(u+\varepsilon v)-g(u)}{\varepsilon}}[\varphi(u+\varepsilon v)-\varphi(u)-\varepsilon v\varphi'(u)-\varepsilon^2\frac{v^2}{2}\varphi''(u)]\widetilde{\Gamma}^0(dv)$$

$$+\varepsilon^{-2}\int_{\mathbf{R}}[e^{vg'(u)} + \varepsilon e^{vg'(u)}\frac{v^2}{2}g''(\theta)][\varepsilon v\varphi'(u) + \varepsilon^2\frac{v^2}{2}\varphi''(u)]\widetilde{\Gamma}^\delta(dv).$$

The subintegral function in the first integral has order ε^3, so when $\varepsilon, \delta \to 0, \varepsilon^{-1}\delta \to 1$, the first integral is $o(\varepsilon, \delta)$. In the second integral, we have:

$$\mathbf{\Gamma}_g^{\varepsilon,\delta}\varphi(u) = \varepsilon^{-1}\varphi'(u)\int_{\mathbf{R}} ve^{vg'(u)}\widetilde{\Gamma}^\delta(dv) + \int_{\mathbf{R}}[\frac{v^2}{2}\varphi''(u) + o(\varepsilon)]\widetilde{\Gamma}^\delta(dv) + o(\varepsilon, \delta).$$

*Again, using the Poisson approximation conditions **PA1, PA2**, we get:*

$$\mathbf{\Gamma}_g^{\varepsilon,\delta}\varphi(u) = \varepsilon^{-1}\delta\varphi'(u)\int_{\mathbf{R}} ve^{vg'}\widetilde{\Gamma}^0(dv) + o(\varepsilon, \delta).$$

Hence, denoting $z = g'$ and putting $\psi(z) = \int_{\mathbf{R}} e^{vz}\widetilde{\Gamma}^0(dv)$, we have a limit generator with control z in the form:

$$\mathbf{\Gamma}_z\varphi(u) = \varphi'(u)\psi'(z).$$

Now put $\varphi' = p, \psi' = q$ and calculate the function $L(q)$ by the formula $(A(p,q) := \mathbf{\Gamma}_z\varphi(u))$:

$$L(q) = \sup_p(A(p,q) - H(p)) = \sup_p(qp - H(p)),$$

where $H(p)$ is determined by the explicit form of the generator [1.27]*:*

$$H(p) := H_\Gamma(\varphi') = ap - \int_{\mathbf{R}}[e^{vp} - 1 - vp]\widetilde{\Gamma}^0(dv).$$

We have:

$$(qp - H(p))'_p = (qp - ap - \int_{\mathbf{R}}[e^{vp} - 1 - vp]\widetilde{\Gamma}^0(dv))'_p$$

$$= q - a - \int_{\mathbf{R}}[ve^{vp} - v]\widetilde{\Gamma}^0(dv) = 0,$$

i.e. p is determined from the relation

$$\int_{\mathbf{R}} ve^{vp}\widetilde{\Gamma}^0(dv) = q - (a - a_0) < \infty.$$

The explicit expression of the function $L(q)$ can be obtained by substituting the value of $p(q)$ in the formula

$$L(q) = qp - H(p).$$

The rate functional will thus have the following form:

$$I(x) = I_0(x(0)) + \int_0^\infty L(x'(s))ds.$$

1.3. Compactness and comparison principle

1.3.1. *Compactness and exponential compactness*

Proving the weak convergence of different types of processes and solving the large deviations problem are associated with proving the compactness, or exponential compactness of the respective processes. Note that all of the studied processes are considered in the space $D_{\mathbf{R}^d}[0,\infty)$, with Skorokhod topology J_1.

DEFINITION 1.7.– *A family of processes $x^\varepsilon(t)$ is relatively compact (dense) if for each $c > 0$, there exists a compact $K_c \in \mathbf{R}^d$, such that*

$$\sup_\varepsilon P(x^\varepsilon(t) \notin K_c) \leq c.$$

DEFINITION 1.8.– *A family of processes $x^\varepsilon(t)$ is exponentially compact if for every $c > 0$, there exists a compact $K_c \in \mathbf{R}^d$, such that*

$$\limsup_{\varepsilon \to 0} \varepsilon \ln P(x^\varepsilon(t) \notin K_c) \leq -c.$$

The compact containment condition (CCC) plays an important role in proving the relative compactness of processes, namely:

CCC: A family of processes $x^\varepsilon(t)$ satisfies the compact containment condition, if for any $T > 0$ and $c > 0$, there exists a compact $K_{c,T} \in \mathbf{R}^d$, such that for $0 < t \leq T$:

$$\limsup_{\varepsilon \to 0} P(x^\varepsilon(t) \notin K_{c,T}) \leq c.$$

According to Feng and Kurtz (2006, Theorem 3.9.1), the necessary and sufficient conditions for the relative compactness of a process are the **CCC** and the condition of the existence of a family of functions $F \subset C(\mathbf{R}^d)$, which is closed with respect to the addition operation and separates points in $\mathbf{R}^d$ and for any $f \in F$, $\{f(x^\varepsilon)\}$ is compact in $D_{\mathbf{R}^d}[0,\infty)$.

A similar result (Schied 1995) concerns exponential compactness. The exponential compact containment condition (ECCC) has the form:

ECCC: A family of processes $x^\varepsilon(t)$ satisfies the exponential compact containment condition, if for each $T > 0$ and $c > 0$ there exists a compact $K_{c,T} \in \mathbf{R}^d$, such that for $0 < t \leq T$:

$$\limsup_{\varepsilon \to 0} \varepsilon \ln P(x^\varepsilon(t) \notin K_{c,T}) \leq -c.$$

The necessary and sufficient conditions for the exponential compactness of a process are the **ECCC** and a condition of the existence of a family of functions $F \subset C(\mathbf{R}^d)$, which is closed with respect to addition and separates points in $\mathbf{R}^d$ and for any $f \in F$, $\{f(x^\varepsilon)\}$ is exponentially compact in $D_{\mathbf{R}^d}[0, \infty)$.

In proving the weak convergence of the processes in Chapter 2, we will prove the relative compactness of pre-limit processes using the **CCC** in a slightly different form, namely

$$\lim_{c \to \infty} \limsup_{\varepsilon \to 0} P(\sup_{t \leq T} |\xi^\varepsilon(t)| \geq c) = 0.$$

Since the proof of relative compactness will require the application of some new methods, we will present these results in the substantive part (see Chapter 2).

Instead, we will discuss issues related to the proof of exponential compactness, i.e. with the program proposed at the beginning of section 1.2.

As in section 1.2.1, in particular in Lemma 1.4, let $\mathbf{L}^\varepsilon$ be a family of generators, corresponding to a Markov processes $x^\varepsilon(t)$, then the nonlinear operator

$$\mathbf{H}^\varepsilon \varphi(x) = e^{-\varphi/\varepsilon} \varepsilon \mathbf{L}^\varepsilon e^{\varphi/\varepsilon}.$$

Introduce the operator $\mathbf{H}^\varepsilon_\dagger$, such that

$$e^{(\varphi(x^\varepsilon(t)) - \varphi(x^\varepsilon(0)) - \int_0^t \mathbf{H}^\varepsilon_\dagger \varphi(x^\varepsilon(s)) ds)/\varepsilon}$$

is a supermartingale for each solution $x^\varepsilon(t)$ of the martingale problem for $\mathbf{L}^\varepsilon$. Obviously, $\mathbf{H}^\varepsilon \subset \mathbf{H}^\varepsilon_\dagger$.

DEFINITION 1.9.– *We will say that a set of functions $F \subset C_b(\mathbf{R}^d)$ approximates the metric q, if for each compact $K \subset \mathbf{R}^d$ and $z \in K$, there exists a sequence $\varphi^\varepsilon \in F$, such that*

$$\lim_{\varepsilon \to 0} \sup_{u \in K} |\varphi^\varepsilon(u) - q(u, z)| = 0.$$

The following proposition holds true:

PROPOSITION 1.5.– *(see Feng and Kurtz (2006, Corollary 4.17)) Let $F \subset C_b(\mathbf{R}^d)$ and $S \subset \mathbf{R}$. Assume that the following conditions are met:*

a) A family of processes $x^\varepsilon(t)$ satisfies the ***ECCC****.*

b) If $S = \mathbf{R}$, then the set of functions F is closed by adding an operation and separating the points in $\mathbf{R}^d$, if $S = (0, \infty)$, then F approximates the metric q.

c) For every $\lambda \in S$ and $\varphi \in F$, there exists a sequence $\varphi^\varepsilon(u)$, such that

$$\mathbf{H}^\varepsilon_\dagger \lambda \varphi^\varepsilon(u) = \mathbf{H}^\varepsilon \varphi^\varepsilon(u), \ \sup_\varepsilon ||\varphi^\varepsilon|| < \infty,$$

and for every compact $K \subset \mathbf{R}^d$

$$\lim_{\varepsilon \to 0} \sup_{u \in K} |\varphi^\varepsilon(u) - \varphi(u)| = 0,$$

and

$$\sup_\varepsilon \sup_{u \in K} \mathbf{H}^\varepsilon \varphi^\varepsilon(u) = C_\lambda(\varphi, K) < \infty.$$

Then, the family of processes $x^\varepsilon(t)$ is exponentially compact.

When using Proposition 1.5, the most difficult part is checking condition a). The classical approach to solving this problem is associated with the use of methods from the stability theory, in particular the Lyapunov function. The following proposition holds true.

PROPOSITION 1.6.– *(see Feng and Kurtz (2006, Lemma 4.22)) Let $x^\varepsilon(t)$ be a solution of the martingale problem for $\mathbf{L}^\varepsilon$. Let K be a compact and $G \supset K$ an open set. We denote the test functions $\varphi^\varepsilon(u)$ and operator $\mathbf{H}^\varepsilon_\dagger \varphi^\varepsilon(u)$ and let the corresponding supermartingale be a continuous case. Also, denote*

$$\beta(K,G) := \liminf_{\varepsilon \to 0}(\inf_{u \in G^c} \varphi^\varepsilon(u) - \sup_{u \in K} \varphi^\varepsilon(u)), \gamma(G) := \limsup_{\varepsilon \to 0} \sup_{u \in G} \mathbf{H}^\varepsilon_\dagger \varphi^\varepsilon(u).$$

Then

$$\limsup_{\varepsilon \to 0} \varepsilon \ln P(\exists t \leq T : x^\varepsilon(t) \notin G)$$

$$\leq \max\{-\beta(K,G) + T\gamma(G), \limsup_{\varepsilon \to 0} \varepsilon \ln P(x^\varepsilon(0) \notin K)\}.$$

Consider the processes for which the large deviations problem will be further analyzed in Chapters 4 and 5, namely, random evolutionary systems with independent increments and switching in the Lévy approximation scheme

$$\xi^\delta_\varepsilon(t) = \xi^\delta_\varepsilon(0) + \int_0^t \eta^\delta_\varepsilon(ds; x(s/\varepsilon^3)), \ t \geq 0,$$

$$\eta^\delta_\varepsilon(t;x) := \varepsilon \eta^\delta(t/\varepsilon^3; x)$$

which is determined by the generator of the two-component process $\xi_\varepsilon^\delta(t), x(t), t \geq 0$

$$\mathbf{L}_\varepsilon^\delta \varphi(u,x) = \varepsilon^{-3} Q\varphi(u,\cdot)(x) + \mathbf{\Gamma}_\varepsilon^\delta(x)\varphi(\cdot,x)(u),$$

where

$$\widetilde{\Gamma}_\varepsilon^\delta(x)\varphi(u) = \varepsilon^{-3}\int_{\mathbf{R}^d} [\varphi(u+\varepsilon v) - \varphi(u)]\widetilde{\Gamma}^\delta(dv;x), \ x \in E, u \in \mathbf{R}^d.$$

The question of the asymptotic behavior of the corresponding nonlinear operator

$$\mathbf{H}^{\varepsilon,\delta}\varphi_\varepsilon^\delta(u,x) = e^{-\varphi_\varepsilon^\delta/\varepsilon}\varepsilon \mathbf{L}_\varepsilon^\delta e^{\varphi_\varepsilon^\delta/\varepsilon} = e^{-\varphi_\varepsilon^\delta/\varepsilon}[\varepsilon^{-2}Q + \varepsilon \mathbf{\Gamma}_\varepsilon^\delta(x)]e^{\varphi_\varepsilon^\delta/\varepsilon}$$

is investigated in Chapter 4.

Taking into account the asymptotic properties of the operator $\mathbf{H}^{\varepsilon,\delta}$, let us put

$$\varphi^\varepsilon(u,x) = \varphi(u) + \varepsilon R_0 b_1(x)\varphi'(u) + \varepsilon^2\varphi_2(x),$$

where the functions $\varphi(u), \varphi_2(u)$ are bounded with their first derivatives and introduce the operator

$$\mathbf{H}_\dagger^{\varepsilon,\delta}\varphi^\varepsilon(u,x) = Q\varphi_2(x) + b_1(x)R_0 b_1^*(x)(\varphi'(u))^2 + b(x)\varphi'(u) + \frac{1}{2}c(x)(\varphi'(u))^2$$
$$+ \int_{\mathbf{R}^d} [e^{v\varphi'(u)} - 1 - v\varphi'(u) - \frac{vv^*}{2}(\varphi'(u))^2]\widetilde{\Gamma}(dv;x) + h^{\varepsilon,\delta}(x), \ \varepsilon,\delta \to 0, \quad [1.42]$$

where the explicit form of the negligible term $h^{\varepsilon,\delta}(x)$ is obtained, for example, by proving Theorem 4.2 (see Chapter 4).

Suppose the following condition: there exists a positive constant L, such that

$$|b(u)| \leq L, \ |b_1(u)| \leq L, \quad \text{and} \quad |c(u)| \leq L,$$

for all real-valued non-negative functions $f(v), v \in \mathbf{R}^d$, such that $\int_{\mathbf{R}^d}(1 + f(v)) \times |v|^3 dv < \infty$, then we have

$$|\Lambda(v;x)| \leq Lf(v),$$

where $\Lambda(v;x)$ is a derivative of the Radon–Nicodemus kernel $\Gamma(B;x)$ in relation to the Lebesgue measure dv in $\mathbf{R}^d$, i.e. $\widetilde{\Gamma}(dv;x) = \Lambda(v;x)dv$.

Then, obviously, the first term in [1.42] is bounded due to the boundedness of the operator Q, and all of the subsequent terms are also bounded by the condition that the derivative $\varphi'(u)$ is bounded by.

Hence, for any bounded set G

$$\gamma(G) \leq \limsup_{\varepsilon \to 0} \sup_{u \in \mathbf{R}^d} \mathbf{H}_{\dagger}^{\varepsilon,\delta} \varphi^{\varepsilon}(u,x) < \infty.$$

Suppose that $K = \{v : |v| \leq c\}$ for $\beta > 0$ and let d satisfy the relation $\varphi(d) = \beta + \varphi(c)$. Then, for $G = \{v : |v| \leq d\}$, the conditions of Proposition 1.6 are satisfied. Therefore, if $\xi_{\varepsilon}^{\delta}(0)$ is exponentially compact, then the **ECCC** takes place for $\xi_{\varepsilon}^{\delta}(t)$.

The fulfillment of conditions b) and c) of Proposition 1.5 obviously follow from the boundedness of the operator $\mathbf{H}_{\dagger}^{\varepsilon,\delta}$ on test functions $\varphi^{\varepsilon}(u,x)$, and hence the family $\xi_{\varepsilon}^{\delta}(t)$ is exponentially compact.

1.3.2. *Comparison principle*

The comparison principle suggests that nonlinear semigroups, which correspond to the pre-limit exponential generators, converge to a unique nonlinear semigroup corresponding to the limit exponential generator.

Thus, provided that the comparison principle is fulfilled, we can use generators instead of the corresponding semigroups to solve the large deviations problem.

It is easier to prove this principle on a compact space E; therefore, we will consider the single-point compaction $E = \mathbf{R}^d \cup \{\infty\}$.

DEFINITION 1.10.– *Let E be a compact metric space and $\mathbf{H} \subset C(E) \times B(E)$. Let us fix a function $h \in C(E)$ and let $\alpha > 0$. Consider a function $f \in B(E)$.*

We denote the function $g = \alpha^{-1}(f - h)$, i.e. $f - \alpha g = h$. Then

a) we will call f the viscosity sub-solution of the equation $(I - \alpha\mathbf{H})f = h$ if and only if f is upper semi-continuous and for all f_0, such that

$$\sup_x (f(v) - f_0(v)) = ||f - f_0||$$

there exists $v_0 \in E$, such that

$$f(v_0) - f_0(v_0) = ||f - f_0||$$

and

$$\alpha^{-1}(f(v_0) - h(v_0)) = \mathbf{H}f(v_0) \leq \mathbf{H}^* f(v_0),$$

where

$$\mathbf{H}^* f(v) = \lim_{\delta\to 0} \sup_{u\in B_\delta(v)} \mathbf{H}f(u).$$

b) we will call f the viscosity super-solution of the equation $(I - \alpha\mathbf{H})f = h$ if and only if f is lower semi-continuous and for all f_0, such that

$$\sup_x (f(v) - f_0(v)) = ||f - f_0||$$

there exists $v_0 \in E$, such that

$$f(v_0) - f_0(v_0) = ||f - f_0||$$

and

$$\alpha^{-1}(f(v_0) - h(v_0)) = \mathbf{H}f(v_0) \geq \mathbf{H}_* f(v_0),$$

where

$$\mathbf{H}_* f(v) = \lim_{\delta\to 0} \inf_{u\in B_\delta(v)} \mathbf{H}f(u).$$

The comparison principle takes place for h, if for any sub-solution $\overline{f}$ and for any super-solution $\underline{f}$

$$\overline{f} \leq \underline{f}.$$

The relationship between the comparison principle and the convergence of the respective semigroups is described in detail in Feng and Kurtz (2006) (see also Crandall and Lions 1983; Tataru 1992).

We use the results obtained in Feng and Kurtz (2006). Let

$$\mathbf{D}(H) \subset \{f \in C(E) : \nabla f \text{be continuous}, \nabla f(\infty) := \lim_{|v|\to\infty} \nabla f(v) = 0\},$$

where $\nabla := \left(\frac{\partial}{\partial x_1}, \ldots, \frac{\partial}{\partial x_n}\right)$.

Let the following conditions be met (Feng and Kurtz (2006, Condition 9.10)):

1) The operator $\mathbf{H}$ satisfies the relations

$$\mathbf{H}^* f(v) = H^*(v, \nabla f(v)),\ f \in \mathcal{D}(\mathbf{H}),$$

$$\mathbf{H}_* f(v) = H_*(v, \nabla f(v)),\ f \in \mathcal{D}(\mathbf{H}).$$

2) For any compact $K \in \mathbf{R}^d$, there exists a sequence $\mu_m \to \infty$ and a map $\omega : (0, \infty) \to [0, \infty]$, for which

$$\lim_{\delta \to 0} \inf_{|\lambda - 1| \leq \delta} \omega(\lambda) = 0,$$

such that for any $\lambda > 0$ and any sequence $\{(x_m, y_m)\} \subset K \times K$, which satisfies the relation $\mu_m |x_m - y_m|^2 \to 0$,

$$\sum_m H_*(y_m, \mu_m(x_m - y_m)) < \infty,$$

and

$$\inf_m H^* \left(y_m, \frac{\mu_m(x_m - y_m)}{\lambda} \right) > -\infty,$$

we have

$$\liminf_{m \to \infty} \left[\lambda H^* \left(y_m, \frac{\mu_m(x_m - y_m)}{\lambda} \right) - H_*(y_m, \mu_m(x_m - y_m)) \right] \leq \omega(\lambda).$$

3) If $x_m \to \infty$ and $|x_m||p_m| \to 0$, then

$$\limsup_{m \to \infty} H(x_m, p_m) \leq 0, \ \liminf_{m \to \infty} H(x_m, p_m) \geq 0.$$

The following proposition takes place:

PROPOSITION 1.7.– *(Feng and Kurtz (2006, Lemma 9.15)) Let* $E = \mathbf{R}^d \cup \{\infty\}$ *and*

$$\mathcal{D}(\mathbf{H}) = \{f \in C(E) : f|_{\mathbf{R}^d} - f(\infty) \in C_c^2(\mathbf{R}^d).\}$$

Denote $\nabla f(\infty) = 0$ *and assume that the operator* $\mathbf{H}$ *can be represented as*

$$\mathbf{H} f(v) = H(v, \nabla f(v)).$$

Let conditions 1)–3) be met, then for $h \in C(E)$ *and* $\alpha > 0$*, there is a comparison principle for the equation*

$$(I - \alpha \mathbf{H}) = h.$$

As shown in Chapters 4 and 5, the limit exponential generators for the considered random evolutionary systems have the asymptotic representation

$$\mathbf{H}\varphi(u) = (\widehat{b} - \widehat{b}_0)\varphi'(u) + \frac{1}{2}\sigma^2(\varphi'(u))^2 + \int_{\mathbf{R}} [e^{v\varphi'(u)} - 1]\widehat{\Gamma}(dv),$$

where

$$\widehat{b} = \Pi b(x) = \int_E \pi(dx)b(x), \widetilde{b}_0 = \Pi b_0(x) = \int_E \pi(dx)b_0(x), b_0(x) = \int_{\mathbf{R}} v\widetilde{\Gamma}(dv; x),$$

$$\widehat{c} = \Pi c(x) = \int_E \pi(dx)c(x), \widetilde{c}_0 = \Pi c_0(x) = \int_E \pi(dx)c_0(x), c_0(x) = \int_{\mathbf{R}} v^2\widetilde{\Gamma}(dv; x),$$

$$\sigma^2 = (\widehat{c} - \widetilde{c}_0) + 2\int_E \pi(dx)b_1(x)R_0 b_1^*(x), \widehat{\Gamma}(v) = \Pi\widetilde{\Gamma}(v; x) = \int_E \pi(dx)\widetilde{\Gamma}(v; x).$$

So, obviously, the generator $\mathbf{H}$ can be represented as

$$\mathbf{H}\varphi(u) = H(\nabla\varphi(u)) = H(p)$$

and it does not depend on u. Therefore, conditions 1)–3) are obviously verified for such a generator, i.e. according to Proposition 1.7, the comparison principle takes place.

REMARK 1.8.– *When the generator depends on u, i.e. it admits a representation*

$$\mathbf{H}\varphi(u) = H(u, \nabla\varphi(u)) = H(u, p)$$

we need to impose additional conditions on the functions $\widehat{b}(u), \widehat{\sigma}(u)$ and the jump measures in the process.

This situation may arise in the study of, for example, a random evolutionary system with locally independent increments (see Chapter 4).

The following proposition indicates a possible variant of such additional conditions:

PROPOSITION 1.8.– *(Feng and Kurtz (2006, Lemma 10.5)) Assume that the function $H(u, p)$ is continuous, and for all $u, p \in \mathbf{R}^d$*

$$\lim_{r\to\infty} H(x, rp) = \infty.$$

Also, suppose that the functions $\widehat{b}(u)$ and $\widehat{\sigma}(u)$ are locally bounded and measurable, grow no faster than linearly, and for each α, the following condition is satisfied for $\alpha > 0$:

$$\sup_u \int_{\mathbf{R}^d} \left|\frac{v}{1+|u|}\right|^2 e^{\alpha|v|/(1+|u|)}\widehat{\Gamma}(u, dv).$$

Then, conditions 1)–3) are satisfied and the comparison principle for the generator $\mathbf{H}$ takes place.

2

Weak Convergence in Poisson and Lévy Approximation Schemes

In this chapter, we propose a method for proving the weak convergence of random evolutionary systems, such as processes with locally independent increments and impulsive recurrent processes, that is significantly different from those used by other authors (Stroock and Varadhan 1979; Ethier and Kurtz 1986; Liptser and Shiryayev 1989; Kushner 1990; Sviridenko 1998; Skorokhod *et al.* 2002; Jacod and Shiryaev 2003; Silvestrov 2004; Korolyuk and Limnios 2005; Skorokhod 2008): the main task is to prove convergence of the predictable characteristics of the semimartingale, which are integral functionals that depend on the switching process. But the problem is that these predictable characteristics also depend on the previous values of the process itself. So, to prove weak convergence of the process, we have to prove weak convergence of its predictable characteristics that are dependent on the process. Conventional methods cannot be used separately in a situation such as this.

We propose to use a combination of two methods. Namely, the approach of R. Liptzer (1994), based on the theory of semimartingales, is combined with the singular perturbation problem solution, which appears to be more effective in this situation in comparison to the classical approach with the use of the ergodic theorem. Thus, we have two stages of proof.

At the first stage, the relative compactness of the semimartingale representation is proved for the family of processes $\xi^\varepsilon(t)$, $t \geq 0$, $\varepsilon > 0$, which follows from the following two statements (see section 1.3.1):

$$\lim_{c\to\infty} \sup_{\varepsilon\leq\varepsilon_0} P\{\sup_{t\leq T} |\xi^\varepsilon(t)| > c\} = 0,$$

well known as a compact containment condition, in the following **CCC**, and

$$\mathbf{E}|\xi^\varepsilon(t) - \xi^\varepsilon(s)|^2 \leq k|t - s|,$$

for some positive constant k, independent of ε.

In a second step using the singular perturbation method (Korolyuk and Limnios 2005), the convergence of the predictable characteristics of the semimartingale is proved, which are integral functionals of the form:

$$\int_0^t a(\xi^\varepsilon(s), x^\varepsilon(s))ds.$$

Special selection of test functions allows us to solve the problem of dependence of the predictable characteristics on the values of the process itself.

Finally, Theorem 2.1 is used, which states a weak convergence of semimartingales under appropriate conditions.

Here are some well-known theorems that play an important role in obtaining the results of this chapter. These theorems allow us to obtain the results on the convergence in $D_{\mathbf{R}^d}[0,\infty)$ for random evolutionary processes under the Poisson approximation scheme.

Let us introduce three classes of functions (we follow Jacod and Shiryaev (2003, p. 395)) $C_2(\mathbf{R}^d), C_1(\mathbf{R}^d), C_3(\mathbf{R}^d)$, defined as:

– $C_2(\mathbf{R}^d)$ is a class of all real continuous bounded functions defined on $\mathbf{R}^d$, equal to 0 in the neighborhood of 0, having a finite limit at infinity;

– $C_1(\mathbf{R}^d)$ is a subclass of $C_2(\mathbf{R}^d)$ with $g_a(x) = (a|x| - 1)^+ \wedge 1$ for all positive rational a, and has the following property: let μ_n, μ be positive measures on $\mathbf{R}^d \setminus \{0\}$, finite on any complement around 0; then, $\mu_n f \to \mu f$ for all $f \in C_1(\mathbf{R}^d)$ will mean the convergence of $\mu_n f \to \mu f$ for all $f \in C_2(\mathbf{R}^d)$. Thus, this class defines the measure;

– $C_3(\mathbf{R}^d)$ is also a class that defines a measure and contains real-valued bounded functions φ, such that

$$\varphi(u)/|u|^2 \to 0, |u| \to 0.$$

The following inclusion takes place:

$$C_1(\mathbf{R}^d) \subset C_2(\mathbf{R}^d) \subset C_3(\mathbf{R}^d).$$

Consider the random measure $\nu = \nu(\omega, dt, dx); \omega \in \Omega$ at $(\mathbf{R}_+ \times E, \mathcal{B}_+ \times \mathcal{E})$ such that $\nu(\{0\} \times E) \equiv 0$. Let R be a measurable function $(\Omega \times \mathbf{R}_+ \times E, \widetilde{O} \times \mathcal{E})$, where $\widetilde{O}$ is an algebra at $\Omega \times \mathbf{R}_+$.

Let us denote the integral process $R * \nu(t, \omega)$ as follows:

$$R * \nu(\omega, t) = \int_{[0,t]\times E} R(\omega; s, x)\nu(\omega, ds, dx),$$

$$\text{if } R * \nu(t, \omega) = \int_{[0,t]\times E} |R(\omega; s, x)|\nu(\omega, ds, dx) < \infty; \text{ else } R * \nu(t, \omega) = 0.$$

For the process M defined on $(\Omega, \mathcal{F}, P)$, we denote the corresponding distribution law by $\widetilde{P} = \mathcal{L}(M)$. That is, it is a mapping of $\widetilde{P} = P \circ M^{-1}$ of the law P after the action of $M: \Omega \to \mathbf{D}$. Note that the family of processes x^ε can also be considered a mapping from Ω to $D_{\mathbf{R}^d}$.

THEOREM 2.1.– *(see Jacod and Shiryaev (2003, Theorem IX.3.27)) Let D be a dense subset of* $\mathbf{R}_+$, *and suppose that* $C_1(\mathbf{R}^d)$ *contains positive and negative components of the following functions:*

$$g_a^i(x) = (x^i - h^i(x))(1 - g_a(x)),$$

$$g_a^{ij}(x) = (x^i x^j - h^i(x)h^j(x))(1 - g_a(x)),$$

where $a \in \mathbb{Q}_+$ *and* $g_a(x) = (a|x| - 1)^+ \wedge 1$.

Suppose also that ν^ε *and* ν *are such that*

$$|x| * \nu^\varepsilon(t) < +\infty, |x| * \nu(t) < +\infty$$

for all $t \geq 0$.

Let the following conditions also be satisfied:

(i) Strong majority hypothesis: there is a continuous deterministic increasing function F, which strongly majorizes the functions $\sum_{i \leq d} Var(B'^i(\alpha))$ *and* $\sum_{i \leq d} Var(\widetilde{C}'^{ii}(\alpha))$.

(ii) Condition for big jumps:

$$\lim_{a\uparrow\infty} \sup_{\alpha\in\Omega} |x|^2 \mathbf{1}_{\{|x|>a\}} * \nu(t; \alpha) = 0.$$

(iii) Unity condition: there is only one probabilistic measure P on $(\Omega, \mathcal{F})$ *such that the canonical process* $x(t), t \geq 0$, *is a semimartingale on* $(\Omega, \mathcal{F}, \mathbf{F}, P)$ *with characteristics* (B, C, ν) *and the initial distribution* η.

(iv) Continuity condition: for any $t \in D, g \in C_1(\mathbf{R}^d)$ *functions* $\omega \mapsto B'(t,\omega)$, $\widetilde{C}'(t,\omega), g * \nu(t,\omega)$ *are continuous in the Skorokhod topology* J_1 *at* $D_{\mathbf{R}^d}$.

(v) $\mathcal{L}(x^\varepsilon(0)) \Rightarrow P_0$, *where* $\mathcal{L}(x^\varepsilon(0))$ *is the law of random variable distribution* $x^\varepsilon(0)$ *and* P_0 *is the initial distribution of the limit process.*

(vi) Convergence in probability takes place

$$g * \nu^\varepsilon(t) - (g * \nu(t)) \circ x^\varepsilon \to 0$$

for all $t \in D, g \in C_1(\mathbf{R}^d)$, *and the following conditions of convergence in probability are met for all* $t \geq 0$:

$$\sup_{s \leq t} |B'^{\varepsilon}(s) - B'(s) \circ x^\varepsilon| \to 0,$$

$$\widehat{C}'^{\varepsilon} - \widehat{C}' \circ x^\varepsilon \to 0,$$

and also condition

$$\lim_{a \uparrow \infty} \limsup_{\varepsilon > 0} P^\varepsilon(|x|^2 \mathbf{1}_{\{|x|>a\}} * \nu^\varepsilon(t) > \varepsilon) = 0,$$

for all $\varepsilon > 0, t \in \mathbf{R}_+$.

Then

$$\mathcal{L}(x^\varepsilon) \Rightarrow P, \varepsilon \to 0,$$

where $\mathcal{L}(x^\varepsilon)$ *is the distribution law of the process* $x^\varepsilon(t), t \geq 0$.

THEOREM 2.2.– *(see Korolyuk and Limnios (2005, Theorem 6.3)) Let* $C_0^2(\mathbf{R} \times E)$ *be a space of real twice continuously differentiable by the first argument functions defined on* $\mathbf{R} \times E$ *and equal to 0 at infinity and* $C(\mathbf{R} \times E)$ *be a space of real continuous bounded functions on* $\mathbf{R} \times E$.

Suppose that the following conditions are met for a family of two-component Markov process $\xi^\varepsilon(t), x^\varepsilon(t), t \geq 0, \varepsilon > 0$:

CD1: *There is a family of test functions* $\varphi^\varepsilon(u,x)$ *in* $C_0^2(\mathbf{R} \times E)$, *such that*

$$\lim_{\varepsilon \to 0} \varphi^\varepsilon(u,x) = \varphi(u),$$

uniformly over u, x.

CD2: *Weak convergence takes place*

$$\lim_{\varepsilon \to 0} \mathbf{L}^\varepsilon \varphi^\varepsilon(u,x) = \mathbf{L}\varphi(u),$$

uniformly over u, x. *The family of functions* $\mathbf{L}^\varepsilon \varphi^\varepsilon, \varepsilon > 0$ *is uniformly bounded* $\mathbf{L}\varphi(u)$ *and* $\mathbf{L}^\varepsilon \varphi^\varepsilon$ *belong to* $C(\mathbf{R} \times E)$.

CD3: *Quadratic characteristics of martingales corresponding to the two-component Markov process* $\xi^\varepsilon(t), x^\varepsilon(t), t \geq 0, \varepsilon > 0$ *have the representation* $\langle \mu^\varepsilon \rangle_t = \int_0^t \zeta^\varepsilon(s) ds$, *where random functions* $\zeta^\varepsilon, \varepsilon > 0$, *satisfy the condition*

$$\sup_{0 \leq s \leq T} \mathbf{E}|\zeta^\varepsilon(s)| \leq c < +\infty.$$

CD4: *There is a convergence of initial conditions and*

$$\sup_{\varepsilon > 0} \mathbf{E}|\zeta^\varepsilon(0)| \leq C < +\infty.$$

Then, the weak convergence becomes true

$$\xi^\varepsilon(t) \Rightarrow \xi(t), \quad \varepsilon \to 0.$$

THEOREM 2.3.– *(see Liptser and Shiryayev (1989, Theorem 8.2.1)) Let* Q^ε *be the probability distribution for the* P^ε*-semimartingale* $\xi^\varepsilon = (\xi^\varepsilon(t), \mathcal{F}_t^\varepsilon)$ *with the triplet of predictable characteristics* $\mathcal{T}^\varepsilon = (B^\varepsilon, C^\varepsilon, \Gamma^\varepsilon)$ *and* Q *be the probability distribution for the semimartingale* $\xi^0 = (\xi^0(t), \mathcal{D}_t^Q)$ *with the triplet* $\mathcal{T}^0 = (B^0, C^0, \Gamma^0)$.

If the following condition holds for the triplet $\mathcal{T}$:

LP1:

$$|\int_E b(\xi(t), x)\pi(dx)| \leq L(1 + \xi^\dagger(t)),$$

$$|\int_E c(\xi(t), x)\pi(dx)| \leq L(1 + (\xi^\dagger(t))^2),$$

and for any non-negative measurable function $f(v) \leq v^2 \wedge 1$

$$\int_E \int_{\mathbf{R}} f(v)\Gamma(\xi(t), dv; x)\pi(dx) \leq L(1 + \xi^\dagger(t)).$$

And for the triplets $\mathcal{T}^\varepsilon$ *for any fixed* $T > 0$:

LP2:

$$\lim_{l \to \infty} \limsup_{\varepsilon \to 0} \int_0^T \int_{|v| > l} \Gamma^\varepsilon(\xi^\varepsilon(s), dv; x_s^\varepsilon) ds = 0,$$

$$\lim_{\delta \to 0} \limsup_{\varepsilon \to 0} \int_0^T \int_{|v| \leq \delta} v^2 \widehat{\Gamma}(\xi^\varepsilon(s), dv) ds = 0.$$

LP3: *For any bounded measurable function* $f(v)$

$$\lim_{\varepsilon\to 0}\sup_{t\leq T}\left|\int_0^t\int_{|v|>\delta} f(v)[\Gamma^\varepsilon(\xi^\varepsilon(s),dv;x_s^\varepsilon)-\widehat{\Gamma}(\xi^\varepsilon(s),dv)]ds\right|=0.$$

LP4:

$$\lim_{\varepsilon\to 0}\sup_{t\leq T}\left|\int_0^t [b^\varepsilon(\xi^\varepsilon(s),x_s^\varepsilon)-\widehat{b}(\xi^\varepsilon(s))]ds\right|=0.$$

LP5:

$$\lim_{\varepsilon\to 0}\sup_{t\leq T}\left|\int_0^t [c^\varepsilon(\xi^\varepsilon(s),x_s^\varepsilon)-\widehat{c}(\xi^\varepsilon(s))]ds\right|=0,$$

$$\lim_{\delta\to 0}\lim_{\varepsilon\to 0}\sup\int_0^T\int_{|v|\leq\delta} v^2\Gamma^\varepsilon(\xi^\varepsilon(s),dv;x_s^\varepsilon)ds=0,$$

then under the compact containment condition – ***CCC***

$$\lim_{c\to\infty}\lim_{\varepsilon\to 0}\sup P^\varepsilon\{|\xi^\varepsilon(0)|\geq c\}=0$$

the family $Q^\varepsilon, \varepsilon>0$ *is relatively compact.*

2.1. Random evolutionary systems with locally independent increments

2.1.1. *Markov switching*

In the following, we consider spaces $\mathbf{R}^d$ with norm $|\cdot|$ ($d\geq 1$), and $(E,\mathcal{E})$, – *standard phase space* (i.e. E – Polish space, and $\mathcal{E}$ – its Borel σ-algebra). For an arbitrary vector $v\in\mathbf{R}^d$ and a matrix $c\in\mathbf{R}^{d\times d}$, we denote by v^* and c^* the corresponding transposed vector and matrix.

Random evolutionary systems with locally independent increments $\xi^\varepsilon(t)$ are defined in section 1.1.8 (see [1.10], [1.11]). The switching process is a Markov process $x(t)$, introduced in section 1.1.3; Poisson approximation conditions for normalized random evolutions

$$\xi^\varepsilon(t)=\xi_0^\varepsilon+\int_0^t \eta_\varepsilon(ds;x(s/\varepsilon)) \qquad [2.1]$$

were introduced in section 1.1.10 (conditions **C1–C3**).

Before formulating the main result, we introduce the concept of the local Lipschitz condition for the functions that determine the process generator with locally independent increments

$$\widehat{\mathbf{\Gamma}}\varphi(u)=\widehat{b}(u)\varphi'(u)+\frac{1}{2}\sigma^2(u)\varphi''(u)+\int_{\mathbf{R}^d}[\varphi(u+v)-\varphi(u)-v\varphi'(u)]\widehat{\Gamma}(u,dv).$$

DEFINITION 2.1.– *We say that the functions* $\widehat{b}(u), \sigma^2(u)$ *and* $\widehat{\Gamma}(u, dv)$ *satisfy the local Lipschitz condition if*

LL: *For an arbitrary* $r > 0$, *there is a constant* l_r *i*

$$|\widehat{b}(u) - \widehat{b}(u')| + |\sigma^2(u) - \sigma^2(u')| + |\widehat{\Gamma}(u, v) - \widehat{\Gamma}(u', v)| \leq l_r |u - u'|$$

if $|u| \leq r, |v| \leq r.$

We formulate the main result in the form of a theorem.

THEOREM 2.4.– *Under the conditions* ***CM, C1–C3*** *a weak convergence in* $D_{\mathbf{R}^d}[0, \infty)$ *holds true*

$$\xi^\varepsilon(t) \Rightarrow \xi^0(t), \quad \varepsilon \to 0.$$

Under the ***LL*** *condition, the limit process* $\xi^0(t), t \geq 0$, *is determined by the generator*

$$\widehat{\mathbf{\Gamma}}\varphi(u) = \widehat{b}(u)\varphi'(u) + \int_{\mathbf{R}^d} [\varphi(u + v) - \varphi(u) - v\varphi'(u)]\widehat{\Gamma}(u, dv), \qquad [2.2]$$

or otherwise

$$\widehat{\mathbf{\Gamma}}\varphi(u) = \sum_{k=1}^{d} \widehat{b}_k(u)\varphi'_k(u) + \int_{\mathbf{R}^d} [\varphi(u + v) - \varphi(u) - \sum_{k=1}^{d} v_k\varphi'_k(u)]\widehat{\Gamma}(u, dv),$$

$$\varphi'_k(u) := \partial\varphi(u)/\partial u_k, 1 \leq k \leq d,$$

where the averaged deterministic shift has the form

$$\widehat{b}(u) = \int_E \pi(dx)b(u; x),$$

and the averaged intensity kernel

$$\widehat{\Gamma}(u, dv) = \int_E \pi(dx)\Gamma(u, dv; x).$$

REMARK 2.1.– *According to Skorokhod* et al. *(2002, Chapter 2), the local Lipschitz condition for the functions that define the limit generator of the Markov process, in particular the process with locally independent increments, together with the growth condition* ***C3*** *(see section 1.1.10) are sufficient conditions for the limit generator to correspond to the limit process of the pre-limit family of processes (see, for example, Skorokhod* et al. *(2002, p. 70, Theorem 2)).*

REMARK 2.2.– *The limit process $\xi^0(t), t \geq 0$ is a process with locally independent increments or a piecewise deterministic Markov process (see, for example, Korolyuk and Limnios (2005)).*

Generator [2.2] can be rewritten in the form

$$\widehat{\mathbf{\Gamma}}\varphi(u) = \widehat{b}_0(u)\varphi'(u) + \int_{\mathbf{R}^d} [\varphi(u+v) - \varphi(u)]\widehat{\Gamma}(u, dv),$$

where $\widehat{b}_0(u) = \widehat{b}(u) - \int_{\mathbf{R}^d} v\widehat{\Gamma}(u, dv)$.

The consequence of the theorem is the result where the limit process is a compound Poisson process.

COROLLARY 2.1.– *In the partial case when the following conditions of Poisson approximation become true:*

***PA1'**:*

$$b_\varepsilon(u; x) = \int_{\mathbf{R}^d} v\Gamma^\varepsilon(u, dv; x) = \varepsilon[b(x) + \theta_b^\varepsilon(u; x)]$$

and

$$c_\varepsilon(u; x) = \int_{\mathbf{R}^d} vv^*\Gamma^\varepsilon(u, dv; x) = \varepsilon[c(u; x) + \theta_c^\varepsilon(u; x)]$$

and additional condition

***PA4**:*

$$\int_{\mathbf{R}^d} v\Gamma^\varepsilon(u, dv) = \int_E \pi(dx)b(u; x), \quad \Gamma^\varepsilon(u, dv) = \int_{\mathbf{R}^d} \pi(dx)\Gamma^\varepsilon(u, dv; x),$$

the limit process $\xi^0(t), t \geq 0$ is a compound Poisson process

$$\xi^0(t) = u + \sum_{k=1}^{\nu(t)} \alpha_k, \quad t \geq 0,$$

defined by the generator

$$\mathbf{\Gamma}\varphi(u) = \int [\varphi(u+v) - \varphi(u)]\Gamma(u, dv)$$

*under the **LL** condition, where*

$$\Gamma(u, dv) = \int_E \pi(dx)\Gamma(u, dv; x), \quad \Gamma_g(u, x) = \int g(v)\Gamma(u, dv; x).$$

The random variables α_k, $k = 1, 2, ...$ are i.i.d variables with the distribution function $\mathbf{P}(\alpha_k \in dv) = \Gamma(u, dv)/\Lambda(u), \Lambda(u) = \Gamma(u, \mathbf{R}^d)$. The homogeneous by time Poisson process $\nu(t), t \geq 0$, is determined by its intensity: $\Lambda(u) > 0$.

PROOF OF THEOREM 2.4.– The proof is based on the semimartingale representation of the process [2.1]. Predictable characteristics of the semimartingale [2.1] have the following form (see Çinlar *et al.* (1980, *Theorems 6.27 and 7.16*)):

– $B^\varepsilon(t) = \varepsilon^{-1} \int_0^t b_\varepsilon(\xi^\varepsilon(s); x_s^\varepsilon)ds = \int_0^t b(\xi^\varepsilon(s); x_s^\varepsilon)ds + \theta_b^\varepsilon$,

– $C^\varepsilon(t) = \varepsilon^{-1} \int_0^t c_\varepsilon(\xi^\varepsilon(s); x_s^\varepsilon)ds = \int_0^t c(\xi^\varepsilon(s); x_s^\varepsilon)ds + \theta_c^\varepsilon$,

– $\Gamma^\varepsilon(t) = \varepsilon^{-1} \int_0^t \int_{\mathbf{R}^d} g(v)\Gamma^\varepsilon(\xi^\varepsilon(s), dv; x_s^\varepsilon)ds$

$= \int_0^t \int_{\mathbf{R}^d} g(v)\Gamma(\xi^\varepsilon(s), dv; x_s^\varepsilon)ds + \theta_g^\varepsilon$,

where $x_t^\varepsilon := x(t/\varepsilon), t \geq 0$, and $\sup_{x \in E} |\theta_\cdot^\varepsilon| \to 0, \quad \varepsilon \to 0$.

The jumping martingale part of semimartingale [2.1] is given by the ratio

$$\mu^\varepsilon(t) = \int_0^t \int_{\mathbf{R}^d} v[\mu^\varepsilon(ds, dv; x_s^\varepsilon) - \Gamma^\varepsilon(\xi^\varepsilon(s), dv; x_s^\varepsilon)ds].$$

Here, $\mu^\varepsilon(ds, dv; x), x \in E$ is a family of counting measures with the characteristics

$$\mathbf{E}\mu^\varepsilon(ds, dv; x) = \Gamma^\varepsilon(u, dv; x)ds.$$

Therefore, it is obvious that the predictable characteristics depend on the process $\xi^\varepsilon(s)$. That is, to prove the convergence of $\xi^\varepsilon(s)$, it is necessary to prove the convergence of functionals that themselves depend on $\xi^\varepsilon(s)$. We solve this problem by a combination of two methods.

We will prove this in two stages.

STAGE 1. We prove the relative compactness of the family of processes $\xi^\varepsilon(t), t \geq 0, \varepsilon > 0$. As soon as the space of all probabilistic measures defined on the standard phase space $(E, \mathcal{E})$ is also a Polish space, the relative compactness is equivalent to the density of a family of measures.

First, we prove the following lemma.

LEMMA 2.1.– *Under conditions **C3, PA1**, a constant $k > 0$, independent of ε but dependent on T exists, such that*

$$\mathbf{E} \sup_{t \leq T} |\xi^\varepsilon(t)|^2 \leq k_T.$$

COROLLARY 2.2.– *Under conditions* **C3, PA1**, *the* **CCC** *condition holds true:*

$$\lim_{c\to\infty}\limsup_{\varepsilon\to 0}\mathbf{P}\{\sup_{t\leq T}|\xi^\varepsilon(t)|>c\}=0.$$

PROOF.– Corollary follows from the Kolmogorov inequality with the additional statement of Lemma 2.1.

REMARK 2.3.– *Another way to prove the* **CCC** *condition is proposed by Feng and Kurtz (2006, Theorem 8.10) (see also Sviridenko (1998); Ethier and Kurtz (1986)). The function $\varphi(u)=\sqrt{1+u^2}$ is used and the corresponding corollary is proved for $\varphi(\xi_t^\varepsilon)$ using martingale characterization of the Markov process.*

PROOF OF LEMMA 2.1.– For an arbitrary process $y(t), t\geq 0$, we denote the process $y^\dagger(t):=\sup_{s\leq t}|y(s)|$.

From the conditions **PA1** and **C3**, it follows that for an arbitrary fixed $t>0$

$$\int_0^t\int_{\mathbf{R}^d} vv^*\Gamma^\varepsilon(\xi^\varepsilon(s),dv;x_s^\varepsilon)ds=\varepsilon\int_0^t c(\xi^\varepsilon(s);x_s^\varepsilon)ds+\varepsilon t\theta_c^\varepsilon(u;x)$$

$$\leq\varepsilon t[L(1+((\xi_t^\varepsilon)^\dagger)^2)+\theta_c^\varepsilon(u;x)]<\infty P-\text{a.s.}$$

Increasing process $\int_0^t\int_{\mathbf{R}^d} vv^*\Gamma^\varepsilon(\xi^\varepsilon(s),dv;x_s^\varepsilon)ds$ is continuous by t; thus, according to Theorem 28 (see Dellacherie and Meyer (1978, Chapter 5), also Theorem 1.6.3 from Liptser and Shiryayev (1989)), this is a compensator $\int_0^t\int_{\mathbf{R}^d} vv^*\mu^\varepsilon(\xi^\varepsilon(s),d(s/\varepsilon),dv;x_s^\varepsilon)$.

Thus, process [2.1] has the following semimartingale representation:

$$\xi^\varepsilon(t)=u+B_t^\varepsilon+M_t^\varepsilon, \qquad [2.3]$$

where $u=\xi^\varepsilon(0)$; B_t^ε is a predictable drift

$$B_t^\varepsilon=\int_0^t\int_{\mathbf{R}^d} v\Gamma(\xi^\varepsilon(s),dv;x_s^\varepsilon)ds+\theta^\varepsilon(t)=\int_0^t b(\xi^\varepsilon(s),x_s^\varepsilon)ds+\theta^\varepsilon(t),$$

M_t^ε is the local quadratically integrated martingale

$$M_t^\varepsilon=\int_0^t\int_{\mathbf{R}^d} v[\mu(ds,dv;x_s^\varepsilon)-\Gamma(\xi^\varepsilon(s),dv;x_s^\varepsilon)ds]+\theta_g^\varepsilon(t)$$

with a quadratic characteristic

$$\langle M^\varepsilon\rangle_t=\int_0^t\int_{\mathbf{R}^d} vv^*\Gamma(\xi^\varepsilon(s),dv;x_s^\varepsilon)ds+\theta_g^\varepsilon(t),$$

and for any bounded $T > 0$

$$\sup_{0\le t\le T} |\theta^\varepsilon_\cdot(t)| \to 0, \varepsilon \to 0.$$

From [2.3], we obtain

$$((\xi^\varepsilon(t))^\dagger)^2 \le 3[u^2 + ((B^\varepsilon(t))^\dagger)^2 + ((M_t^\varepsilon)^\dagger)^2]. \qquad [2.4]$$

By **C3**:

$$(B^\varepsilon(t))^\dagger = \int_0^t b(\xi^\varepsilon(s), x_s^\varepsilon)ds \le L\int_0^t (1 + (\xi^\varepsilon(s))^\dagger)ds, \qquad [2.5]$$

$$|\langle M^\varepsilon\rangle_t| = |\int_0^t \int_{\mathbf{R}^d} vv^*\Gamma(\xi^\varepsilon(s), dv; x_s^\varepsilon)ds| \le Lr_1 \int_0^t [1 + ((\xi^\varepsilon(s))^\dagger)^2]ds, \qquad [2.6]$$

where $r_1 = \int_{\mathbf{R}^d} |v|^2 f(v)dv$.

Inequalities [2.4]–[2.6], the Doob inequality (e.g. Theorem 1.9.2 from Liptser and Shiryayev (1989))

$$\mathbf{E}((M_t^\varepsilon)^\dagger)^2 \le 4|\mathbf{E}\langle M^\varepsilon\rangle_t|,$$

and the Cauchy–Bunyakovsky–Schwarz inequality ($[\int_0^t \varphi(s)ds]^2 \le t\int_0^t \varphi^2(s)ds$) give the following result:

$$\mathbf{E}((\xi^\varepsilon(t))^\dagger)^2 \le k_1 + k_2\int_0^t \mathbf{E}[((\xi^\varepsilon(s))^\dagger)^2 ds],$$

where k_1 and k_2 are positive constants independent of ε.

Now applying the Gronwall–Bellman inequality (e.g. Ethier and Kurtz 1986, p. 498), we have

$$\mathbf{E}((\xi^\varepsilon(t))^\dagger)^2 \le k_1 \exp(k_2 t).$$

Thus, for the process $\xi^\varepsilon(t)$, the following inequality holds true:

$$\mathbf{E}\sup_{t\le T} |\xi^\varepsilon(t)|^2 \le k_T.$$

Lemma 2.1 is proved.

LEMMA 2.2.– *Under* **C3**, *there is a constant $k > 0$, independent of ε and such that*

$$\mathbf{E}|\xi^\varepsilon(t) - \xi^\varepsilon(s)|^2 \le k|t - s|, \forall s, t \le T.$$

PROOF.– As in the case of inequality [2.4], we can write

$$|\xi^\varepsilon(t) - \xi^\varepsilon(s)|^2 \leq 2|B_t^\varepsilon - B_s^\varepsilon|^2 + 2|M_t^\varepsilon - M_s^\varepsilon|^2.$$

According to the Doob inequality, we have

$$\mathbf{E}|\xi^\varepsilon(t) - \xi^\varepsilon(s)|^2 \leq 2\mathbf{E}\{|B_t^\varepsilon - B_s^\varepsilon|^2 + 8|\langle M^\varepsilon\rangle_t - \langle M^\varepsilon\rangle_s|\}.$$

Inequalities [2.5], [2.6] and condition **C3** lead to the estimation

$$|B_t^\varepsilon - B_s^\varepsilon|^2 + 8|\langle M^\varepsilon\rangle_t - \langle M^\varepsilon\rangle_s| \leq k_3[1 + ((\xi^\varepsilon(T))^\dagger)^2]|t - s|,$$

where k_3 is a positive constant independent of ε.

Applying the result of Lemma 2.1 to the last estimate, we obtain the desired result.

The lemma is proved.

The conditions proved in Corollary 2.2 and Lemma 2.2 are necessary and sufficient conditions for the compactness of the family of processes $\xi^\varepsilon(t), t \geq 0, \varepsilon > 0$.

We can now use Theorem 2.3. Obviously, condition **LP1** is a direct consequence of condition **C3**. Condition **LP2** and the second part of condition **LP5** are fulfilled due to the limited jump measures of the process. Finally, conditions **LP4**, **LP4** and the second part of condition **LP5** are the result of the convergence of the predictable characteristics of the process.

STAGE 2. We will prove convergence of the predictable characteristics. At the same time, we will use the methods described in Korolyuk and Limnios (2005) (see sections 3.2 and 3.3).

In the following, we denote by $A^\varepsilon(t)$ the main part of any of the predictable characteristics $B^\varepsilon(t), C^\varepsilon(t), \Gamma^\varepsilon(t)$.

Consider a three-component Markov process $A^\varepsilon(t), \xi^\varepsilon(t), x_t^\varepsilon, t \geq 0$, which can be characterized by a martingale

$$\mu_t^\varepsilon = \varphi(A^\varepsilon(t), \xi^\varepsilon(t), x_t^\varepsilon) - \int_0^t \mathbf{L}^\varepsilon \varphi(A^\varepsilon(s), \xi^\varepsilon(s), x_t^\varepsilon) ds,$$

where the generator $\mathbf{L}^\varepsilon$ has a representation (see [1.12])

$$\mathbf{L}^\varepsilon \varphi(u, v; x) = \varepsilon^{-1} Q\varphi(\cdot, \cdot; x) + \mathbf{\Gamma}^\varepsilon(x)\varphi(u, \cdot; \cdot) + A^\varepsilon(u; x)\varphi(\cdot, v; \cdot), \qquad [2.7]$$

here

$$\mathbf{\Gamma}^\varepsilon(x)\varphi(u) = \varepsilon^{-1} \int_{\mathbf{R}^d} [\varphi(u + v) - \varphi(u)]\Gamma^\varepsilon(u, dv; x), x \in E,$$

$$Q\varphi(x) = q(x)\int_E P(x,dy)[\varphi(y) - \varphi(x)],$$

$$A^\varepsilon(u;x)\varphi(v) = a_\varepsilon(u;x)\varphi'(v) = a(u;x)\varphi'(v) \\ + \theta_a^\varepsilon\varphi'(v) =: A(u;x)\varphi(v) + \theta_a^\varepsilon\varphi'(v).$$

Given the compactness of the process $\xi^\varepsilon(t)$, to prove the convergence of the predictable characteristics, it is enough to consider the action of the generator $\mathbf{L}^\varepsilon$ on the test function with two variables $\varphi^\varepsilon(v;x) = \varphi(v) + \varepsilon\varphi_1(v;x)$.

Thus, the generator has the form

$$\mathbf{L}^\varepsilon\varphi^\varepsilon(v;x) = [\varepsilon^{-1}Q + A^\varepsilon(u;x)][\varphi(v) + \varepsilon\varphi_1(v;x)] = \varepsilon^{-1}Q\varphi(v) \\ + [Q\varphi_1(v;x) + A(u;x)\varphi(v)] + \varepsilon A^\varepsilon(u;x)\varphi_1(v;x) + \theta_a^\varepsilon\varphi'(v). \tag{2.8}$$

Therefore, the asymptotic representation of the generator [2.8] can be obtained, solving the singular perturbation problem given by the equations

$$Q\varphi(v) = 0,$$

$$Q\varphi_1(v;x) + A(u;x)\varphi(v) = \mathbf{L}\varphi(v).$$

From the first equation, it is obvious $\varphi(v) \in \mathcal{N}_Q$, and from the solvability condition for the second equation, we have:

$$\mathbf{L}\varphi(v) = \widehat{A}(u)\varphi(v) := \Pi A(u;x)\Pi\varphi(v). \tag{2.9}$$

Thus, applying Theorem 2.2, we can state a weak convergence of the predictable characteristics of the process.

The last stage of the proof is the application of Theorem 2.1. Indeed, all the conditions of the theorem are satisfied.

The strong majorization hypothesis is performed with the majorizing function given in condition **C3**. Condition **C2** is a condition of large jumps for a predictable jump measures, moreover, provided **C2** follows the last condition of Theorem 2.1. Conditions (iv) and (v) are obviously met.

Weak convergence of predictable characteristics follows from the solution of the singular perturbation problem for the generator [2.8].

Thus, there is a weak convergence in $D_{\mathbf{R}^d}[0,\infty)$ to the process, uniquely determined by the limit predictable characteristics.

Substituting the operators $B(u;x), C(u;x), \Gamma(u;x)$ instead of $A(u;x)$ in [2.9], we see that the limit Markov process has the following predictable characteristics

$$B^0(t) = \int_0^t \widehat{b}(\xi^0(s))ds, \quad C^0(t) = \int_0^t \widehat{c}(\xi^0(s))ds, \quad \Gamma_g^0(t) = \int_0^t \widehat{\Gamma}_g(\xi^0(s))ds.$$

Since the weak convergence of the process is proved, the limit generator can be obtained from the asymptotic representation of the generator [2.7] on the test functions of the form $\varphi^\varepsilon(u;x) = \varphi(u) + \varepsilon\varphi_1(u;x)$.

Namely,

$$\mathbf{L}^\varepsilon\varphi^\varepsilon(u;x) = [\varepsilon^{-1}Q + \mathbf{\Gamma}^\varepsilon(x)][\varphi(u) + \varepsilon\varphi_1(u;x)]$$

$$= \varepsilon^{-1}Q\varphi(u) + [Q\varphi_1(u;x) + \mathbf{\Gamma}^\varepsilon(x)\varphi(u)] + \varepsilon\mathbf{\Gamma}^\varepsilon(x)\varphi_1(u;x).$$

Applying Lemma 1.1, we can write the following asymptotic representation:

$$\mathbf{L}^\varepsilon\varphi^\varepsilon(u;x) = \varepsilon^{-1}Q\varphi(u) + [Q\varphi_1(u;x) + \mathbf{\Gamma}(x)\varphi(u)] + \varepsilon\mathbf{\Gamma}^\varepsilon(x)\varphi_1(u;x) + \theta^\varepsilon\varphi(u),$$

where

$$\mathbf{\Gamma}(x)\varphi(u) := b(u;x)\varphi'(u) + \int_{\mathbf{R}^d} [\varphi(u+v) - \varphi(u) - v\varphi'(u)]\Gamma(u, dv; x).$$

Thus, the singular perturbation problem is reduced to a system of equations

$$Q\varphi(u) = 0,$$

$$Q\varphi_1(u;x) + \mathbf{\Gamma}(x)\varphi(u) = \widehat{\mathbf{\Gamma}}\varphi(u).$$

From the first equation, it is obvious that $\varphi(u) \in \mathcal{N}_Q$, and by the solvability condition for the second equation, we have:

$$\widehat{\mathbf{\Gamma}}\varphi(u) = \Pi\mathbf{\Gamma}(x)\Pi\varphi(u)$$

$$= \widehat{b}(u)\varphi'(u) + \int_{\mathbf{R}^d} [\varphi(u+v) - \varphi(u) - v\varphi'(u)]\widehat{\Gamma}(u, dv).$$

Using condition **LL** and the results of Skorokhod *et al.* (2002, Chapter 2), we can say that the limit generator really determines the limit process $\xi^0(t)$.

Theorem 2.4 is proved.

Now consider the normalized random evolutionary systems

$$\xi^\varepsilon(t) = \xi_0^\varepsilon + \int_0^t \eta_\varepsilon(ds; x(s/\varepsilon^2)) \qquad [2.10]$$

under Lévy approximation conditions, which are introduced in section 1.1.10 (conditions **C1’, C2–C3**).

We formulate the main result in the form of a theorem.

THEOREM 2.5.– *Under the conditions* **CM, C1', C2–C3**, *a weak convergence in* $D_{\mathbf{R}^d}[0,\infty)$ *holds true:*

$$\xi^\varepsilon(t) \Rightarrow \xi^0(t), \quad \varepsilon \to 0.$$

The limit process $\xi^0(t), t \geq 0$ *is a Lévy process, determined by the generator*

$$\widehat{\mathbf{\Gamma}}\varphi(u) = (\widehat{b}(u) - \widehat{b}_0(u) + \widehat{b}_1(u))\varphi'(u) + \frac{1}{2}\sigma^2(u)\varphi''(u)$$

$$+\lambda(u)\int_{\mathbf{R}^d}[\varphi(u+v) - \varphi(u)]\Gamma(u, dv),$$

or otherwise

$$\widehat{\mathbf{\Gamma}}\varphi(u) = \sum_{k=1}^{d}(\widehat{b}_k(u) - \widehat{b}_{0k}(u) + \widehat{b}_{1k}(u))\varphi'_k(u) + \frac{1}{2}\sum_{k,r=1}^{d}\sigma^2_{kr}(u)\varphi''_{kr}(u)$$

$$+\int_{\mathbf{R}^d}[\varphi(u+v) - \varphi(u)]\widehat{\Gamma}(u, dv),$$

$$\varphi'_k(u) := \partial\varphi(u)/\partial u_k, \varphi'_{kr}(u) := \partial^2\varphi(u)/\partial u_k\partial u_r, 1 \leq k, r \leq d$$

under additional **LL**, *here* $\widehat{b}(u) = \int_E \pi(dx)b(u;x), \widehat{b}_0(u) = \int_{\mathbf{R}^d} v\widehat{\Gamma}(u, dv)$,

$$\widehat{b}_1(u) = \int_E \pi(dx)b_1(u;x)R_0 b'^*_1(u;x),$$

$$\widehat{\Gamma}(u, dv) = \int_E \pi(dx)\Gamma(u, dv;x), c_0(u;x) = \int_{\mathbf{R}^d} vv^*\Gamma(u, dv;x),$$

$$\sigma^2(u) = 2\int_E \pi(dx)\{b_1(u;x)R_0 b_1^*(u;x) + \frac{1}{2}[c(u;x) - c_0(u;x)]\}, \quad \sigma^2(u) > 0,$$

where R_0 *is the potential operator of the switching Markov process,* $QR_0 = R_0Q = \Pi - I$,

$$\lambda(u) = \widehat{\Gamma}(u, \mathbf{R}^d), \qquad \Gamma^0(u, dv) = \widehat{\Gamma}(u, dv)/\lambda(u).$$

REMARK 2.4.– *The assertion of the positivity of the characteristic*

$$\sigma^2(u) = \frac{1}{2}\int_E \pi(dx)[c(u;x) - c_0(u;x)]$$

for random evolutionary systems with switching was investigated and proved in Pinsky (1991, pp. 24–26).

REMARK 2.5.– *The limit Lévy process contains three components: the deterministic drift, the diffusion part and the Poisson jumping part. Comparing the obtained result with Example 1.2 and the Lévy approximation conditions from section 1.1.10, we draw the following obvious conclusion: the function that determines the deterministic drift is conditioned by frequent small jumps of the process; function* $\sigma^2(u)$*, defining the diffusion component, is determined by frequent large jumps of the process; finally, the rare extremes of the limit process are set by the measure of rare large jumps of the pre-limit process.*

There are three possible cases:

1) if the condition $\varepsilon^{-2}\int_{|v|>\delta} v^2\Gamma^\varepsilon(dv;x) \to 0, \varepsilon \to 0, \forall\delta > 0$ *is met (see Jacod and Shiryaev (2003, p. 558, Theorem 4.21)), then the limit process has no Poisson component;*

2) if $\widehat{b}(u) - \widehat{b}_0(u) + \widehat{b}_1(u) = 0$*, then the limit process does not contain the deterministic drift;*

3) if $\sigma^2(u) = 0$*, then the limit process has no diffusion component. A possible variant of this case occurs when* $c(u;x) = c_0(u;x)$*, hence* $b_1(u;x) = 0$ *and we fall into conditions of Poisson approximation after renormalization* $\varepsilon^2 = \widetilde{\varepsilon}$ *(see, for example, Korolyuk and Limnios (2005, Chapter 7)).*

REMARK 2.6.– *The presence of the term* $\widehat{b}_{1k}(u)$ *in the part of the limit generator that determines the deterministic drift significantly distinguishes the results obtained in our book for processes with local independent increments, from the results of Korolyuk and Limnios (2005) for processes with independent increments. This is due to the fact that in the case where* $b_1(x)$ *does not depend on* u *derivative in the term*

$$\widehat{b}_1(u) = \int_E \pi(dx)b_1(u;x)R_0{b'}_1^*(u;x)$$

equals 0.

In addition, it should be noted that the asymptotics of the second moment in condition ***L1*** *(see section 1.1.10) contains the second modified characteristic* $c(u;x)$ *(see Jacod and Shiryaev (2003, p. 555, relation (4.2))). In the limit, this characteristic consists of two parts: the second moment of the Poisson component and the dispersion of the diffusion component, namely* $c(u;x) = c_0(u;x) + \sigma^2(u;x)$*. Under Poisson approximation conditions,* $c(u;x) = c_0(u;x)$*, hence* $\sigma^2 = 0$ *and the limit process* $\xi^0(t)$ *does not contain a diffusion component.*

PROOF OF THEOREM 2.5.– The proof is based on the semimartingale representation of process [2.10]. Predictable characteristics of semimartingale [2.10] have the following form (see Çinlar *et al.* (1980, *Theorems 6.27 and 7.16*)):

– $B^\varepsilon(t) = \varepsilon^{-2}\int_0^t b_\varepsilon(\xi^\varepsilon(s); x_s^\varepsilon)ds = \varepsilon^{-1}\int_0^t b_1(\xi^\varepsilon(s); x_s^\varepsilon)ds + \int_0^t b(\xi^\varepsilon(s); x_s^\varepsilon)ds$ $+ \theta_b^\varepsilon$,

– $C^\varepsilon(t) = \varepsilon^{-2}\int_0^t c_\varepsilon(\xi^\varepsilon(s); x_s^\varepsilon)ds = \int_0^t c(\xi^\varepsilon(s); x_s^\varepsilon)ds + \theta_c^\varepsilon$,

– $\Gamma^\varepsilon(t) = \varepsilon^{-2}\int_0^t \int_{\mathbf{R}^d} g(v)\Gamma^\varepsilon(\xi^\varepsilon(s), dv; x_s^\varepsilon)ds$
$= \int_0^t \int_{\mathbf{R}^d} g(v)\Gamma(\xi^\varepsilon(s), dv; x_s^\varepsilon)ds + \theta_g^\varepsilon$,

where $x_t^\varepsilon := x(t/\varepsilon^2), t \geq 0$, and $\sup_{x\in E} |\theta_\cdot^\varepsilon| \to 0, \quad \varepsilon \to 0$.

The jumping martingale part of semimartingale [2.10] is given by the ratio

$$\mu^\varepsilon(t) = \int_0^t \int_{\mathbf{R}^d} v[\mu^\varepsilon(ds, dv; x_s^\varepsilon) - \Gamma^\varepsilon(\xi^\varepsilon(s), dv; x_s^\varepsilon)ds].$$

Here, $\mu^\varepsilon(ds, dv; x), x \in E$ is the family of counting measures with the characteristics

$$\mathbf{E}\mu^\varepsilon(ds, dv; x) = \Gamma^\varepsilon(u, dv; x)ds.$$

The proof is carried out in two stages.

STAGE 1. In terms of the Lévy approximation, it is more difficult to prove the compactness of the process, in particular the lemma, similar to Lemma 2.1.

LEMMA 2.3.– *Given* **C3**, **PA1**, *there is a constant* $k > 0$, *independent of* ε *but dependent on* T, *such that*

$$\mathbf{E} \sup_{t\leq T} |\xi^\varepsilon(t)|^2 \leq k_T.$$

PROOF.– Semimartingale [2.10] also has a representation

$$\xi^\varepsilon(t) = u + B_t^\varepsilon + M_t^\varepsilon,$$

where $u = \xi^\varepsilon(0)$; B_t^ε is the predictable drift:

$$B_t^\varepsilon = \int_0^t \int_{\mathbf{R}^d} v\Gamma(\xi^\varepsilon(s), dv; x_s^\varepsilon)ds + \theta_b^\varepsilon(t) = \varepsilon^{-1}\int_0^t b_1(\xi^\varepsilon(s), x_s^\varepsilon)ds$$

$$+ \int_0^t b(\xi^\varepsilon(s), x_s^\varepsilon)ds + \theta_b^\varepsilon(t) =: \widetilde{B}_1^\varepsilon(t) + \widetilde{B}^\varepsilon(t) + \theta_b^\varepsilon(t),$$

and M_t^ε is a local quadratically integrated martingale

$$M_t^\varepsilon = \int_0^t \int_{\mathbf{R}^d} v[\mu(ds, dv; x_s^\varepsilon) - \Gamma(\xi^\varepsilon(s), dv; x_s^\varepsilon)ds] + \theta_c^\varepsilon(t),$$

and for each finite $T > 0$

$$\sup_{0\leq t\leq T} |\theta^{\varepsilon}_{\cdot}(t)| \to 0, \varepsilon \to 0.$$

Divide the process $\xi^{\varepsilon}(t)$ into two components.

The first part is of the order ε^{-1}

$$\widetilde{B}_1^{\varepsilon}(t) = \varepsilon^{-1} \int_0^t b_1(\xi^{\varepsilon}(s), x_s^{\varepsilon}) ds,$$

its martingale characterization has the form

$$\widetilde{\mu}_t^{\varepsilon} = \varphi^{\varepsilon}(\widetilde{B}_1^{\varepsilon}(t), x_t^{\varepsilon}) + \varphi^{\varepsilon}(\widetilde{B}_1^{\varepsilon}(0), x_0^{\varepsilon}) - \int_0^t \mathbf{L}^{\varepsilon}\varphi^{\varepsilon}(\widetilde{B}_1^{\varepsilon}(s), x_s^{\varepsilon}) ds, \qquad [2.11]$$

where the generator

$$\mathbf{L}^{\varepsilon}\varphi(u;x) = \varepsilon^{-2} q(x)[\widetilde{\mathbf{B}}_1^{\varepsilon}(x) - I]\varphi(u;x),$$

here $\widetilde{\mathbf{B}}_1^{\varepsilon}(x)\varphi(u) = \varphi(u + \varepsilon b_1(u;x)) = \varepsilon b_1(u;x)\varphi'(u) + \varepsilon\theta^{\varepsilon}(x)\varphi(u)$.

The last generator can be written as:

$$\mathbf{L}^{\varepsilon} = \varepsilon^{-2} Q + \varepsilon^{-1}\widetilde{\mathbf{B}}_1(x)P + \theta^{\varepsilon},$$

here $\widetilde{\mathbf{B}}_1(x)\varphi(u) = \varepsilon b_1(u;x)\varphi'(u)$.

Thus (see Korolyuk and Limnios (2005, Remark 1.2.2 and Theorem 1.2)), the quadratic characteristic of martingale [2.11] has the form

$$< \widetilde{\mu}^{\varepsilon} >_t = \int_0^t \left[\mathbf{L}^{\varepsilon}(\varphi^{\varepsilon}(\widetilde{B}_1^{\varepsilon}(s), x_t^{\varepsilon}))^2 - 2\varphi^{\varepsilon}(\widetilde{B}_1^{\varepsilon}(s), x_s^{\varepsilon})\mathbf{L}^{\varepsilon}\varphi^{\varepsilon}(\widetilde{B}_1^{\varepsilon}(s), x_s^{\varepsilon})\right] ds.$$

Consider the action of the generator $\mathbf{L}^{\varepsilon} = \varepsilon^{-2}Q + \varepsilon^{-1}\widetilde{\mathbf{B}}_1(x)P + \theta^{\varepsilon}$ on test functions

$$\varphi^{\varepsilon}(u;x) = \varphi(u) + \varepsilon\varphi_1(u;x), \varphi(u) \in C_3(\mathbf{R}^d).$$

We obtain the following form of the integrand:

$$Q\varphi_1^2 - 2\varphi_1 Q\varphi_1 + \theta^{\varepsilon}\varphi^{\varepsilon}.$$

Thus, the integrand is bounded. Due to boundedness of the quadratic characteristic, the compactness of the martingale $\widetilde{\mu}_t^{\varepsilon}$ follows. Thus, the functional $\varphi(\widetilde{B}_1^{\varepsilon}(t))$ is also compact and bounded uniformly by ε. According to Ethier and Kurtz (1986), it follows the compactness of $\widetilde{B}_1^\varepsilon(t)$ and uniform by ε boundedness of

$$\sup_{t\leq T} \widetilde{B}_1^\varepsilon(t) < \sqrt{k_3}, \qquad [2.12]$$

because the test functions $\varphi(u)$ belong to the class that determines measure.

Now consider the other part of the process $\xi^\varepsilon(t)$.

For an arbitrary process $y(t), t \geq 0$, we denote $y_t^\dagger = \sup_{s\leq t} |y(s)|$, so we have

$$((\xi_t^\varepsilon)^\dagger)^2 \leq 4[u^2 + ((\widetilde{B}_1^\varepsilon)^\dagger(t))^2 + ((\widetilde{B}^\varepsilon)^\dagger(t))^2 + ((M_t^\varepsilon)^\dagger)^2]. \qquad [2.13]$$

By **C3**:

$$(\widetilde{B}_t^\varepsilon)^\dagger = \int_0^t b(\xi^\varepsilon(s), x_s^\varepsilon)ds \leq L\int_0^t (1 + (\xi_s^\varepsilon)^\dagger)ds. \qquad [2.14]$$

and by **C3**, we obtain

$$|\langle M^\varepsilon\rangle_t| = |\int_0^t \int_{\mathbf{R}^d} vv^*\Gamma(\xi^\varepsilon(s), dv; x_s^\varepsilon)ds| \leq Lr_1 \int_0^t [1 + ((\xi_s^\varepsilon)^\dagger)^2]ds, \qquad [2.15]$$

where $r_1 = \int_{\mathbf{R}^d} |v|^2 f(v)dv$.

According to inequalities [2.12]–[2.15], the Doob inequality

$$\mathbf{E}((M_t^\varepsilon)^\dagger)^2 \leq 4|\mathbf{E}\langle M^\varepsilon\rangle_t|,$$

and the Cauchy–Bunyakovsky–Schwarz inequality ($[\int_0^t \varphi(s)ds]^2 \leq t\int_0^t \varphi^2(s)ds$), we get an estimate

$$\mathbf{E}((\xi_t^\varepsilon)^\dagger)^2 \leq k_1 + 4k_3 + k_2\int_0^t \mathbf{E}((\xi_s^\varepsilon)^\dagger)^2 ds,$$

where k_1, k_2 and k_3 are positive constants independent of ε.

According to the Gronwall–Bellman inequality (see, for example, Ethier and Kurtz (1986, p. 498), we have

$$\mathbf{E}((\xi_t^\varepsilon)^\dagger)^2 \leq (k_1 + 4k_3)\exp(k_2 t).$$

The lemma is proved.

Proof of Corollary 2.2 and Lemma 2.2 in the case of the Lévy approximation does not differ from the case of the Poisson approximation.

Therefore, the conditions proved in Corollary 2.2 and Lemma 2.2 are necessary and sufficient conditions for the compactness of the family of processes $\xi^\varepsilon(t)$, $t \geq 0, \varepsilon > 0$.

STAGE 2. In the following, we denote by $A^\varepsilon(t)$ the main part of any of the predictable characteristics $B^\varepsilon(t), C^\varepsilon(t), \Gamma^\varepsilon(t)$.

Consider a three-component Markov process $A^\varepsilon(t), \xi^\varepsilon(t), x_t^\varepsilon, t \geq 0$, which can be characterized by a martingale

$$\mu_t^\varepsilon = \varphi(A^\varepsilon(t), \xi^\varepsilon(t), x_t^\varepsilon) - \int_0^t \mathbf{L}^\varepsilon \varphi(A^\varepsilon(s), \xi^\varepsilon(s), x_t^\varepsilon) ds,$$

where the generator $\mathbf{L}^\varepsilon$ has a representation (see [1.12])

$$\mathbf{L}^\varepsilon \varphi(u, v; x) = \varepsilon^{-2} Q\varphi(\cdot, \cdot; x) + \mathbf{\Gamma}^\varepsilon(x)\varphi(u, \cdot; \cdot) + A^\varepsilon(u; x)\varphi(\cdot, v; \cdot), \qquad [2.16]$$

here

$$\mathbf{\Gamma}^\varepsilon(x)\varphi(u) = \varepsilon^{-2} \int_{\mathbf{R}^d} [\varphi(u+v) - \varphi(u)]\Gamma^\varepsilon(u, dv; x), x \in E,$$

$$Q\varphi(x) = q(x) \int_E P(x, dy)[\varphi(y) - \varphi(x)],$$

$$\begin{aligned} A^\varepsilon(u; x)\varphi(v) &= a_\varepsilon(u; x)\varphi'(v) = a(u; x)\varphi'(v) \\ &+ \theta_a^\varepsilon \varphi'(v) =: A(u; x)\varphi(v) + \theta_a^\varepsilon \varphi'(v). \end{aligned}$$

To prove the convergence of the predictable characteristics, it suffices to consider the action of the generator $\mathbf{L}^\varepsilon$ on the test function of two variables $\varphi^\varepsilon(v; x) = \varphi(v) + \varepsilon^2 \varphi_2(v; x)$.

Thus, the generator has the form

$$\begin{aligned} \mathbf{L}^\varepsilon \varphi^\varepsilon(v; x) &= [\varepsilon^{-2} Q + A^\varepsilon(u; x)][\varphi(v) + \varepsilon^2 \varphi_2(v; x)] \\ &= \varepsilon^{-2} Q\varphi(v) + [Q\varphi_2(v; x) + A(u; x)\varphi(v)] + \varepsilon^2 A^\varepsilon(u; x)\varphi_2(v; x) + \theta_a^\varepsilon \varphi'(v). \end{aligned}$$

Therefore, the asymptotic representation of the last generator can be obtained by solving the singular perturbation problem given by the equations

$$\begin{aligned} &Q\varphi(v) = 0, \\ &Q\varphi_2(v; x) + A(u; x)\varphi(v) = \mathbf{L}\varphi(v). \end{aligned}$$

From the first equation, it is obvious $\varphi(v) \in \mathcal{N}_Q$, and from the solvability condition for the second equation, we have:

$$\mathbf{L}\varphi(v) = \widehat{A}(u)\varphi(v) := \Pi A(u;x)\Pi\varphi(v).$$

Thus, applying Theorem 2.2, we can state a weak convergence of the predictable characteristics of the process.

Having received result about weak convergence of the process, the form of the limit generator can be obtained from the asymptotic representation of the generator [2.16] on the test functions of the form $\varphi^\varepsilon(u;x) = \varphi(u) + \varepsilon\varphi_1(u;x) + \varepsilon^2\varphi_2(u;x)$.

The generator has the form

$$\mathbf{L}^\varepsilon\varphi^\varepsilon(u;x) = [\varepsilon^{-2}Q + \mathbf{\Gamma}^\varepsilon(x)][\varphi(u) + \varepsilon\varphi_1(u;x) + \varepsilon^2\varphi_2(u;x)]$$

$$= \varepsilon^{-2}Q\varphi(u) + \varepsilon^{-1}Q\varphi_1(u;x) + [Q\varphi_2(u;x) + \mathbf{\Gamma}^\varepsilon(x)\varphi(u)]$$

$$+\varepsilon\mathbf{\Gamma}^\varepsilon(x)\varphi_1(u;x) + \varepsilon^2\mathbf{\Gamma}^\varepsilon(x)\varphi_2(u;x).$$

Using Lemma 1.2, we can write the following asymptotic representation:

$$\mathbf{L}^\varepsilon\varphi^\varepsilon(u;x) = \varepsilon^{-2}Q\varphi(u) + \varepsilon^{-1}[Q\varphi_1(u;x) + b_1(u;x)\varphi'(u)] + [Q\varphi_2(u,x)$$

$$+\mathbf{\Gamma}(x)\varphi(u) + b_1(u;x)\varphi_1'(u;x)] + \varepsilon\mathbf{\Gamma}(x)\varphi_1(u;x) + \varepsilon^2\mathbf{\Gamma}^\varepsilon(x)\varphi_2(u;x),$$

where

$$\mathbf{\Gamma}(x)\varphi(u) := (b(u;x) - b_0(u;x))\varphi'(u) + \frac{1}{2}(c(u;x) - c_0(u;x))\varphi''(u)$$

$$+ \int_{\mathbf{R}^d}[\varphi(u+v) - \varphi(u)]\Gamma(u,dv;x).$$

Thus, the singular perturbation problem is reduced to a system of equations

$$Q\varphi(u) = 0,$$

$$Q\varphi_1(u;x) + b_1(u;x)\varphi'(u) = 0,$$

$$Q\varphi_2(u,x) + \mathbf{\Gamma}(x)\varphi(u) + b_1(u;x)\varphi_1'(u;x) = \widehat{\mathbf{\Gamma}}\varphi(u).$$

From the first equation, it is obvious that $\varphi(u) \in \mathcal{N}_Q$, and from the solvability condition for the second equation and balance condition **L3**, we have:

$$\widehat{b}_1(u;x)\varphi'(u) = 0, \ \varphi_1(u;x) = R_0 b_1^*(u;x)\varphi'(u).$$

From the solvability condition for the last equation, we have:

$$\begin{aligned}\widehat{\mathbf{\Gamma}}\varphi(u) &= \Pi\mathbf{\Gamma}(x)\Pi\varphi(u) + \Pi b_1(u;x)R_0 b_1'^{*}(u;x)\varphi'(u)\\ &\quad + \Pi b_1(u;x)R_0 b_1^{*}(u;x)\varphi''(u)\\ &= (\widehat{b}(u) - \widehat{b}_0(u) + \widehat{b}_1(u))\varphi'(u) + \frac{1}{2}\sigma^2(u)\varphi''(u)\\ &\quad + \lambda(u)\int_{\mathbf{R}^d}[\varphi(u+v) - \varphi(u)]\Gamma^0(u,dv).\end{aligned}$$

Using condition **LL** and the results from Skorokhod *et al.* (2002, Chapter 2), we can say that the limit generator really determines the limit process $\xi^0(t)$.

Theorem 2.5 is proved.

2.1.2. *Semi-Markov switching*

In the case of a semi-Markov switching process (see section 1.1.4), the compensating operator plays an important role (see Definition 1.5 in section 1.1.8, formula [1.13]) and its representation [1.14].

An additional condition for the distribution function of jump moments of the switching process is Cramer's condition:

C4

$$\sup_{x\in E}\int_0^{\infty} e^{ht}F_x(dt) \leq H < +\infty, h > 0.$$

Therefore, under the Poisson approximation conditions, the following theorem holds true.

THEOREM 2.6.– *Under the conditions* ***CM, C1', C2–C4*** *and normalization* [2.1], *a weak convergence in* $D_{\mathbf{R}^d}[0,\infty)$ *holds true*

$$\xi^{\varepsilon}(t) \Rightarrow \xi^0(t), \quad \varepsilon \to 0.$$

The limit process $\xi^0(t), t \geq 0$ *is determined by the generator*

$$\widehat{\mathbf{\Gamma}}\varphi(u) = \widehat{b}(u)\varphi'(u) + \int_{\mathbf{R}^d}[\varphi(u+v) - \varphi(u) - v\varphi'(u)]\widehat{\Gamma}(u,dv) \qquad [2.17]$$

provided ***LL****, where the averaged deterministic drift*

$$\widehat{b}(u) = \int_E \pi(dx)b(u;x),$$

and the averaged intensity kernel

$$\widehat{\Gamma}(u, dv) = \int_E \pi(dx)\Gamma(u, dv; x).$$

The first stage of proof coincides with the case of Markov switching.

In the second stage, we denote by $A^\varepsilon(t)$ the main part of any of the predictable characteristics $B^\varepsilon(t), C^\varepsilon(t), \Gamma^\varepsilon(t)$.

The extended Markov renewal process is a four-component Markov chain

$$A_n^\varepsilon = A_0^\varepsilon(\tau_n^\varepsilon), \xi_n^\varepsilon, x_n^\varepsilon, \tau_n^\varepsilon, n \geq 0,$$

where $x_n^\varepsilon = x^\varepsilon(\tau_n^\varepsilon), x^\varepsilon(t) := x(t/\varepsilon), \xi_n^\varepsilon = \xi^\varepsilon(\tau_n^\varepsilon)$ and $\tau_{n+1}^\varepsilon = \tau_n^\varepsilon + \varepsilon\theta_n^\varepsilon, n \geq 0$, and

$$P(\theta_{n+1}^\varepsilon \leq t | x_n^\varepsilon = x) = F_x(t) = P(\theta_x \leq t).$$

According to Lemma 7.5 of Korolyuk and Limnios (2005), the compensating operator of the extended Markov renewal process defined by formula [1.13] has a representation [1.14] (see also Korolyuk and Limnios (2005, section 2.8)).

$$\mathbb{L}^\varepsilon\varphi(u, v; x, t) = \varepsilon^{-1}q(x)$$
$$\times \left[\int_0^\infty F_x(ds)A_{\varepsilon s}(u; x)\Gamma_{\varepsilon s}(x) \int_E P(x, dy)\varphi(u, v; y, t+\varepsilon s) - \varphi(u, v; x, t)\right],$$

where $\Gamma_{\varepsilon s}(x), s \geq 0, x \in E$, is a family of semigroups corresponding to the generator $\mathbf{\Gamma}^\varepsilon(x), x \in E$; $A_{\varepsilon s}(u; x), s \geq 0, x \in E, u \in \mathbf{R}^d$ is a family of semigroups with the generator $A(u; x)\varphi(v) = a(u; x)\varphi'(v)$.

According to Lemma 7.9 from Korolyuk and Limnios (2005) on test functions $\varphi \in C^2(\mathbf{R}^d \times \mathbf{R}^d) \times \mathbf{B}(E)$, the last generator has an asymptotic representation

$$\mathbb{L}^\varepsilon\varphi(u, v; x) = \varepsilon^{-1}Q\varphi(\cdot, \cdot; x) + A(u, x)P\varphi(\cdot, v; \cdot)$$
$$+ \mathbf{\Gamma}^\varepsilon(x)P\varphi(u, \cdot; \cdot) + \varepsilon\theta^\varepsilon(x)P\varphi(u, v; x), \quad [2.18]$$

where P is the transition operator associated with $P(x, B)$,

$$Q\varphi(x) = q(x)\int_E P(x, dy)[\varphi(\cdot; y) - \varphi(\cdot; x)].$$

To prove the convergence of the predictable characteristics, it suffices to consider the action of the generator $\mathbb{L}^\varepsilon$ on test functions of two variables $\varphi(v, x)$.

Thus, it has a representation

$$\mathbb{L}^\varepsilon\varphi(v; x) = [\varepsilon^{-1}Q + A(u, x)P + \varepsilon\theta^\varepsilon(x)P]\varphi(v; x).$$

Solution of the singular perturbation problem in the form $\mathbb{L}^\varepsilon\varphi^\varepsilon = \widehat{\mathbb{L}}\varphi + \theta^\varepsilon\varphi$ on test functions of the form $\varphi^\varepsilon(v;x) = \varphi(v) + \varepsilon\varphi_1(v;x)$ is similar to the case of Markov switching and has the form

$$\widehat{\mathbb{L}} = \widehat{A}(u), \; \widehat{A}(u) := \Pi A(u;x)\Pi, \qquad [2.19]$$

because $P\Pi = \Pi$.

We can rewrite $\widehat{A}(u) = \int_E \pi(dx)A(u,x)$.

So,

$$\widehat{A}(u)\varphi(v) = \widehat{a}(u)\varphi'(v),$$

where $\widehat{a}(u) := \int_E \pi(dx)a(u,x)$.

Applying Theorem 2.2, we can state a weak convergence of the predictable characteristics of the process.

The last stage of the proof is the application of Theorem 2.1. Indeed, all the conditions of the theorem are satisfied.

The strong majorization hypothesis is performed with the majorizing function given in condition **C3**. Condition **C2** is a condition of large jumps for the predictable jump measures; moreover, the last condition of Theorem 2.1 follows from condition **C2**. Conditions (iv) and (v) are obviously fulfilled.

Weak convergence of predictable characteristics follows from the solution of the singular perturbation problem.

Thus, there is a weak convergence in $D_{\mathbf{R}^d}[0,\infty)$ to the process, which is uniquely determined by the limit predictable characteristics.

Substituting the operators $B(u;x), C(u;x), \Gamma(u;x)$ instead of $A(u;x)$ in [2.19], we see that the limit Markov process has the following predictable characteristics:

$$B^0(t) = \int_0^t \widehat{b}(\xi^0(s))ds, \quad C^0(t) = \int_0^t \widehat{c}(\xi^0(s))ds, \quad \Gamma_g^0(t) = \int_0^t \widehat{\Gamma}_g(\xi^0(s))ds.$$

Since the weak convergence of the process is proved, the form of the limit operator can be obtained from the asymptotic representation of the generator [2.18] on the test functions of the form $\varphi^\varepsilon(u;x) = \varphi(u) + \varepsilon\varphi_1(u;x)$.

The generator has the form

$$\mathbb{L}^\varepsilon\varphi^\varepsilon(u;x) = [\varepsilon^{-1}Q + \mathbf{\Gamma}^\varepsilon(x)P][\varphi(u) + \varepsilon\varphi_1(u;x)]$$

$$= \varepsilon^{-1}Q\varphi(u) + [Q\varphi_1(u;x) + \mathbf{\Gamma}^\varepsilon(x)P\varphi(u)] + \varepsilon\mathbf{\Gamma}^\varepsilon(x)P\varphi_1(u;x).$$

Using Lemma 1.1, we can write the following asymptotic representation:

$$\mathbb{L}^\varepsilon \varphi^\varepsilon(u;x) = \varepsilon^{-1}Q\varphi(u) + [Q\varphi_1(u;x) + \mathbf{\Gamma}(x)P\varphi(u)] + \varepsilon\mathbf{\Gamma}^\varepsilon(x)P\varphi_1(u;x) + \theta^\varepsilon\varphi(u).$$

Thus, the singular perturbation problem is reduced to a system of equations

$$Q\varphi(u) = 0,$$

$$Q\varphi_1(u;x) + \mathbf{\Gamma}(x)P\varphi(u) = \widehat{\mathbf{\Gamma}}\varphi(u).$$

From the first equation, it is obvious that $\varphi(u) \in \mathcal{N}_Q$, and from the solvability condition for the second equation, we have:

$$\widehat{\mathbf{\Gamma}}\varphi(u) = \Pi\mathbf{\Gamma}(x)P\Pi\varphi(u)$$

$$= \widehat{b}(u)\varphi'(u) + \int_{\mathbf{R}^d} [\varphi(u+v) - \varphi(u) - v\varphi'(u)]\widehat{\Gamma}(u,dv).$$

Using condition **LL** and the results from Skorokhod *et al.* (2002, Chapter 2), we can say that the limit generator really determines the limit process $\xi^0(t)$.

Theorem 2.6 is proved.

The following theorem holds in the Lévy approximation scheme.

THEOREM 2.7.– *Under the conditions **CSM, C1', C2–C4** and normalization* [2.10], *there is a weak convergence in* $D_{\mathbf{R}^d}[0,\infty)$

$$\xi^\varepsilon(t) \Rightarrow \xi^0(t), \quad \varepsilon \to 0.$$

The limit process $\xi^0(t), t \geq 0$ *is a Lévy process, which under condition **LL** is determined by the generator*

$$\widehat{\mathbf{\Gamma}}\varphi(u) = (\widehat{b}(u) - \widehat{b}_0(u) + \widehat{b}_1(u))\varphi'(u) + \frac{1}{2}\sigma^2(u)\varphi''(u) + \lambda(u)\int_{\mathbf{R}^d}[\varphi(u+v) - \varphi(u)]\Gamma^0(u,dv),$$

where:

$$\widehat{b}(u) = \int_E \pi(dx)b(u;x), \widehat{b}_0(u) = \int_E v\widehat{\Gamma}(u,dv),$$

$$\widehat{b}_1(u) = \int_E \pi(dx)b_1(u;x)\widetilde{R}_0 b_1'^*(u;x) - \int_E \pi(dx)b_1(u;x)b_1'^*(u;x),$$

$$\widehat{\Gamma}(u,dv) = \int_E \pi(dx)\Gamma(u,dv;x),$$

$$\widehat{b_1^2}(u) = \int_E \pi(dx) b_1(u;x) b_1^*(u;x), c_0(u;x) = \int_E vv^*\Gamma(u, dv; x),$$

$$\sigma^2(u) = 2[\int_E \pi(dx)\{b_1(u;x)\widetilde{R}_0 b_1^*(u;x) + \frac{1}{2}[c(u;x) - c_0(u;x)]\} - \widehat{b_1^2}(u)],$$

$$\sigma^2(u) > 0, \lambda(u) = \Gamma(u, \mathbf{R}^d), \qquad \Gamma^0(u, dv) = \widehat{\Gamma}(u, dv)/\lambda(u).$$

Here $\widetilde{R}_0$ is the potential operator of the embedded Markov chain (see Korolyuk and Limnios 2005).

By using the potential operator of the embedded Markov chain $\widetilde{R}_0$ instead of the potential operator of the Markov process R_0, it was possible to simplify the expression for $\sigma^2(u)$ significantly and show that it really corresponds to the well-known results (see Theorem 2.5 and Remark 2.5).

At the first stage, the proof completely corresponds to the case of Markov switching in the Lévy approximation scheme (section 2.1.1).

The proof of weak convergence of predictable characteristics is carried out similarly to the case of semi-Markov switching in the Poisson approximation scheme.

Since the weak convergence of the process is proved, the form of the limit operator can be obtained from the asymptotic representation of the generator

$$\mathbb{L}^\varepsilon \varphi(u, v; x) = \varepsilon^{-2} Q\varphi(\cdot, \cdot; x) + A(u, x)P\varphi(\cdot, v; \cdot) + \mathbf{\Gamma}^\varepsilon(x)P\varphi(u, \cdot; \cdot)$$
$$+ \varepsilon\theta^\varepsilon(x)P\varphi(u, v; x),$$

on test functions of the form $\varphi^\varepsilon(u;x) = \varphi(u) + \varepsilon\varphi_1(u;x) + \varepsilon^2\varphi_2(u;x)$.

The generator has the form

$$\mathbb{L}^\varepsilon \varphi^\varepsilon(u;x) = [\varepsilon^{-2}Q + \mathbf{\Gamma}^\varepsilon(x)P][\varphi(u) + \varepsilon\varphi_1(u;x) + \varepsilon^2\varphi_2(u;x)]$$
$$= \varepsilon^{-2}Q\varphi(u) + \varepsilon^{-1}Q\varphi_1(u;x) + [Q\varphi_2(u;x) + \mathbf{\Gamma}^\varepsilon(x)P\varphi(u)]$$
$$+ \varepsilon\mathbf{\Gamma}^\varepsilon(x)P\varphi_1(u;x) + \varepsilon^2\mathbf{\Gamma}^\varepsilon(x)P\varphi_2(u;x).$$

Using Lemma 1.2, we can write the following asymptotic representation:

$$\mathbb{L}^\varepsilon \varphi^\varepsilon(u;x) = \varepsilon^{-2}Q\varphi(u) + \varepsilon^{-1}[Q\varphi_1(u;x) + b_1(u,x)P\varphi'(u)]$$
$$+ [Q\varphi_2(u;x) + \mathbf{\Gamma}(x)P\varphi(u) + b_1(u,x)P\varphi_1'(u,x)]$$
$$+ \varepsilon\mathbf{\Gamma}(x)P\varphi_1(u;x) + \varepsilon^2\mathbf{\Gamma}^\varepsilon(x)P\varphi_2(u;x) + \theta^\varepsilon\varphi(u).$$

Thus, the singular perturbation problem is reduced to a system of equations

$$Q\varphi(u) = 0,$$

$$Q\varphi_1(u;x) + b_1(u,x)P\varphi'(u) = 0,$$

$$Q\varphi_2(u;x) + \mathbf{\Gamma}(x)P\varphi(u) + b_1(u,x)P\varphi_1'(u,x) = \widehat{\mathbf{\Gamma}}\varphi(u).$$

To simplify the formulas, we rewrite these equations for the embedded Markov chain. The corresponding operator

$$\widetilde{Q} := P - I,$$

and the potential operator has the property

$$\widetilde{R}_0(P - I) = \widetilde{\Pi} - I.$$

Thus,

$$\widetilde{Q}\varphi(u) = 0,$$

$$\widetilde{Q}\varphi_1(u;x) + b_1(u,x)P\varphi'(u) = 0,$$

$$\widetilde{Q}\varphi_2(u;x) + \mathbf{\Gamma}(x)P\varphi(u) + b_1(u,x)P\varphi_1'(u,x) = \widehat{\mathbf{\Gamma}}\varphi(u).$$

From the first equation, it is obvious that $\varphi(u) \in \mathcal{N}_{\widetilde{Q}}$, and from the solvability condition for the second equations and balance conditions **L3**, we have:

$$-\widehat{b}_1(u,x)P\varphi'(u) = 0, \varphi_1(u;x) = \widetilde{R}_0 b_1(u,x)P\varphi'(u)$$

Substitute the resulting expression into the last equation:

$$\widetilde{Q}\varphi_2(u;x) + \mathbf{\Gamma}(x)P\varphi(u) + b_1(u,x)P\widetilde{R}_0 b_1'(u,x)P\varphi'(u)$$

$$+b_1(u,x)P\widetilde{R}_0 b_1(u,x)P\varphi''(u) = \widehat{\mathbf{\Gamma}}\varphi(u).$$

Using the solvability condition and taking into account an equality $P\widetilde{R}_0 = \widetilde{R}_0 + \widetilde{\Pi} - I$, we finally obtain:

$$\begin{aligned}\widehat{\mathbf{\Gamma}}\varphi(u) &= \widetilde{\Pi}[\mathbf{\Gamma}(x)P\varphi(u) + b_1(u,x)P\widetilde{R}_0 b_1'(u,x)P\varphi'(u)\\ &\quad +b_1(u,x)P\widetilde{R}_0 b_1(u,x)P\varphi''(u)]\widetilde{\Pi}\\ &= (\widehat{b}(u) - \widehat{b}_0(u) + \widehat{b}_1(u))\varphi'(u) + \frac{1}{2}\sigma^2(u)\varphi''(u)\\ &\quad + \int_{\mathbf{R}^d}[\varphi(u+v) - \varphi(u)]\widehat{\Gamma}(u,dv).\end{aligned}$$

Using condition **LL** and the results from Skorokhod *et al.* (2002, Chapter 2), we can say that the limit generator really determines the limit process $\xi^0(t)$.

Theorem 2.7 is proved.

2.1.3. *Unbounded jump measures*

Consider random evolutionary system $\xi^\varepsilon(t), t \geq 0, \varepsilon > 0$, in $\mathbf{R}^d$ in normalization

$$\xi^\varepsilon(t) = \xi_0^\varepsilon + \int_0^t \eta_\varepsilon(ds; x(s/\varepsilon)).$$

A family of processes with locally independent increments $\eta^\varepsilon(t;x), t \geq 0, x \in E$ in $\mathbf{R}^d$, is set by generators (see Bertoin (1996, section I.2) and Korolyuk and Limnios (2005, section 1.2.4))

$$\begin{aligned}\mathbf{\Gamma}^\varepsilon(x)\varphi(u) &= b_\varepsilon(u;x)\varphi'(u) \\ &+ \int_{\mathbf{R}^d} [\varphi(u+v) - \varphi(u) - v\varphi'(u)\mathbf{1}_{(|v|\leq 1)}]\Gamma^\varepsilon(u, dv; x), \qquad [2.20]\end{aligned}$$

where $\varphi(u)$ is a real twice differentiable function in $\mathbf{R}^d$ equal to 0 at infinity, with sup norm $||\varphi|| = C_0^2(\mathbf{R}^d)$, $b_\varepsilon(u;x) = \int_{\mathbf{R}^d} v\Gamma^\varepsilon(u, dv; x)$ and $\Gamma^\varepsilon(u, dv; x)$ is the intensity kernel that satisfies the condition

$$\Gamma^\varepsilon(u, \{0\}; x) = 0.$$

THEOREM 2.8.– *Under the conditions* ***CM, C1–C3****, a weak convergence holds true*

$$\xi^\varepsilon(t) \Rightarrow \xi^0(t), \quad \varepsilon \to 0.$$

The limit process $\xi^0(t), t \geq 0$ under condition ***LL*** *is determined by the generator*

$$\widehat{\mathbf{\Gamma}}\varphi(u) = \widehat{b}(u)\varphi'(u) + \int_{\mathbf{R}^d} [\varphi(u+v) - \varphi(u) - v\varphi'(u)\mathbf{1}_{(|v|\leq 1)}]\widehat{\Gamma}(u, dv), \qquad [2.21]$$

where the averaged deterministic drift

$$\widehat{b}(u) = \Pi b(u;x) = \int_E \pi(dx) b(u;x),$$

and the averaged intensity kernel

$$\widehat{\Gamma}(u, dv) = \Pi\Gamma(u, dv; x) = \int_E \pi(dx)\Gamma(u, dv; x).$$

PROOF.– In general, the proof repeats the scheme described in section 2.1.1 for the bounded jump measures. However, the proof of a lemma similar to Lemma 2.1 changes slightly.

LEMMA 2.4.– *Under the conditions* **C3, PA1**, *there is a constant* $k_T > 0$, *independent of* ε *and dependent on* T, *such that*

$$\mathbf{E}\sup_{t\le T}|\xi^\varepsilon(t)|^2 \le k_T.$$

PROOF.– For an arbitrary process $y(t), t \ge 0$, we denote the process $y_t^\dagger = \sup_{s\le t}|y(s)|$.

From conditions **PA1** and **C4**, it follows that for an arbitrary fixed $t > 0$

$$\int_0^t \int_{\mathbf{R}^d} vv^*\Gamma^\varepsilon(\xi^\varepsilon(s), dv; x_s^\varepsilon)ds = \varepsilon\int_0^t c(\xi^\varepsilon(s); x_s^\varepsilon)ds + \varepsilon t\theta_c^\varepsilon(u; x)$$

$$\le \varepsilon t[L(1 + ((\xi_t^\varepsilon)^\dagger)^2) + \theta_c^\varepsilon(u; x)] < \infty P - \text{a.s.}$$

Increasing process $\int_0^t \int_{\mathbf{R}^d\setminus\{0\}} vv^*\Gamma^\varepsilon(\xi^\varepsilon(s), dv; x_s^\varepsilon)ds$ is continuous by t and therefore according to Dellacherie and Meyer (1978, Chapter 5, Theorem 28) (see also Liptser and Shiryayev (1989, Theorem 1.6.3)) is a compensator

$$\int_0^t \int_{\mathbf{R}^d\setminus\{0\}} vv^*\mu^\varepsilon(\xi^\varepsilon(s), d(s/\varepsilon), dv; x_s^\varepsilon).$$

Therefore, the process is a special semimartingale of the form

$$\xi^\varepsilon(t) = u + B_t^\varepsilon + M_t^\varepsilon, \qquad [2.22]$$

where $u = \xi^\varepsilon(0)$; B_t^ε is a predictable drift (see Ethier and Kurtz (1986)):

$$B_t^\varepsilon = \int_0^t b(\xi^\varepsilon(s), x_s^\varepsilon)ds + \int_0^t \int_{|v|>1} v\Gamma(\xi^\varepsilon(s), dv; x_s^\varepsilon)ds + \theta_b^\varepsilon(t),$$

and M_t^ε is a local quadratically integrated martingale

$$M_t^\varepsilon = \int_0^t \int_{\mathbf{R}^d\setminus\{0\}} v[\mu(\xi^\varepsilon(s), ds, dv; x_s^\varepsilon) - \Gamma(\xi^\varepsilon(s), dv; x_s^\varepsilon)ds] + \theta_m^\varepsilon(t),$$

and for an arbitrary finite $T > 0$

$$\sup_{0\le t\le T}|\theta_\cdot^\varepsilon(t)| \to 0, \varepsilon \to 0.$$

From [2.22], we have

$$((\xi_t^\varepsilon)^\dagger)^2 \le 3[u^2 + ((B_t^\varepsilon)^\dagger)^2 + ((M_t^\varepsilon)^\dagger)^2]. \qquad [2.23]$$

From condition **C3**, it follows the inequality

$$(B_t^\varepsilon)^\dagger \le L\int_0^t (1+(\xi_s^\varepsilon)^\dagger)ds + L\int_0^t\int_{|v|>1} |v|f(v)(1+(\xi_s^\varepsilon)^\dagger)dvds$$

$$\le L(1+r_1)\int_0^t (1+(\xi_s^\varepsilon)^\dagger)ds, \qquad [2.24]$$

where $r_1 = \int_{\mathbf{R}^d\setminus\{0\}} vv^* f(v)dv$.

According to the Doob inequality

$$\mathbf{E}((M_t^\varepsilon)^\dagger)^2 \le 4|\mathbf{E}\langle M^\varepsilon\rangle_t|,$$

where under condition **C3**, we have

$$|\langle M^\varepsilon\rangle_t| = |\int_0^t\int_{\mathbf{R}^d\setminus\{0\}} vv^*\Gamma(\xi^\varepsilon(s), dv; x_s^\varepsilon)ds|$$

$$\le Lr_1\int_0^t [1+((\xi_s^\varepsilon)^\dagger)^2]ds. \qquad [2.25]$$

From inequalities [2.23]–[2.26] and the Cauchy–Bunyakovsky–Schwarz inequality ($[\int_0^t \varphi(s)ds]^2 \le t\int_0^t \varphi^2(s)ds$), it follows:

$$\mathbf{E}((\xi_t^\varepsilon)^\dagger)^2 \le k_1 + k_2\int_0^t \mathbf{E}((\xi_s^\varepsilon)^\dagger)^2 ds,$$

where k_1 and k_2 are positive constants independent of ε.

According to the Gronwall–Bellman inequality, we have

$$\mathbf{E}((\xi_t^\varepsilon)^\dagger)^2 \le k_1\exp(k_2 t).$$

The lemma is proved.

Lemma 2.2 and Corollary 2.2 are proved as in the case of a bounded measure.

We can now apply Theorem 2.3, which establishes the exponential compactness of semimartingales.

LEMMA 2.5.– *Under the conditions **C1–C3**, the family of processes $\xi^\varepsilon(t)$ is relatively compact.*

PROOF.– To prove the relative compactness, it is necessary to check the fulfillment of the conditions **LP1–LP5** of Theorem 2.3.

It is easy to see that condition **LP1** is a direct consequence of condition **C3**.

Let us show that under condition **CCC**, which is a consequence of Lemma 2.4 (see Corollary 2.2), condition **LP2** and the second part of condition **LP5** holds true. Note that in the case of a bounded measure, their implementation is obvious. Indeed, with **C3** and the definition of $\Gamma(u, v; x)$ on the set $\{\sup_{t\le T} |\xi^\varepsilon(t) \le c|\}$, we have for the function $g(v) = \begin{cases} 0, |v| \le l \\ 1, |v| > l \end{cases}$:

$$\int_0^T \int_{|v|>l} \Gamma^\varepsilon(\xi^\varepsilon(s), dv; x_s^\varepsilon)ds = \int_0^T \int_{\mathbf{R}^d} g(v)\Gamma^\varepsilon(\xi^\varepsilon(s), dv; x_s^\varepsilon)ds$$

$$= \varepsilon \int_0^T \int_{\mathbf{R}^d} g(v)\Gamma(\xi^\varepsilon(s), dv; x_s^\varepsilon)ds + \varepsilon T\theta_g^\varepsilon \le \varepsilon TL(1+c) \int_{|v|>l} f(v)dv + \varepsilon T\theta_g^\varepsilon$$

$$\le \varepsilon \frac{TL(1+c)}{l} \int_{\mathbf{R}^d} vv^* f(v)dv + \varepsilon T\theta_g^\varepsilon \to 0, l \to \infty, \varepsilon \to 0.$$

The use of condition **PA2** in this case is due to the fact that $g(v) \in C_2(\mathbf{R}^d) \subset C_3(\mathbf{R}^d)$.

By analogy,

$$\int_0^T \int_{|v|\le\delta} vv^*\widehat{\Gamma}(\xi^\varepsilon(s), dv)ds = \int_0^T \int_{|v|\le\delta} \int_E vv^*\Gamma(\xi^\varepsilon(s), dv; x)\pi(dx)ds$$

$$\le TL(1+c) \int_{|v|\le\delta} vv^* f(v)dv \to 0, \delta \to 0,$$

and under the conditions $\Gamma^\varepsilon(u, \{0\}; x) = 0$, **PA1** and **C3** when $\varepsilon \to 0, \delta \to 0$

$$\int_0^T \int_{|v|\le\delta} vv^*\Gamma^\varepsilon(\xi^\varepsilon(s), dv; x_s^\varepsilon)ds \le \int_0^T \int_{\mathbf{R}^d} vv^*\Gamma^\varepsilon(\xi^\varepsilon(s), dv; x_s^\varepsilon)ds$$

$$= \varepsilon \int_0^T c(\xi^\varepsilon(s); x_s^\varepsilon)ds + \varepsilon T\theta_c^\varepsilon(u; x) \le \varepsilon TL(1 + c^2) + \varepsilon T\theta_c^\varepsilon(u; x) \to 0.$$

It is obvious that conditions **LP3**, **LP4** and the first part of condition **LP5** are the consequence of weak convergence of predictable characteristics. Therefore, the last step of proving is to verify such weak convergence.

Note that proof of weak convergence and obtaining the form of the limit generator [2.21] completely repeat the reasoning given in section 2.1.1 regarding the bounded jump measures.

The lemma is proved.

The theorem is proved.

2.2. Impulsive recurrent process

2.2.1. *Markov switching*

In Korolyuk and Limnios (2005), an impulsive process with Markov switching under Poisson approximation conditions was studied

$$\widetilde{\xi}^{\varepsilon}(t) = \widetilde{\xi}^{\varepsilon}_0 + \sum_{k=1}^{\nu(t/\varepsilon)} \alpha^{\varepsilon}_k(x^{\varepsilon}_{k-1}), \quad t \geq 0. \qquad [2.26]$$

Consider the generalization of problem [2.26], namely the impulsive recurrent process under Lévy approximation conditions:

$$\widetilde{\xi}^{\varepsilon}(t) = \widetilde{\xi}^{\varepsilon}_0 + \sum_{k=1}^{\nu(t/\varepsilon^2)} \alpha^{\varepsilon}_k(\widetilde{\xi}^{\varepsilon}_{k-1}, x^{\varepsilon}_{k-1}), \quad t \geq 0. \qquad [2.27]$$

Here is the switching Markov process $x^{\varepsilon}(t) := x(t/\varepsilon^2), t \geq 0$, the corresponding embedded Markov renewal process $x^{\varepsilon}_k, \tau^{\varepsilon}_k, k \geq 0$, and the counting process of jumps $\nu^{\varepsilon}(t) = \nu(t/\varepsilon^2), t \geq 0$.

Thus, the moments $\tau^{\varepsilon}_k, k \geq 0$ are the moments of process jumps, and

$$x^{\varepsilon}_k := x^{\varepsilon}(\tau^{\varepsilon}_k)$$

and

$$\nu^{\varepsilon}(t) := \max\{k \geq 0 : \tau^{\varepsilon}_k \leq t\}.$$

Now we can denote

$$\widetilde{\xi}^{\varepsilon}_n := \widetilde{\xi}(\varepsilon^2 \tau_n) = \widetilde{\xi}^{\varepsilon}_0 + \sum_{k=1}^{n} \alpha^{\varepsilon}_k(\widetilde{\xi}^{\varepsilon}_{k-1}, x^{\varepsilon}_{k-1}).$$

It should be noted that the pair $\widetilde{\xi}^{\varepsilon}(t), x^{\varepsilon}(t), t \geq 0$, is an additive Markov process (see, for example, Korolyuk and Limnios (2005, section 2.5)).

Note that in this setting, random variables $\alpha^{\varepsilon}_k(u, x), k \geq 1$ depend on the values of the process $\widetilde{\xi}^{\varepsilon}(t)$. This fact significantly complicates the task because, as in the case of continuous evolution, the predictable characteristics of the process recurrently depend on the values of the process itself.

This model describes, in particular, a recurrent equation

$$\widetilde{\xi}^{\varepsilon}_n = \widetilde{\xi}^{\varepsilon}_{n-1} + a^{\varepsilon}_k(\widetilde{\xi}^{\varepsilon}_{n-1}, x_n),$$

with a random stop, namely at the moment of the n-th jump. This equation has many applications to insurance, finance, reliability problems etc.

We impose the following condition on the counting process $\nu(t)$:

$$\int_0^t \mathbf{E}[\varphi(s)d\nu(s)] < l_1 \int_0^t \mathbf{E}(\varphi(s))ds \qquad [2.28]$$

for any non-negative increasing function $\varphi(s)$ and $l_1 > 0$.

Also recall the Lévy approximation conditions:

C1' A family of impulsive recurrent processes $\widetilde{\xi}^\varepsilon(t), t \geq 0$, satisfies the Lévy approximation conditions:

L1 *Approximation of means:*

$$b_\varepsilon(u;x) = \int_{\mathbf{R}^d} v G^\varepsilon_{u,x}(dv) = \varepsilon b_1(u;x) + \varepsilon^2[b(u;x) + \theta^\varepsilon_b(u;x)],$$

and

$$c_\varepsilon(u;x) = \int_{\mathbf{R}^d} vv^* G^\varepsilon_{u,x}(dv) = \varepsilon^2[c(u;x) + \theta^\varepsilon_c(u;x)].$$

L2 *Condition of the intensity kernel:*

$$G^\varepsilon_g(u;x) = \int_{\mathbf{R}^d} g(v) G^\varepsilon_{u,x}(dv) = \varepsilon^2[G_g(u;x) + \theta^\varepsilon_g(u;x)]$$

for all $g \in C_3(\mathbf{R}^d)$, and the kernel $G_g(u;x)$ is bounded for all $g \in C_3(\mathbf{R}^d)$, i.e.

$$|G_g(u;x)| \leq G_g \quad \text{(constant, dependent on } g\text{)}.$$

Here

$$G_g(u;x) = \int_{\mathbf{R}^d} g(v) G_{u,x}(dv), \quad g \in C_3(\mathbf{R}^d).$$

Negligible terms $\theta^\varepsilon_b, \theta^\varepsilon_c, \theta^\varepsilon_g$ satisfy the condition

$$\sup_{x \in E} |\theta^\varepsilon_\cdot(u;x)| \to 0, \quad \varepsilon \to 0.$$

L3 *Balance condition* (see the definition of $\rho(x)$ in condition **CSM**):

$$\int_E \rho(dx) b_1(u;x) = 0.$$

L4 *Conditions for initial values:*

$$\sup_{\varepsilon>0} E|\widetilde{\xi}_0^\varepsilon| \leq C < \infty$$

and

$$\widetilde{\xi}_0^\varepsilon \Rightarrow \widetilde{\xi}_0, \varepsilon \to 0.$$

C2 *Uniform quadratic integrity:*

$$\lim_{c\to\infty} \sup_{x\in E} \int_{|v|>c} vv^* G_{u,x}(dv) = 0.$$

C3 *Growth condition:* there is a positive constant L such that

$$|b(u;x)| \leq L(1+|u|), \quad \text{and} \quad |c(u;x)| \leq L(1+|u|^2),$$

and for all real-valued non-negative functions $f(x), x \in \mathbf{R}^d$, such that $\int_{\mathbf{R}^d}(1+f(x))|x|^2 dx < \infty$, we have

$$|G_{u,x}(v)| \leq Lf(v)(1+|u|).$$

The following theorem holds.

THEOREM 2.9.– *Under the conditions* ***CM, C1'–C3****, a weak convergence in* $D_{\mathbf{R}^d}[0,\infty)$ *holds true*

$$\widetilde{\xi}^\varepsilon(t) \Rightarrow \widetilde{\xi}^0(t), \quad \varepsilon \to 0.$$

The limit process $\widetilde{\xi}^0(t), t \geq 0$ *is a Lévy process, and under condition* ***LL****, it is determined by the generator*

$$\widehat{\mathbf{L}}\varphi(u) = (\widehat{b}(u) - \widehat{b}_0(u) + \widehat{b}_1(u))\varphi'(u) + \frac{1}{2}\sigma^2(u)\varphi''(u) \qquad [2.29]$$
$$+ \lambda(u) \int_{\mathbf{R}^d} [\varphi(u+v) - \varphi(u)] G_u^0(dv),$$

where $\widetilde{b}_1(u;x) := q(x) \int_E P(x,dy) b_1(u;y), G_u(dv) = q \int_E \rho(dx) G_{u,x}(dv),$

$$\sigma^2(u) = 2q \int_E \rho(dx)[\widetilde{b}_1(u;x) R_0 \widetilde{b}_1^*(u;x) + \frac{1}{2}(c(u;x) - c_0(u;x))], \sigma^2(u) > 0,$$

$$\lambda(u) = qG_u(\mathbf{R}^d), \qquad G_u^0(dv) = G_u(dv)/G_u(\mathbf{R}^d).$$

The proof of the theorem is based on the semimartingale representation of the process [2.27]. Predictable characteristics of semimartingale [2.27] are as follows (see Çinlar *et al.* (1980, Theorems 6.27 and 7.16)):

$$\diamond\, B^\varepsilon(t) = \sum_{k=1}^{\nu(t/\varepsilon^2)} b_\varepsilon(\widetilde{\xi}^\varepsilon_{k-1}; x^\varepsilon_{k-1}) = \varepsilon \sum_{k=1}^{\nu(t/\varepsilon^2)} b_1(\widetilde{\xi}^\varepsilon_{k-1}; x^\varepsilon_{k-1})$$

$$+ \varepsilon^2 \sum_{k=1}^{\nu(t/\varepsilon^2)} b(\widetilde{\xi}^\varepsilon_{k-1}; x^\varepsilon_{k-1}) + \theta^\varepsilon_b,$$

$$\diamond\, C^\varepsilon(t) = \sum_{k=1}^{\nu(t/\varepsilon^2)} c_\varepsilon(\widetilde{\xi}^\varepsilon_{k-1}; x^\varepsilon_{k-1}) = \varepsilon^2 \sum_{k=1}^{\nu(t/\varepsilon^2)} c(\widetilde{\xi}^\varepsilon_{k-1}; x^\varepsilon_{k-1}) + \theta^\varepsilon_c,$$

$$\diamond\, G^\varepsilon(t) = \sum_{k=1}^{\nu(t/\varepsilon^2)} \int_{\mathbf{R}^d} g(v) G^\varepsilon(\widetilde{\xi}^\varepsilon_{k-1}, dv; x^\varepsilon_{k-1})$$

$$= \varepsilon^2 \sum_{k=1}^{\nu(t/\varepsilon^2)} \int_{\mathbf{R}^d} g(v) G(\widetilde{\xi}^\varepsilon_{k-1}, dv; x^\varepsilon_{k-1}) + \theta^\varepsilon_g,$$

where $\sup_{x\in E} |\theta^\varepsilon_\cdot| \to 0, \quad \varepsilon \to 0.$

For convenience of notation, we write $G(u, dv; x)$ instead of $G_{u,x}(dv)$, and $G^\varepsilon(u, dv; x)$ instead of $G^\varepsilon_{u,x}(dv)$.

The proof is separated into two stages.

STAGE 1. The proof of the compactness of the family of processes $\widetilde{\xi}^\varepsilon(t)$, as in the case of continuous evolution, is based on the following lemma.

LEMMA 2.6.– *Given* **C3** *there is a constant* $k > 0$, *independent of* ε *and dependent on* T, *such that*

$$\mathbf{E} \sup_{t\le T} |\widetilde{\xi}^\varepsilon(t)|^2 \le k_T.$$

PROOF.– Semimartingale [2.27] has a representation

$$\widetilde{\xi}^\varepsilon(t) = u + B^\varepsilon_t + M^\varepsilon_t,$$

where $u = \widetilde{\xi}^\varepsilon(0)$; B^ε_t is the predictable drift:

$$B^\varepsilon_t = \sum_{k=1}^{\nu(t/\varepsilon^2)} b_\varepsilon(\widetilde{\xi}^\varepsilon_{k-1}, x^\varepsilon_{k-1}) = \widetilde{B}^\varepsilon_1(t) + \widetilde{B}^\varepsilon(t) + \theta^\varepsilon_b(t),$$

here

$$\widetilde{B}_1^\varepsilon(t) := \varepsilon \sum_{k=1}^{\nu(t/\varepsilon^2)} b_1(\widetilde{\xi}_{k-1}^\varepsilon, x_{k-1}^\varepsilon), \widetilde{B}^\varepsilon(t) := \varepsilon^2 \sum_{k=1}^{\nu(t/\varepsilon^2)} b(\widetilde{\xi}_{k-1}^\varepsilon, x_{k-1}^\varepsilon).$$

M_t^ε is the local quadratically integrated martingale with the quadratic characteristic

$$\langle M^\varepsilon \rangle_t = \varepsilon^2 \sum_{k=1}^{\nu(t/\varepsilon^2)} \int_{\mathbf{R}^d} vv^* G(\widetilde{\xi}_{k-1}^\varepsilon, dv; x_{k-1}^\varepsilon).$$

Let us separate the process $\widetilde{\xi}^\varepsilon(t)$ into two components.

The first part is of the order ε

$$\widetilde{B}_1^\varepsilon(t) = \varepsilon \sum_{k=1}^{\nu(t/\varepsilon^2)} b_1(\xi_{k-1}^\varepsilon; x_{k-1}^\varepsilon),$$

its martingale characterization has the form

$$\widetilde{\mu}_t^\varepsilon = \varphi^\varepsilon(\widetilde{B}_1^\varepsilon(t)) + \varphi^\varepsilon(\widetilde{B}_1^\varepsilon(0)) - \int_0^t \mathbf{L}^\varepsilon \varphi^\varepsilon(\widetilde{B}_1^\varepsilon(s))ds, \qquad [2.30]$$

where the generator

$$\mathbf{L}^\varepsilon \varphi(u; x) = \varepsilon^{-2} q(x)[\widetilde{\mathbf{B}}_1^\varepsilon(x) - I]\varphi(u; x),$$

here $(\widetilde{\mathbf{B}}_1^\varepsilon(x) - I)\varphi(u) = \varphi(u + \varepsilon b_1(u; x)) - \varphi(u) = \varepsilon b_1(u; x)\varphi'(u) + \varepsilon\theta^\varepsilon(x)\varphi(u)$.

The generator $\mathbf{L}^\varepsilon$ can be written as:

$$\mathbf{L}^\varepsilon = \varepsilon^{-2} Q + \varepsilon^{-1} Q_0 \widetilde{\mathbf{B}}_1(x) + \theta^\varepsilon,$$

here $\widetilde{\mathbf{B}}_1(x)\varphi(u) = \varepsilon b_1(u; x)\varphi'(u), Q_0\varphi(x) := q(x) \int_E P(x, dy)\varphi(y)$.

Thus (see, for example, Korolyuk and Limnios (2005, Theorem 1.2)), the quadratic characteristic of martingale [2.30] has the form

$$< \widetilde{\mu}^\varepsilon >_t = \int_0^t \left[\mathbf{L}^\varepsilon (\varphi^\varepsilon(\widetilde{B}_1^\varepsilon(s)))^2 - 2\varphi^\varepsilon(\widetilde{B}_1^\varepsilon(s)) \mathbf{L}^\varepsilon \varphi^\varepsilon(\widetilde{B}_1^\varepsilon(s)) \right] ds.$$

Consider the action of the generator $\mathbf{L}^\varepsilon = \varepsilon^{-2} Q + \varepsilon^{-1} Q_0 \widetilde{\mathbf{B}}_1(x) + \theta^\varepsilon$ on test functions

$$\varphi^\varepsilon(u; x) = \varphi(u) + \varepsilon\varphi_1(u; x), \varphi(u) \in C_3(\mathbf{R}^d).$$

We obtain the following form of the integrand:

$$Q\varphi_1^2 - 2\varphi_1 Q\varphi_1 + \theta^\varepsilon \varphi^\varepsilon.$$

Thus, the integrand is bounded. The compactness of the martingale $\widetilde{\mu}_t^\varepsilon$ follows from the boundedness of the quadratic characteristic (see Liptser and Shiryayev (1989); Jacod and Shiryaev (2003)). Thus, the functional $\varphi(\widetilde{B}_1^\varepsilon(t))$ is also compact and bounded uniformly by ε. According to Ethier and Kurtz (1986), this leads to compactness of $\widetilde{B}_1^\varepsilon(t)$ and uniform by ε boundedness of

$$\sup_{t\le T} \widetilde{B}_1^\varepsilon(t) < \sqrt{k_3}, \qquad [2.31]$$

because the test functions $\varphi(u)$ belong to the class that defines a measure.

Now consider the other part of the process $\widetilde{\xi}^\varepsilon(t)$.

For an arbitrary process $y(t), t \ge 0$, we denote $y_t^\dagger = \sup_{s\le t} |y(s)|$, so we have

$$((\xi_t^\varepsilon)^\dagger)^2 \le 4[u^2 + ((\widetilde{B}_1^\varepsilon)^\dagger(t))^2 + ((\widetilde{B}^\varepsilon)^\dagger(t))^2 + ((M_t^\varepsilon)^\dagger)^2]. \qquad [2.32]$$

We apply the result from Korolyuk and Limnios (2005, section 2.3), namely

$$\sum_{k=1}^{\nu(t)} a(\widetilde{\xi}_{k-1}^\varepsilon, x_{k-1}^\varepsilon) = \int_0^t a(\widetilde{\xi}^\varepsilon(s), x^\varepsilon(s))d\nu(s).$$

By condition **C3**:

$$(\widetilde{B}^\varepsilon(t))^\dagger = \varepsilon^2 \int_0^{t/\varepsilon^2} b(\xi^\varepsilon(s), x^\varepsilon(s))d\nu(s)$$

$$\le L\varepsilon^2 \int_0^{t/\varepsilon^2} (1 + (\widetilde{\xi}^\varepsilon(s))^\dagger)d\nu(s). \qquad [2.33]$$

According to the Doob inequality

$$\mathbf{E}((M_t^\varepsilon)^\dagger)^2 \le 4|\mathbf{E}\langle M^\varepsilon\rangle_t|,$$

and from condition **C3**, we have

$$|\langle M^\varepsilon\rangle_t| = \left|\varepsilon^2 \int_0^{t/\varepsilon^2} \int_{\mathbf{R}^d} vv^* G(\widetilde{\xi}^\varepsilon(s), dv; x_s^\varepsilon)d\nu(s)\right|$$

$$\le Lr_1\varepsilon^2 \int_0^{t/\varepsilon^2} [1 + ((\widetilde{\xi}^\varepsilon(s))^\dagger)^2]d\nu(s), \qquad [2.34]$$

where $r_1 = \int_{\mathbf{R}^d} |v|^2 f(v)dv$.

From inequalities [2.31]–[2.34], condition [2.28] and the Cauchy–Bunyakovsky–Schwarz inequality ($[\int_0^t \varphi(s)ds]^2 \leq t\int_0^t \varphi^2(s)ds$), the estimate follows:

$$\mathbf{E}((\widetilde{\xi}^\varepsilon(t))^\dagger)^2 \leq k_1 + k_3 + k_2\varepsilon^2 \int_0^{t/\varepsilon^2} \mathbf{E}[((\widetilde{\xi}^\varepsilon(s))^\dagger)^2 d\nu(s)]$$

$$\leq k_1 + k_3 + k_2 l_1 \varepsilon^2 \int_0^{t/\varepsilon^2} \mathbf{E}((\widetilde{\xi}^\varepsilon(s))^\dagger)^2 ds$$

$$= k_1 + k_3 + k_2 l_1 \int_0^t \mathbf{E}((\widetilde{\xi}^\varepsilon(s))^\dagger)^2 ds,$$

where k_1, k_2, k_3 and l_1 are positive constants independent of ε.

According to the Gronwall–Bellman inequality (see, for example, Ethier and Kurtz (1986, p. 498)), we have

$$\mathbf{E}((\widetilde{\xi}^\varepsilon(t))^\dagger)^2 \leq (k_1 + k_3)\exp(k_2 l_1 t).$$

Therefore, both components of the process $\widetilde{\xi}^\varepsilon(t)$ are compact and bounded

$$\mathbf{E}\sup_{t\leq T} |\xi^\varepsilon(t)|^2 \leq k_T.$$

The lemma is proved.

The following corollary and lemma hold true; the proof does not differ from the case of continuous evolution.

COROLLARY 2.3.– *Under condition* **C3**, *condition* ***CCC*** *takes place:*

$$\lim_{c\to\infty} \sup_{\varepsilon\leq\varepsilon_0} \mathbf{P}\{\sup_{t\leq T} |\widetilde{\xi}^\varepsilon(t)| > c\} = 0.$$

LEMMA 2.7.– *Under condition* **C3**, *there exists a constant* $k > 0$, *independent of* ε *and such that*

$$\mathbf{E}|\widetilde{\xi}^\varepsilon(t) - \widetilde{\xi}^\varepsilon(s)|^2 \leq k|t-s|, \forall s, t \leq T.$$

Thus, from the conditions proved in Corollary 2.3 and Lemma 2.7, compactness of the family of the processes $\widetilde{\xi}^\varepsilon(t), t \geq 0, \varepsilon > 0$, follows.

STAGE 2. At the second stage, the convergence of the predictable characteristics is proved by analogy with the continuous case.

Next, we denote by $A^\varepsilon(t)$ the main part of any of the predictive characteristics $B^\varepsilon(t), C^\varepsilon(t), \Gamma^\varepsilon(t)$.

Consider a three-component Markov process $A^\varepsilon(t), \widetilde{\xi}^\varepsilon(t), x_t^\varepsilon$, $t \geq 0$, which is characterized by a martingale

$$\mu_t^\varepsilon = \varphi(\widetilde{\xi}^\varepsilon(t), A^\varepsilon(t), x_t^\varepsilon) - \int_0^t \mathbf{L}^\varepsilon \varphi(\widetilde{\xi}^\varepsilon(t), A^\varepsilon(s), x_s^\varepsilon) ds,$$

where the generator $\mathbf{L}^\varepsilon$ has a representation (see Korolyuk and Limnios (2005, Chapters 2 and 7))

$$\mathbf{L}^\varepsilon \varphi(u, v; x) = \varepsilon^{-2} Q\varphi(\cdot, \cdot; x) + Q_0 A^\varepsilon(u; x)\varphi(\cdot, v; x) \\ + Q_0 G^\varepsilon(x)\varphi(u, \cdot; x), \qquad [2.35]$$

here

$$Q_0\varphi(\cdot, \cdot; x) := q(x) \int_E P(x, dy)\varphi(\cdot, \cdot; y),$$

$$G^\varepsilon(x)\varphi(u, \cdot; \cdot) := \varepsilon^{-2} \int_{\mathbf{R}^d} G_{u,x}^\varepsilon(dv)[\varphi(u + v) - \varphi(u)],$$

$$A^\varepsilon(u; x)\varphi(\cdot, v; \cdot) = \varepsilon^{-2}[\varphi(\cdot, v + \varepsilon^2 a(u; x); \cdot) - \varphi(\cdot, v; \cdot)],$$

and

$$A(u; x)\varphi(\cdot, v; \cdot) = a(u, x)\varphi'(\cdot, v; \cdot).$$

Considering the generator [2.35] on the test functions $\varphi^\varepsilon(v; x) = \varphi(v) + \varepsilon\varphi_1(v; x) + \varepsilon^2\varphi_2(v; x)$, similar to the continuous case, we obtain the result about the convergence of the predictable characteristics.

Thus, taking into account the compactness of the process, the result about weak convergence of the process will follow from Theorem 3.1.

It remains to find the form of the limit operator; to do this, we regard the action of the generator [2.35] on the test function $\varphi^\varepsilon(u; x) = \varphi(u) + \varepsilon\varphi_1(u; x) + \varepsilon^2\varphi_2(u; x)$.

We have:

$$\mathbf{L}^\varepsilon \varphi^\varepsilon(u; x) = [\varepsilon^{-2} Q + Q_0 G^\varepsilon(x)][\varphi(u) + \varepsilon\varphi_1(u; x) + \varepsilon^2\varphi_2(u; x)].$$

Similar to Lemma 1.2, we can write an asymptotic representation

$$G^\varepsilon(x)\varphi(u) = [\varepsilon^{-1} B_1(u; x) + G(u; x)]\varphi(u) + \theta^\varepsilon := \varepsilon^{-1} b_1(u; x)\varphi'(u)$$

$$+(b(u; x) - b_0(u; x))\varphi'(u) + \frac{1}{2}(c(u; x) - c_0(u; x))\varphi''(u)$$

$$+ \int_{\mathbf{R}^d} [\varphi(u + v) - \varphi(u)] G_{u,x}(dv) + \theta^\varepsilon,$$

where $b_0(u; x) := \int_{\mathbf{R}^d} v G_{u,x}(dv)$, $c_0(u; x) := \int_{\mathbf{R}^d} vv^* G_{u,x}(dv)$.

Thus:

$$\mathbf{L}^\varepsilon \varphi^\varepsilon(u;x) = [\varepsilon^{-2}Q + \varepsilon^{-1}Q_0B_1(u;x) + Q_0G(u;x)][\varphi(u) + \varepsilon\varphi_1(u;x)$$

$$+\varepsilon^2\varphi_2(u;x)] + \theta^\varepsilon\varphi^\varepsilon = \varepsilon^{-2}Q\varphi(u) + \varepsilon^{-1}[Q\varphi_1(u;x) + Q_0b_1(u;x)\varphi'(u)]$$

$$+[Q\varphi_2(u;x)+Q_0b_1(u;x)\varphi_1'(u;x)+Q_0G(u;x)\varphi(u)]+\varepsilon Q_0B_1(u;x)\varphi_2(u;x)$$

$$+\varepsilon Q_0G(u;x)\varphi_1(u;x) + \varepsilon^2 Q_0G(u;x)\varphi_2(u;x) + \theta^\varepsilon\varphi^\varepsilon.$$

So, the singular perturbation problem is reduced to a system of equations:

$$Q\varphi(u) = 0,$$

$$Q\varphi_1(v;x) + Q_0b_1(u;x)\varphi'(u) = 0,$$

$$Q\varphi_2(v;x) + Q_0b_1(u;x)\varphi_1'(u;x) + Q_0G(u;x)\varphi(u) = \widehat{\mathbf{L}}\varphi(u).$$

Therefore, $\varphi(u) \in \mathcal{N}_Q$, and from balance condition **L3** and the second equation

$$Q\varphi_1(u;x) = -Q_0b_1(u;x)\varphi'(u), \varphi_1(u;x) = R_0Q_0b_1(u;x)\varphi'(u).$$

Substituting into the last equation, we obtain

$$Q\varphi_2(u;x)+Q_0b_1(u;x)R_0Q_0b_1'(u;x)\varphi'(u)+Q_0b_1(u;x)R_0Q_0b_1(u;x)\varphi''(u)$$

$$+Q_0G(u;x)\varphi(u) = \widehat{\mathbf{L}}\varphi(u).$$

Applying the solvability condition to the last equation, we finally have the generator [2.29].

Using condition **LL** and the results from Skorokhod *et al.* (2002, Chapter 2), we can say that the limit generator really defines the limit process $\widetilde{\xi}^0(t)$.

The theorem is proved.

2.2.2. *Semi-Markov switching*

The next generalization of the previous model is an impulsive recurrent process with semi-Markov switching (see sections 1.1.4, 1.1.9 and 1.1.10)

$$\widetilde{\xi}^\varepsilon(t) = \widetilde{\xi}_0^\varepsilon + \sum_{k=1}^{\nu(t/\varepsilon)} \alpha_k^\varepsilon(\widetilde{\xi}_{k-1}^\varepsilon, x_{k-1}^\varepsilon), \quad t \geq 0. \qquad [2.36]$$

For such a process, we consider both approximation schemes: Poisson and Lévy.

Recall that the switching semi-Markov process $x(t), t \geq 0$ is determined by the semi-Markov kernel

$$Q(x, B, t) = P(x, B)F_x(t), x \in E, B \in \mathcal{E}, t \geq 0,$$

which also defines the associated Markov renewal process $x_n, \tau_n, n \geq 0$:

$$Q(x, B, t) = \mathbf{P}(x_{n+1} \in B, \theta_{n+1} \leq t | x_n = x)$$

$$= \mathbf{P}(x_{n+1} \in B | x_n = x)\mathbf{P}(\theta_{n+1} \leq t | x_n = x).$$

In addition, $\widetilde{\xi}_n^\varepsilon$ in [2.36] is defined as:

$$\widetilde{\xi}_n^\varepsilon := \widetilde{\xi}(\varepsilon\tau_n) = \widetilde{\xi}_0^\varepsilon + \sum_{k=1}^{n} \alpha_k^\varepsilon(\widetilde{\xi}_{k-1}^\varepsilon, x_{k-1}^\varepsilon).$$

Condition **CSM**, formulated in section 1.1.4, and condition [2.28] are fulfilled.

Also recall the Poisson approximation conditions for the impulsive recurrent process, formulated in section 1.1.10.

C1 A family of impulse processes $\widetilde{\xi}^\varepsilon(t), t \geq 0$ satisfies Poisson approximation conditions.

PA1 *Approximation of means:*

$$b_\varepsilon(u; x) = \int_{\mathbf{R}^d} v G_{u,x}^\varepsilon(dv) = \varepsilon[b(u; x) + \theta_b^\varepsilon(u; x)],$$

and

$$c_\varepsilon(u; x) = \int_{\mathbf{R}^d} v v^* G_{u,x}^\varepsilon(dv) = \varepsilon[c(u; x) + \theta_c^\varepsilon(u; x)].$$

PA2 *Condition for the distribution function:*

$$G_{u,x}^\varepsilon(g) = \int_{\mathbf{R}^d} g(v) G_{u,x}^\varepsilon(dv) = \varepsilon[G_{u,x}(g) + \theta_g^\varepsilon(u; x)]$$

for all $g \in C_3(\mathbf{R}^d)$, and the measure $G_g(u; x)$ is bounded for all $g \in C_3(\mathbf{R}^d)$, such that,

$$|G_{u,x}(g)| \leq G_g \quad \text{(constant, dependent on } g\text{)},$$

where the measure $G_{u,x}(dv)$ is defined on the class $C_3(\mathbf{R}^d)$ by the ratio

$$G_{u,x}(g) = \int_{\mathbf{R}^d} g(v) G_{u,x}(dv), \quad g \in C_3(\mathbf{R}^d).$$

Negligible terms $\theta_b^\varepsilon, \theta_c^\varepsilon, \theta_g^\varepsilon$ satisfy the condition

$$\sup_{x \in E} |\theta_\cdot^\varepsilon(u; x)| \to 0, \quad \varepsilon \to 0.$$

PA3 *Conditions for initial values*:

$$\sup_{\varepsilon > 0} E|\widetilde{\xi}_0^\varepsilon| \leq C < \infty$$

and

$$\widetilde{\xi}_0^\varepsilon \Rightarrow \widetilde{\xi}_0, \varepsilon \to 0.$$

In addition, the following conditions are met:

C2 *Uniform quadratic integrability*:

$$\lim_{c \to \infty} \sup_{x \in E} \int_{|v| > c} vv^* G_{u,x}(dv) = 0.$$

C3 *Growth condition*: there is a positive constant L, such that

$$|b(u; x)| \leq L(1 + |u|), \quad \text{and} \quad |c(u; x)| \leq L(1 + |u|^2),$$

and for any real-valued non-negative function $f(v), v \in \mathbf{R}^d$, such that $\int_{\mathbf{R}^d}(1 + f(v))|v|^2 dv < \infty$, we have

$$G_{u,x}(v) \leq Lf(v)(1 + |u|).$$

The following theorem holds.

THEOREM 2.10.– *Under the conditions* **CSM, C1–C4**, *the impulsive recurrent process* [2.36] *weakly converges in $D_{\mathbf{R}^d}[0, \infty)$ to a compound Poisson process, which under condition* **LL** *is determined by the generator*

$$\widehat{\mathbf{L}}\varphi(u) = b(u)\varphi'(u) + \int_{\mathbf{R}^d} [\varphi(u + v) - \varphi(u)]G_u(dv), \quad [2.37]$$

where

$$G_u(dv) := \int_E \rho(dx)G_{u,x}(dv).$$

The drift parameter $b(u)$ is determined by the ratio

$$b(u) := qb_0(u), b_0(u) = \widehat{b}(u) - b^0(u),$$

$$\widehat{b}(u) := \int_E \rho(dx)b(u; x), \quad b^0(u) := \int_{\mathbf{R}^d} vG_u(dv).$$

The proof of the theorem is based on the semimartingale representation of process [2.36]. Predictable characteristics of semimartingale [2.36] are as follows (see Çinlar *et al.* (1980, Theorems 6.27 and 7.16)):

– $B^\varepsilon(t) = \sum_{k=1}^{\nu(t/\varepsilon)} b_\varepsilon(\widetilde{\xi}_{k-1}^\varepsilon; x_{k-1}^\varepsilon) = \varepsilon \sum_{k=1}^{\nu(t/\varepsilon)} b(\widetilde{\xi}_{k-1}^\varepsilon; x_{k-1}^\varepsilon) + \theta_b^\varepsilon$,

– $C^\varepsilon(t) = \sum_{k=1}^{\nu(t/\varepsilon)} c_\varepsilon(\widetilde{\xi}_{k-1}^\varepsilon; x_{k-1}^\varepsilon) = \varepsilon \sum_{k=1}^{\nu(t/\varepsilon)} c(\widetilde{\xi}_{k-1}^\varepsilon; x_{k-1}^\varepsilon) + \theta_c^\varepsilon$,

– $G^\varepsilon(t) = \sum_{k=1}^{\nu(t/\varepsilon)} \int_{\mathbf{R}^d} g(v) G^\varepsilon(\widetilde{\xi}_{k-1}^\varepsilon, dv; x_{k-1}^\varepsilon)$
$= \varepsilon \sum_{k=1}^{\nu(t/\varepsilon)} \int_{\mathbf{R}^d} g(v) G(\widetilde{\xi}_{k-1}^\varepsilon, dv; x_{k-1}^\varepsilon) + \theta_g^\varepsilon$,

where $\sup_{x \in E} |\theta_\cdot^\varepsilon| \to 0, \quad \varepsilon \to 0$.

For convenience of notation, we write $G(u, dv; x)$ instead of $G_{u,x}(dv)$, and $G^\varepsilon(u, dv; x)$ instead of $G_{u,x}^\varepsilon(dv)$.

The proof is separated into two stages.

STAGE 1. We prove the relative compactness of the family of processes $\widetilde{\xi}^\varepsilon(t), t \geq 0, \varepsilon > 0$. The corresponding lemmas are formulated analogously to Lemmas 2.6, 2.7 and Corollary 2.3. We present only a proof of a lemma similar to Lemma 2.6, since it has certain differences.

PROOF.– The impulse process [2.36] has the following semimartingale representation:

$$\widetilde{\xi}^\varepsilon(t) = u + B_t^\varepsilon + M_t^\varepsilon, \qquad [2.38]$$

where $u = \widetilde{\xi}_0^\varepsilon$; B_t^ε is a predictable drift

$$B_t^\varepsilon = \sum_{k=1}^{\nu(t/\varepsilon)} b_\varepsilon(\widetilde{\xi}_{k-1}^\varepsilon, x_{k-1}^\varepsilon) = \varepsilon \sum_{k=1}^{\nu(t/\varepsilon)} b(\widetilde{\xi}_{k-1}^\varepsilon, x_{k-1}^\varepsilon) + \theta_a^\varepsilon,$$

and M_t^ε is a local quadratically integrated martingale with the quadratic characteristic

$$\langle M^\varepsilon \rangle_t = \varepsilon \sum_{k=1}^{\nu(t/\varepsilon)} \int_{\mathbf{R}^d} vv^* G(\widetilde{\xi}_{k-1}^\varepsilon, dv; x_{k-1}^\varepsilon) + \theta_c^\varepsilon, \qquad [2.39]$$

and for any finite $T > 0$

$$\sup_{0 \leq t \leq T} |\theta_\cdot^\varepsilon(t)| \to 0, \varepsilon \to 0.$$

For an arbitrary process $y(t), t \geq 0$, we denote the process $y^\dagger(t) := \sup_{s \leq t} |y(s)|$; then, from [2.38], we have

$$((\widetilde{\xi}^\varepsilon(t))^\dagger)^2 \leq 3[u^2 + ((B^\varepsilon(t))^\dagger)^2 + ((M_t^\varepsilon)^\dagger)^2]. \qquad [2.40]$$

Rewriting, according to Korolyuk and Limnios (2005, section 2.3),

$$\sum_{k=1}^{\nu(t)} b(\widetilde{\xi}^{\varepsilon}_{k-1}, x^{\varepsilon}_{k-1}) = \int_0^t b(\widetilde{\xi}^{\varepsilon}(s), x^{\varepsilon}(s))d\nu(s),$$

we have by condition **C3** for a sufficiently large ε

$$(B^{\varepsilon}(t))^{\dagger} = \varepsilon \int_0^{t/\varepsilon} b(\widetilde{\xi}^{\varepsilon}(s), x^{\varepsilon}(s))d\nu(s) \leq L\varepsilon \int_0^{t/\varepsilon} (1 + (\widetilde{\xi}^{\varepsilon}(s))^{\dagger})d\nu(s). \qquad [2.41]$$

According to equation [2.39] and condition **C3**, we obtain

$$|\langle M^{\varepsilon}\rangle_t| = |\varepsilon \int_0^{t/\varepsilon} \int_{\mathbf{R}^d} vv^* G(\widetilde{\xi}^{\varepsilon}(s), dv; x^{\varepsilon}_s)d\nu(s)|$$

$$\leq Lr_1\varepsilon \int_0^{t/\varepsilon} [1 + ((\widetilde{\xi}^{\varepsilon}(s))^{\dagger})^2]d\nu(s), \qquad [2.42]$$

where $r_1 = \int_{\mathbf{R}^d} |v|^2 f(v)dv$.

Inequalities [2.40]–[2.42], the Doob inequality (e.g. Liptser and Shiryayev (1989, Theorem 1.9.2))

$$\mathbf{E}((M^{\varepsilon}_t)^{\dagger})^2 \leq 4|\mathbf{E}\langle M^{\varepsilon}\rangle_t|,$$

condition [2.28] and the Cauchy–Bunyakovsky–Schwarz inequality ($[\int_0^t \varphi(s)ds]^2 \leq t\int_0^t \varphi^2(s)ds$) give the following result

$$\mathbf{E}((\widetilde{\xi}^{\varepsilon}(t))^{\dagger})^2 \leq k_1 + k_2\varepsilon \int_0^{t/\varepsilon} \mathbf{E}[((\widetilde{\xi}^{\varepsilon}(s))^{\dagger})^2 d\nu(s)] \leq k_1$$

$$+k_2 l_1 \varepsilon \int_0^{t/\varepsilon} \mathbf{E}((\widetilde{\xi}^{\varepsilon}(s))^{\dagger})^2 ds = k_1 + k_2 l_1 \int_0^t \mathbf{E}((\widetilde{\xi}^{\varepsilon}(s))^{\dagger})^2 ds,$$

where k_1, k_2 and $l_1$are positive constants independent of ε.

Now applying the Gronwall–Bellman inequality (e.g. Ethier and Kurtz (1986, p. 498)), we have

$$\mathbf{E}((\widetilde{\xi}^{\varepsilon}(t))^{\dagger})^2 \leq k_1 \exp(k_2 l_1 t).$$

Thus, for the process $\widetilde{\xi}^{\varepsilon}(t)$, the inequality holds

$$\mathbf{E}\sup_{t\leq T} |\widetilde{\xi}^{\varepsilon}(t)|^2 \leq k_T.$$

The lemma is proved.

STAGE 2. In proving the convergence of the predictable characteristics, we use the methods described in Korolyuk and Limnios (2005, sections 3.2 and 3.3) and Theorem 2.1.

Note that we will further denote by $A^\varepsilon(t)$ the main part of any of the predictable characteristics $B^\varepsilon(t), C^\varepsilon(t), G^\varepsilon(t)$.

The extended Markov renewal process is considered a four-component Markov chain

$$A_n^\varepsilon = A^\varepsilon(\tau_n^\varepsilon), \quad \xi_n^\varepsilon, \quad x_n^\varepsilon, \quad \tau_n^\varepsilon, \quad n \geq 0,$$

where $x_n^\varepsilon = x^\varepsilon(\tau_n^\varepsilon)$, $x^\varepsilon(t) := x(t/\varepsilon)$, $\xi_n^\varepsilon = \xi^\varepsilon(\tau_n^\varepsilon)$ and $\tau_{n+1}^\varepsilon = \tau_n^\varepsilon + \varepsilon\theta_n^\varepsilon, n \geq 0$, and

$$P(\theta_{n+1}^\varepsilon \leq t | x_n^\varepsilon = x) = F_x(t) = P(\theta_x \leq t).$$

The compensating operator of the extended Markov renewal process can be set by equality (see also Korolyuk and Limnios (2005, section 2.8))

$$\mathbb{L}^\varepsilon\varphi(u, v; x) = \varepsilon^{-1}q(x)\left[\int_E P(x, dy)\int_{\mathbf{R}^d} G_{u,y}^\varepsilon(dz)A^\varepsilon(u; y)\varphi(u + z, v; y)\right.$$

$$\left. -\varphi(u, v; x)\right],$$

where $A^\varepsilon(u; y), y \in E, u \in \mathbf{R}^d$ is a family of generators with the limit generator

$$A(u; y)\varphi(v) = a(u; y)\varphi'(v).$$

According to Korolyuk and Limnios (2005, Chapter 3), the compensating operator of the extended Markov renewal process has the following asymptotic representation on the test functions $\varphi \in \overline{C}^2(\mathbf{R}^d \times \mathbf{R}^d \times E)$ (measurable functions denoted at $\mathbf{R}^d \times \mathbf{R}^d \times E$, twice continuously differentiable by the first two arguments and bounded by the third argument):

$$\begin{aligned}\mathbb{L}^\varepsilon\varphi(u, v; x) &= \varepsilon^{-1}Q\varphi(\cdot, \cdot; x) + Q_0 A(u, x)\varphi(\cdot, v; \cdot) \\ &\quad + \varepsilon^{-1}Q_0\mathbf{G}_{u,x}^\varepsilon\varphi(u, \cdot; \cdot) + \varepsilon\theta^\varepsilon(x)\varphi(u, v; x),\end{aligned} \qquad [2.43]$$

where:

$$Q_0\varphi(x) := q(x)\int_E P(x, dy)\varphi(y),$$

$$\mathbf{G}_{u,x}^\varepsilon\varphi(u) := \int_{\mathbf{R}^d}[\varphi(u + v) - \varphi(u)]G_{u,x}^\varepsilon(dv).$$

To prove the convergence of the predictable characteristics, it suffices to consider the action of the generator $\mathbf{L}^\varepsilon$ on the test function of two variables $\varphi(v, x)$.

So, it has a representation

$$\mathbb{L}^\varepsilon \varphi(v; x) = [\varepsilon^{-1} Q + Q_0 A(u, x) + \varepsilon \theta^\varepsilon(x)] \varphi(v; x).$$

Solution of the singular perturbation problem on test functions

$$\varphi^\varepsilon(v; x) = \varphi(v) + \varepsilon \varphi_1(v; x)$$

is similar to the continuous case and has the form

$$\widehat{\mathbb{L}} = \widehat{A}(u),$$

where $\widehat{A}(u) = q \int_E \rho(dx) A(u, x)$.

Thus,

$$\widehat{A}(u)\varphi(v) = \widehat{a}(u)\varphi'(v),$$

where $\widehat{a}(u) := q \int_E \rho(dx) a(u, x)$.

We can now apply Theorem 2.2 and Theorem 2.1, from which follows a weak convergence in $\mathbf{D}_{\mathbf{R}^d}[0, \infty)$ to a process, uniquely defined by the limit predictable characteristics.

Since the weak convergence of the process is proved, the form of the limit generator can be obtained from the asymptotic representation of the generator [2.43] on the test functions of the form $\varphi^\varepsilon(u; x) = \varphi(u) + \varepsilon \varphi_1(u; x)$.

The generator has the form

$$\mathbb{L}^\varepsilon \varphi^\varepsilon(u; x) = [\varepsilon^{-1} Q + \varepsilon^{-1} Q_0 \mathbf{G}^\varepsilon_{u,x}][\varphi(u) + \varepsilon \varphi_1(u; x)]$$

$$= \varepsilon^{-1} Q\varphi(u) + [Q\varphi_1(u; x) + \varepsilon^{-1} Q_0 \mathbf{G}^\varepsilon_{u,x} \varphi(u)] + Q_0 \mathbf{G}^\varepsilon_{u,x} \varphi_1(u; x).$$

Using an analogue of Lemma 1.1, we can write the following asymptotic representation:

$$\mathbb{L}^\varepsilon \varphi^\varepsilon(u; x) = \varepsilon^{-1} Q\varphi(u) + [Q\varphi_1(u; x) + Q_0 \mathbf{G}_{u,x} \varphi(u)]$$
$$+ \varepsilon Q_0 \mathbf{G}_{u,x} \varphi_1(u; x) + \theta^\varepsilon \varphi(u).$$

Thus, the singular perturbation problem is reduced to a system of equations

$$Q\varphi(u) = 0,$$

$$Q\varphi_1(u; x) + Q_0 \mathbf{G}_{u,x} \varphi(u) = \widehat{\mathbf{L}} \varphi(u).$$

From the first equation, it is obvious that $\varphi(u) \in \mathcal{N}_Q$, and from the solvability condition for the second equation, we have the generator:

$$\widehat{\mathbf{L}}\varphi(u) = \Pi Q_0 \mathbf{G}_{u,x} \Pi \varphi(u)$$

$$= \widehat{b}(u)\varphi'(u) + \int_{\mathbf{R}^d} [\varphi(u+v) - \varphi(u) - v\varphi'(u)] G_u(dv),$$

which can also be written as [2.37].

Using condition **LL** and the results of Skorokhod *et al.* (2002, Chapter 2), we can say that the limit generator really determines the limit process $\xi^0(t)$.

Theorem 2.10 is proved.

In the Lévy approximation scheme for impulsive processes with semi-Markov switching (see sections 1.1.4, 1.1.9 and 1.1.10), we have the following normalization:

$$\widetilde{\xi}^\varepsilon(t) = \widetilde{\xi}_0^\varepsilon + \sum_{k=1}^{\nu(t/\varepsilon^2)} \alpha_k^\varepsilon(\widetilde{\xi}_{k-1}^\varepsilon, x_{k-1}^\varepsilon), \quad t \geq 0. \qquad [2.44]$$

Here

$$\widetilde{\xi}_n^\varepsilon := \widetilde{\xi}(\varepsilon^2 \tau_n) = \widetilde{\xi}_0^\varepsilon + \sum_{k=1}^{n} \alpha_k^\varepsilon(\widetilde{\xi}_{k-1}^\varepsilon, x_{k-1}^\varepsilon).$$

Suppose that condition **CSM** and the Lévy approximation conditions **C1'**, **C2** and **C3** for the impulsive process are satisfied.

THEOREM 2.11.– *Under the conditions* **CSM, C1–C4**, *the impulsive recurrent process* [2.44] *converges weakly in* $D_{\mathbf{R}^d}[0,\infty)$ *to a Lévy process* $\xi^0(t), t \geq 0$, *which under condition* **LL** *is determined by the generator*

$$\widehat{\mathbf{L}}\varphi(u) = (\widehat{b}(u) - \widehat{b}_0(u) + \widehat{b}_1(u))\varphi'(u) + \frac{1}{2}\sigma^2(u)\varphi''(u)$$

$$+ \lambda(u) \int_{\mathbf{R}^d} [\varphi(u+v) - \varphi(u)] G_u^0(dv), \qquad [2.45]$$

where $\widehat{b}(u) = q\int_E \rho(dx) b(u;x), \widehat{b}_0(u) = \int_E v G_u(dv)$,

$$\widehat{b}_1(u) = q \int_E \rho(dx) b_1(u;x) \widetilde{R}_0 b_1'^*(u;x) - q \int_E \rho(dx) b_1(u;x) b_1'^*(u;x),$$

$$G_u(dv) = q \int_E \rho(dx) G_{u,x}(dv), \widehat{b_1^2}(u) = q \int_E \rho(dx) b_1(u;x) b_1^*(u;x),$$

$$\sigma^2(u) = 2q \int_E \rho(dx)[b_1(u;x)\widetilde{R}_0 b_1^*(u;x) + \frac{1}{2}(c(u;x)$$

$$- c_0(u;x))] - \widehat{b_1^2}(u), \sigma^2(u) \geq 0,$$

$$\lambda(u) = qG_u(\mathbf{R}^d), \qquad G_u^0(dv) = G_u(dv)/G_u(\mathbf{R}^d).$$

Here, $\widetilde{R}_0$ is the potential operator of the embedded Markov chain.

REMARK 2.7.– *If we investigate a semimartingale without switching (see, for example, Jacod and Shiryaev (2003)), a similar result for the second modified characteristic has the form $\sigma^2 = E(\alpha_k^\varepsilon)^2 - (E(\alpha_k^\varepsilon))^2 = (c - c_0) - b_1^2$. This is obvious from Theorem 2.2.3, while from Korolyuk and Limnios (2005, Theorem 9.3) for the impulsive process with Markov switching, the explicit form of the second modified characteristic does not follow obviously.*

The proof of the theorem is again carried out in two stages.

STAGE 1. Completely repeat the proof of Lemma 2.6, from which Corollary 2.3 follows. Lemma 2.7 is also obviously true.

STAGE 2. In proving the convergence of the predictable characteristics, we use the methods described in Korolyuk and Limnios (2005, sections 3.2 and 3.3) and Theorem 2.1.

Note that we will further denote by $A^\varepsilon(t)$ the main part of any of the predictive characteristics $B^\varepsilon(t), C^\varepsilon(t), G^\varepsilon(t)$.

The extended Markov renewal process is considered a four-component Markov chain

$$A_n^\varepsilon = A^\varepsilon(\tau_n^\varepsilon), \quad \widetilde{\xi}_n^\varepsilon, \quad x_n^\varepsilon, \quad \tau_n^\varepsilon, \quad n \geq 0,$$

where $x_n^\varepsilon = x^\varepsilon(\tau_n^\varepsilon)$, $x^\varepsilon(t) := x(t/\varepsilon^2)$, $\widetilde{\xi}_n^\varepsilon = \widetilde{\xi}^\varepsilon(\tau_n^\varepsilon)$ and $\tau_{n+1}^\varepsilon = \tau_n^\varepsilon + \varepsilon^2\theta_n^\varepsilon, n \geq 0$, and

$$P(\theta_{n+1}^\varepsilon \leq t | x_n^\varepsilon = x) = F_x(t) = P(\theta_x \leq t).$$

The compensating operator of the extended Markov renewal process can be set by equality (see also Korolyuk and Limnios (2005, section 2.8))

$$\mathbb{L}^\varepsilon\varphi(u,v;x) = \varepsilon^{-2}q(x)\left[\int_E P(x,dy)\int_{\mathbf{R}^d} G_{u,y}^\varepsilon(dz)A^\varepsilon(u;y)\varphi(u+z,v;y)\right.$$

$$\left.-\varphi(u,v;x)\right],$$

where $A^\varepsilon(u;y), y \in E, u \in \mathbf{R}^d$ is a family of generators with the limit generator

$$A(u;y)\varphi(v) = a(u;y)\varphi'(v).$$

According to Korolyuk and Limnios (2005, Chapter 3 and Lemma 9.2), the compensating operator of the extended Markov renewal process has the following asymptotic representation on the test functions $\varphi \in \overline{C}^2(\mathbf{R}^d \times \mathbf{R}^d \times E)$ (measurable functions denoted as $\mathbf{R}^d \times \mathbf{R}^d \times E$, twice continuously differentiable by the first two arguments and bounded by the third argument):

$$\begin{aligned}\mathbb{L}^\varepsilon\varphi(u,v;x) &= \varepsilon^{-2}Q\varphi(\cdot,\cdot;x) + Q_0A(u,x)\varphi(\cdot,v;\cdot)\\ &\quad + \varepsilon^{-2}Q_0\mathbf{G}^\varepsilon_{u,x}\varphi(u,\cdot;\cdot) + \varepsilon\theta^\varepsilon(x)\varphi(u,v;x),\end{aligned} \qquad [2.46]$$

where:

$$Q_0\varphi(x) := q(x)\int_E P(x,dy)\varphi(y),$$

$$\mathbf{G}^\varepsilon_{u,x}\varphi(u) := \int_{\mathbf{R}^d}[\varphi(u+v) - \varphi(u)]G^\varepsilon_{u,x}(dv).$$

Considering the generator [2.46] on the test functions $\varphi^\varepsilon(v;x) = \varphi(v) + \varepsilon\varphi_1(v;x) + \varepsilon^2\varphi_2(v;x)$ similar to the continuous case and that of Markov switching for the impulsive process, we obtain the result about convergence of the predictable characteristics.

Thus, taking into account the compactness of the process, the result about weak convergence of the process will follow from Theorem 2.1.

The generator [2.46] on the test functions $\varphi^\varepsilon(u;x) = \varphi(u) + \varepsilon\varphi_1(u;x) + \varepsilon^2\varphi_2(u;x)$ is of the form

$$\mathbb{L}^\varepsilon\varphi^\varepsilon(u;x) = [\varepsilon^{-2}Q + \varepsilon^{-2}Q_0\mathbf{G}^\varepsilon_{u,x}][\varphi(u) + \varepsilon\varphi_1(u;x) + \varepsilon^2\varphi_2(u;x)].$$

Similar to Lemma 2.2, we can write an asymptotic representation

$$\varepsilon^{-2}\mathbf{G}^\varepsilon_{u,x}\varphi(u) = [\varepsilon^{-1}B_1(u;x) + G(u;x)]\varphi(u) + \theta^\varepsilon := \varepsilon^{-1}b_1(u;x)\varphi'(u)$$

$$+(b(u;x) - b_0(u;x))\varphi'(u) + \frac{1}{2}(c(u;x) - c_0(u;x))\varphi''(u)$$

$$+\int_{\mathbf{R}^d}[\varphi(u+v) - \varphi(u)]G_{u,x}(dv) + \theta^\varepsilon,$$

where $b_0(u;x) := \int_{\mathbf{R}^d} vG_{u,x}(dv), c_0(u;x) := \int_{\mathbf{R}^d} vv^*G_{u,x}(dv)$.

So:

$$\mathbb{L}^\varepsilon\varphi^\varepsilon(u;x) = [\varepsilon^{-2}Q + \varepsilon^{-1}Q_0B_1(u;x) + Q_0G(u;x)][\varphi(u) + \varepsilon\varphi_1(u;x)$$

$$+\varepsilon^2\varphi_2(u;x)] + \theta^\varepsilon\varphi^\varepsilon = \varepsilon^{-2}Q\varphi(u) + \varepsilon^{-1}[Q\varphi_1(u;x) + Q_0b_1(u;x)\varphi'(u)]$$

$$+[Q\varphi_2(u;x)+Q_0b_1(u;x)\varphi_1'(u;x)+Q_0G(u;x)\varphi(u)]+\varepsilon Q_0B_1(u;x)\varphi_2(u;x)$$

$$+\varepsilon Q_0G(u;x)\varphi_1(u;x) + \varepsilon^2Q_0G(u;x)\varphi_2(u;x) + \theta^\varepsilon\varphi^\varepsilon.$$

Thus, the singular perturbation problem is reduced to a system of equations:

$$Q\varphi(u) = 0,$$

$$Q\varphi_1(v;x) + Q_0b_1(u;x)\varphi'(u) = 0,$$

$$Q\varphi_2(v;x) + Q_0b_1(u;x)\varphi_1'(u;x) + Q_0G(u;x)\varphi(u) = \widehat{\mathbf{L}}\varphi(u).$$

To simplify the formulas, we rewrite these equations for the embedded Markov chain. The corresponding operator

$$\widetilde{Q} := P - I,$$

and the potential operator has the property

$$\widetilde{R}_0(P - I) = \widetilde{\Pi} - I.$$

Thus

$$\widetilde{Q}\varphi(u) = 0,$$

$$\widetilde{Q}\varphi_1(v;x) + Q_0b_1(u;x)\varphi'(u) = 0,$$

$$\widetilde{Q}\varphi_2(v;x) + Q_0b_1(u;x)\varphi_1'(u;x) + Q_0G(u;x)\varphi(u) = \widehat{\mathbf{L}}\varphi(u).$$

So, $\varphi(u) \in \mathcal{N}_{\widetilde{Q}}$, and from balance condition **L3** and the second equation

$$\widetilde{Q}\varphi_1(u;x) = -Q_0b_1(u;x)\varphi'(u), \varphi_1(u;x) = \widetilde{R}_0Q_0b_1(u;x)\varphi'(u).$$

Substituting into the last equation, we obtain

$$\widetilde{Q}\varphi_2(u;x)+Q_0b_1(u;x)\widetilde{R}_0Q_0b_1'(u;x)\varphi'(u)+Q_0b_1(u;x)\widetilde{R}_0Q_0b_1(u;x)\varphi''(u)$$

$$+Q_0G(u;x)\varphi(u) = \widehat{\mathbf{L}}\varphi(u).$$

Applying the solvability condition to the last equation and taking into account the relation $\widetilde{R}_0P = \widetilde{R}_0 + \widetilde{\Pi} - I$, we finally obtain the generator [2.45].

Using condition **LL** and the results from Skorokhod *et al.* (2002, Chapter 2), we can say that the limit generator really defines the limit process $\widetilde{\xi}^0(t)$.

The theorem is proved.

3

Large Deviations in the Scheme of Asymptotically Small Diffusion

This chapter studies the asymptotic analysis of the large deviations problem for random evolutionary systems in the scheme of asymptotically small diffusion. The theory of large deviations deals with the asymptotic estimation of probabilities of rare events. The method, used in the majority of classical works, is based on the change of measure and the application of the variational formula to the cumulant of the process under study. The method of asymptotic analysis of the exponential generator of the Markov process is used. The limit exponential generators are calculated for random evolutionary systems with ergodic Markov switching. The method proposed here may have applications for finite-dimensional models, such as random evolutionary systems in $\mathbf{R}^d$, queuing theory, etc.

3.1. Statement of the problem

The theory of large deviations was presented in the work of Cramér (1938), which deals with the asymptotic estimation of probabilities of rare events. The main problem in the large deviations theory is the construction of the rate functional to estimate the probabilities of rare events. Different aspects and applications of this problem were studied by many mathematicians. We discuss the Markov processes with independent increments, so it is natural to refer to the fundamental works of Donsker and Varadhan (1975a, 1975b, 1976), Ventsel (1976a, 1976b, 1979) and Freidlin and Ventzell (2012).

Another approach was presented in the works of Hopf (1950) and Cole (1951), which was applied to the large deviations problem in Fleming and Souganidis (1986). It is based on the asymptotic analysis of the nonlinear Hamilton–Jacobi equation corresponding to the process under study. The solution of the limit nonlinear Hamilton–Jacobi equation is given by the variational formula that defines the rate functional of the pre-limit process. The main problem here is to prove the uniqueness of the solution of the limit nonlinear equation.

All the technical problems connected with the application of the last method to different classes of Markov processes are solved in the monograph (Feng and Kurtz 2006). The main idea of this book is the following.

Let $\eta(t)$, $t \geq 0$, be a Markov process in Euclidean space R, defined by its linear generator L

$$L\varphi(u) := \lim_{\Delta \to 0} \frac{1}{\Delta} E[\varphi(u + \Delta\eta(t)) - \varphi(u) | \eta(t) = u],$$

$$\Delta\eta(t) := \eta(t + \Delta) - \eta(t).$$

The classical martingale characterization of Markov processes is defined by the martingale μ_t:

$$\mu_t = \varphi(\eta(t)) - \varphi(\eta(0)) - \int_0^t \mathbf{L}\varphi(\eta(s))ds.$$

The large deviations theory is based on the exponential martingale characterization

$$\widetilde{\mu}_t = \exp\{\varphi(\eta(t)) - \varphi(\eta(0)) - \int_0^t \mathbf{H}\varphi(\eta(s))ds\},$$

which is a martingale.

The exponential (nonlinear) operator $\mathbf{H}$ is connected with the linear generator $\mathbf{L}$ of the Markov process $\eta(t)$, with continuous time $t \geq 0$, as follows:

$$\mathbf{H}\varphi(u) = e^{-\varphi(u)}\mathbf{L}e^{\varphi(u)}, e^{\varphi(u)} \in \mathcal{D}(\mathbf{L}).$$

The large deviations problem can be formulated as a limit theorem in the series scheme with a small parameter $\varepsilon \to 0$ ($\varepsilon > 0$) (see Feng and Kurtz (2006, Chapter 1)):

$$\mathbf{H}^\varepsilon\varphi^\varepsilon \to \mathbf{H}\varphi, \varphi^\varepsilon \to \varphi, \varepsilon \to 0.$$

Here, by definition,

$$\mathbf{H}^\varepsilon\varphi(u) := e^{-\varphi(u)/\varepsilon}\varepsilon\mathbf{L}^\varepsilon e^{\varphi(u)/\varepsilon}.$$

The generator $\mathbf{L}^\varepsilon, \varepsilon > 0$ defines the Markov process $\eta^\varepsilon(t), t \geq 0, \varepsilon > 0$ in the scheme of series under some scaling transform.

EXAMPLE 3.1.– *The asymptotically small diffusion process is given by* $\sqrt{\varepsilon}\sigma w(t)$, $t \geq 0$ *with the standard Brownian motion process* $w(t), t \geq 0$. *The generator of such a process is the following:*

$$\mathbf{L}^{\varepsilon}\varphi(u) = \varepsilon\frac{1}{2}B\varphi''(u), B = \sigma^2, \varphi''(u) := \partial^2\varphi(u)/\partial u^2.$$

The exponential generator of the asymptotically small diffusion process can be easily calculated:

$$\mathbf{H}^{\varepsilon}\varphi(u) = \frac{1}{2}B[\varphi'(u)]^2 + \varepsilon B\varphi''(u).$$

Hence, the limit exponential operator is represented as follows:

$$\mathbf{H}\varphi(u) = \frac{1}{2}B[\varphi'(u)]^2. \qquad [3.1]$$

REMARK 3.1.– *The exponential operator* [3.1] *in Euclidean space* $\mathbf{R}^d, d \geq 2$ *is represented by the quadratic form*

$$\mathbf{H}\varphi(u) = \frac{1}{2}\sum_{k,r=1}^{d} B_{kr}\varphi_k'(u)\varphi_r'(u), \varphi_k'(u) := \partial\varphi(u)/\partial u_k,$$

where $B = [B_{kr}; 1 \leq k, r \leq d]$ *is the variance matrix.*

To simplify the notations, we present all the following results in $\mathbf{R}$.

To start with, the large deviations problem is considered for the processes with locally independent increments under the scaling (Mogulskii 1993):

$$\eta^{\varepsilon}(t) = \varepsilon^2\eta(t/\varepsilon^3), t \geq 0, \varepsilon > 0. \qquad [3.2]$$

The generator of the Markov process [3.2] is given by

$$\Gamma^{\varepsilon}\varphi(u) = \varepsilon^{-3}\int_{\mathbf{R}}[\varphi(u+\varepsilon^2 v) - \varphi(u)]\Gamma(u, dv), u \in \mathbf{R}, \varphi(u) \in \mathcal{B}_{\mathbf{R}}. \qquad [3.3]$$

The Lévy measure $\Gamma(u, dv)$ satisfies the condition

$$\int_{\mathbf{R}} e^{av}\Gamma(u, dv) < \infty, a > 0, u \in \mathbf{R}. \qquad [3.4]$$

In section 3.3, the large deviations problem is considered for the random evolutionary system with Markov switching (see Korolyuk and Limnios (2005, Chapter 2)). The scheme of asymptotically small diffusion is considered under additional balance conditions (local and total).

The large deviations problem in the scheme of phase merging is investigated in Chapter 5.

3.2. Processes with locally independent increments

For simplicity, compound Poisson processes are supposed to be defined by the generator [3.3] under the condition [3.4].

The balance condition (local) is:

LB : $b(u) := \int_{\mathbf{R}} v\Gamma(u, dv) \equiv 0.$

The main part of the asymptotic representation of the generator [3.3] on the smooth enough test functions is the following:

$$\Gamma^\varepsilon \varphi(u) = \varepsilon \frac{1}{2} B(u)\varphi''(u) + \varepsilon\delta^\varepsilon(u)\varphi(u)$$

$$B(u) = \int_{\mathbf{R}} v^2 \Gamma(u, dv)$$

and the negligible term converges uniformly by u on the functions $\varphi(u) \in C^3(\mathbf{R})$:

$$|\delta^\varepsilon(u)\varphi(u)| \to 0, \varepsilon \to 0. \qquad [3.5]$$

The large deviations problem for the processes [3.2] can be solved using the limit approximation of the exponential generator (see Feng and Kurtz (2006, Part 1)):

$$\mathbf{H}^\varepsilon \varphi(u) = e^{-\varphi(u)/\varepsilon} \varepsilon \Gamma^\varepsilon e^{\varphi(u)/\varepsilon} = \varepsilon^{-2} \int_{\mathbf{R}} [e^{\Delta^\varepsilon \varphi} - 1]\Gamma(u, dv),$$

$$\Delta^\varepsilon \varphi := \varepsilon^{-1}[\varphi(u + \varepsilon^2 v) - \varphi(u)] = \varepsilon v \varphi'(u) + \varepsilon\delta^\varepsilon \varphi(u).$$

Hence, due to the **LB** condition,

$$\mathbf{H}^\varepsilon \varphi(u) = \varepsilon^{-2} \int_{\mathbf{R}} [\varepsilon v \varphi'(u) + \varepsilon^2 \frac{1}{2} v^2 [\varphi'(u)]^2]\Gamma(u, dv) + \delta^\varepsilon(u)\varphi(u) \qquad [3.6]$$

$$= \frac{1}{2} B(u)[\varphi'(u)]^2 + \delta^\varepsilon(u)\varphi(u) \qquad [3.7]$$

with the negligible term [3.5].

CONCLUSION.– The limit exponential operator for the processes with locally independent increments in the scheme of asymptotically small diffusion is given by

$$\mathbf{H}\varphi(u) = \frac{1}{2} B(u)[\varphi'(u)]^2. \qquad [3.8]$$

3.3. Random evolutionary systems in the scheme of ergodic phase merging

In this section, we investigate random evolutionary systems with locally independent increments, and switching, so we should note that random evolutionary systems with switching are also studied in Chapter 11 of Feng and Kurtz (2006), using the classical methods of averaging and homogenization. This approach involves perturbed PDE operators and perturbed test functions, as presented in the works of Kurtz (1973) and Papanicolaou *et al.* (1977). Recent monographs (Skorokhod *et al.* 2002; Pavliotis and Stuart 2008) include large bibliographic reference sections on this problem. Application of this method to the nonlinear case can also be found in the work of Evans (1989). This approach is important for infinite-dimensional state space models, such as interacting particles or stochastic PDEs. However, in this case, a large number of additional problems appear: the correct description of the functional space for solutions, the domain of infinitesimal operators, etc.

We use the generators of Markov processes with a locally compact vector state space (see Korolyuk and Limnios (2005) for more details). This simplifies the analysis because the generators are defined for all bounded measurable functions. We lose generality, but may present obvious algorithms for the verification of convergence conditions and the calculation of the limit generators. This approach is important for finite-dimensional models, such as random evolutionary systems in $\mathbf{R}^d$.

The stochastic additive functional (see Korolyuk and Limnios (2005, Chapter 2))

$$\xi(t) = \xi_0 + \int_0^t \eta(ds; x(s)),\ t \geq 0, \qquad [3.9]$$

is considered as a random evolutionary system with independent increments.

The Markov switching process $x(t), t \geq 0$, on a standard phase space $(E, \mathcal{E})$ is given by the generator

$$Q\varphi(x) = q(x) \int_E [\varphi(y) - \varphi(x)] P(x, dy),\ x \in E.$$

Markov processes with independent increments $\eta(t; x), t \geq 0, x \in E$, with values in the real line R, are defined by the generators

$$\Gamma(x)\varphi(u) = \int_R [\varphi(u+z) - \varphi(u)]\Gamma(dz; x), x \in E, u \in R^d.$$

The random evolutionary system [3.9] can be determined by the generator of the two-component Markov process $\xi(t), x(t), t \geq 0$ (see Korolyuk and Limnios (2005, Chapter 2))

$$L\varphi(u, x) = Q\varphi(\cdot, x) + \Gamma(x)\varphi(u, \cdot).$$

3.3.1. *Large deviations problem under the balance condition (total)*

The random evolutionary system [3.9] is considered in the scheme of series with a small series parameter $\varepsilon \to 0(\varepsilon > 0)$ in the following scaling:

$$\xi^\varepsilon(t) = \varepsilon^2\xi(t/\varepsilon^3),\ x(t/\varepsilon^2),\ t \geq 0. \qquad [3.10]$$

The main conditions are uniform ergodicity of the switching Markov process **CM** and the balance condition (total):

$$b(x) := \int_{R^d} z\Gamma(dz;x) \equiv 0. \qquad [3.11]$$

THEOREM 3.1.– *The large deviations problem for a random evolutionary system with independent increments* [3.9] *is realized by the exponential generator of a small diffusion*

$$H\varphi(u) = \frac{1}{2}B[\varphi'(u)]^2,$$

$$B = \int_E \pi(dx)B(x),\ B(x) = \int_R z^2\Gamma(dz;x).$$

PROOF.– The random evolutionary system [3.10] is defined by the generators

$$L^\varepsilon\varphi(u,x) = [\varepsilon^{-2}Q + \varepsilon^{-3}\Gamma_\varepsilon(x)]\varphi(u,x), \qquad [3.12]$$

$$\Gamma_\varepsilon(x)\varphi(u) = \int_R [\varphi(u+\varepsilon^2 z) - \varphi(u)]\Gamma(dz;x),\ x \in E. \qquad [3.13]$$

The exponential generator of the random evolutionary system [3.10], given by the generators [3.12], is considered on perturbed test functions (see Feng and Kurtz (2006, Chapter 1))

$$\varphi^\varepsilon(u,x) = \varphi(u) + \varepsilon\ln[1 + \varepsilon\varphi_1(u,x)].$$

That is,

$$H^\varepsilon\varphi^\varepsilon(u,x) = e^{-\varphi/\varepsilon}[1+\varepsilon\varphi_1]^{-1}\varepsilon L^\varepsilon[1+\varepsilon\varphi_1]e^{\varphi/\varepsilon}.$$

LEMMA 3.1.– *The exponential generator of the random evolutionary system* [3.10] *has the following asymptotic representation:*

$$H^\varepsilon\varphi^\varepsilon(u,x) = Q\varphi_1 + \frac{1}{2}B(x)[\varphi'(u)]^2 + h^\varepsilon(x)\varphi(u) \qquad [3.14]$$

with a negligible term

$$|h^\varepsilon(x)\varphi(u)| \to 0,\ \varepsilon \to 0,\ \varphi(u) \in C^3(R). \qquad [3.15]$$

Now the solution of the singular perturbation problem for the reducible-invertible operator Q gives (see Korolyuk and Limnios (2005, Chapter 5))

$$Q\varphi_1 + \frac{1}{2}B(x)[\varphi'(u)]^2 = \frac{1}{2}B[\varphi'(u)]^2. \qquad [3.16]$$

Representation [3.14]–[3.16] completes the proof of Theorem 3.1.

REMARK 3.2.– *The negligible term in* [3.14] *is calculated by using (see Korolyuk and Limnios (2005, Chapter 5))*

$$\varphi_1(u,x) = \frac{1}{2}R_0\widetilde{B}(x)[\varphi'(u)]^2, \ \widetilde{B}(x) := B(x) - B.$$

The potential operator R_0 is defined by the solution of the equation

$$QR_0 = R_0Q = \Pi - I, \ \Pi\varphi(x) = \int_E \pi(dx)\varphi(x).$$

3.3.2. *Large deviations problem under the (local) balance condition*

Let us consider the (local) balance condition

$$b = \int_E \pi(dx)b(x), \ \ b(x) = \int_R z\Gamma(dz;x), \ \ \widetilde{b}(x) := b(x) - b.$$

The random evolutionary system with independent increments is considered in the series scheme and the scaling:

$$\xi^\varepsilon(t) = \varepsilon^2\xi(t/\varepsilon^3) - bt/\varepsilon, \ x(t/\varepsilon^3), \ t \geq 0. \qquad [3.17]$$

The generator of the random evolutionary system [3.17] is given by the relation

$$L^\varepsilon\varphi(u,x) = [\varepsilon^{-3}Q + \Gamma^\varepsilon_{\widetilde{b}}(x)]\varphi(u,x), \qquad [3.18]$$

$$\Gamma^\varepsilon_{\widetilde{b}}(x)\varphi(u) = \varepsilon^{-1}\widetilde{b}(x)\varphi'(u) + \varepsilon^{-3}\int_R [\varphi(u+\varepsilon^2 z) - \varphi(u) - \varepsilon^2 z\varphi'(u)]\Gamma(dz;x). \qquad [3.19]$$

Note that the function $\widetilde{b}(x) = b(x) - b$ satisfies the balance condition

$$\Pi\widetilde{b}(x) = \int_E \pi(dx)\widetilde{b}(x) = 0. \qquad [3.20]$$

THEOREM 3.2.– *The large deviations problem for the random evolutionary system with independent increments* [3.17], *given by the generators* [3.20] *under the balance condition (local), is realized by the exponential generator of a small diffusion*

$$H\varphi(u) = \frac{1}{2}B[\varphi'(u)]^2, \qquad [3.21]$$

$$B = B_1 + B_2,\ B_k = \int_E \pi(dx)B_k(x),\ k = 1, 2, \qquad [3.22]$$

$$B_1(x) = \int_R z^2\Gamma(dz; x),\ B_2(x) = 2\widetilde{b}(x)R_0\widetilde{b}(x). \qquad [3.23]$$

REMARK 3.3.– *The variance of the small diffusion B contains two terms. The first term B_1 is defined by the second moments of the random evolutionary system. The second term B_2 is defined by the fluctuation of the first moments.*

PROOF.– The exponential generator of the random evolutionary system [3.17]–[3.20] is considered on perturbed test functions

$$\varphi^\varepsilon(u, x) = \varphi(u) + \varepsilon \ln[1 + \varepsilon\varphi_1(u, x) + \varepsilon^2\varphi_2(u, x)],$$

that is,

$$H^\varepsilon\varphi^\varepsilon(u, x) = e^{-\varphi/\varepsilon}[1 + \varepsilon\varphi_1 + \varepsilon^2\varphi_2]^{-1}\varepsilon L^\varepsilon[1 + \varepsilon\varphi_1 + \varepsilon^2\varphi_2]e^{\varphi/\varepsilon}.$$

LEMMA 3.2.– *The exponential generator of the random evolutionary system* [3.17]–[3.20] *has the following asymptotic representation:*

$$H^\varepsilon\varphi^\varepsilon = \varepsilon^{-1}[Q\varphi_1 + \widetilde{b}(x)\varphi'] + [Q\varphi_2 - \varphi_1 Q\varphi_1 + \frac{1}{2}B_1(x)[\varphi']^2] + h^\varepsilon(x)\varphi(u),$$

with a negligible term:

$$|h^\varepsilon(x)\varphi(u)| \to 0,\ \varepsilon \to 0,\ \varphi(u) \in C^3(R). \qquad [3.24]$$

Now we consider the singular perturbation problems for the reducible-invertible operator Q (see Korolyuk and Limnios (2005, Chapter 5))

$$Q\varphi_1 + \widetilde{b}(x)\varphi'(u) = 0,\ \Pi\widetilde{b}(x) = 0.$$

Hence:

$$\varphi_1(u, x) = R_0\widetilde{b}(x)\varphi'(u),\ Q\varphi_1 = -\widetilde{b}(x)\varphi'(u).$$

Moreover

$$Q\varphi_2 - \varphi_1 Q\varphi_1 + \frac{1}{2}B_1(x)[\varphi'(u)]^2 = \frac{1}{2}B[\varphi'(u)]^2. \qquad [3.25]$$

where

$$-\varphi_1 Q\varphi_1 = \frac{1}{2}B_2(x)[\varphi'(u)]^2.$$

The right-hand side of [3.25] is calculated by using the solvability condition for equation [3.25].

CONCLUSION.– The exponential generator of the random evolutionary system [3.17]–[3.20] is represented as follows:

$$H^\varepsilon \varphi^\varepsilon(u,x) = \frac{1}{2}B[\varphi'(u)]^2 + h^\varepsilon(x)\varphi(u),$$

with the negligible term [3.24].

The limit exponential generator is given by the generator of the small diffusion [3.21].

3.4. Markov integral functional

The large deviations problem for a Markov integral functional is considered in the following scaling form:

$$\alpha^\varepsilon(t) = \alpha_0 + \int_0^t a^\varepsilon(x(s/\varepsilon^3))ds, \tag{3.26}$$

$$a^\varepsilon(x) = \varepsilon^{-1}a(x) + b(x). \tag{3.27}$$

The first term in [3.27] satisfies the balance condition:

$$\Pi a(x) := \int_E \pi(dx)a(x) = 0. \tag{3.28}$$

The generator of the Markov integral functional [3.26]–[3.27] is defined by

$$L^\varepsilon \varphi(u,x) = [\varepsilon^{-3}Q + \varepsilon^{-1}\mathbf{A}(x) + \mathbf{B}(x)]\varphi(u,x), \tag{3.29}$$

where, by definition,

$$\mathbf{A}(x)\varphi(u) := a(x)\varphi'(u), \; \mathbf{B}(x)\varphi(u) := b(x)\varphi'(u).$$

THEOREM 3.3.– *The large deviations problem for the Markov integral functional* [3.26]–[3.27] *under the balance condition* [3.28] *is realized by the exponential generator of a small diffusion with drift:*

$$H_b\varphi(u) = \frac{1}{2}B[\varphi'(u)]^2 + b\varphi'(u), \tag{3.30}$$

$$B = \int_E \pi(dx)B(x), \; B(x) = 2a(x)R_0a(x), \tag{3.31}$$

$$b = \int_E \pi(dx)b(x). \tag{3.32}$$

PROOF.– The exponential generator of the Markov integral functional [3.26]–[3.28] is considered on perturbed test functions

$$\varphi^\varepsilon(u,x) = \varphi(u) + \varepsilon \ln[1 + \varepsilon\varphi_1(u,x) + \varepsilon^2\varphi_2(u,x)],$$

that is,

$$H^\varepsilon\varphi^\varepsilon = e^{-\varphi/\varepsilon}[1 + \varepsilon\varphi_1 + \varepsilon^2\varphi_2]^{-1}\varepsilon L^\varepsilon[1 + \varepsilon\varphi_1 + \varepsilon^2\varphi_2]e^{\varphi/\varepsilon}.$$

LEMMA 3.3.– *The exponential generator of the Markov integral functional* [3.26]–[3.27] *under the balance condition* [3.28] *has the following asymptotic representation:*

$$H^\varepsilon\varphi^\varepsilon = \varepsilon^{-1}[Q\varphi_1 + a(x)\varphi'(u)] + [Q\varphi_2 - \varphi_1 Q\varphi_1 + b(x)\varphi'(u)] + h^\varepsilon(x)\varphi(u),$$

with a negligible term

$$|h^\varepsilon(x)\varphi(u)| \to 0,\ \varepsilon \to 0,\ \varphi(u) \in C^3(R).$$

Now the singular perturbation problems are considered (Korolyuk and Limnios 2005, Chapter 5)

$$Q\varphi_1 + a(x)\varphi'(u) = 0,\ \ \Pi a(x) = 0.$$

Hence

$$\varphi_1(u,x) = R_0 a(x)\varphi'(u),\ Q\varphi_1 = -a(x)\varphi'(u),$$

and next

$$Q\varphi_2 - \varphi_1 Q\varphi_1 + b(x)\varphi'(u) = H_b\varphi(u),$$

with the exponential generator [3.30] of a small diffusion with drift. Note that

$$-\varphi_1 Q\varphi_1 = \frac{1}{2}B(x)[\varphi'(u)]^2,$$

$$B(x) = 2a(x)R_0 a(x)$$

and

$$b = \Pi b(x) = \int_E \pi(dx)b(x).$$

REMARK 3.4.– *The exponential generator of a small diffusion with drift* [3.30] *is single-valued and generates the function*

$$H_b(p) = \frac{1}{2}Bp^2 + bp,$$

which is convex in $p \in R$.

The Fenchel–Legendre transform defines the reversed function (see Feng and Kurtz (2006, Chapter 8))

$$\mathcal{L}(q) = \sup_{p \in R} \{p \cdot q - H_b(p)\},$$

that is,

$$\mathcal{L}(q) = \frac{1}{2} B^{-1}(q - b).$$

The rate functional on the bounded time interval $0 \leq t \leq T$, *for a random evolutionary system with independent increment is defined as follows (see Feng and Kurtz (2006, Chapter 8), compare with Freidlin and Ventzell (2012)):*

$$I(u) = \int_0^T \mathcal{L}(\dot{u}(s)) ds,$$

with an absolutely continuous function $u(s), 0 \leq s \leq T$, $u(0) = 0$, *and with a square integrable derivative* $\dot{u}(s) := \partial u(s)/\partial s$.

REMARK 3.5.– *The large deviations problem for a random evolutionary system with locally independent increments (see Korolyuk and Limnios (2005, Chapter 2)), in the asymptotically small diffusion scheme, can be investigated using the algorithms of solution of singular perturbation problems, represented in Korolyuk and Limnios (2005), inserted in the asymptotic representation of the exponential generator.*

4

Large Deviations of Systems in Poisson and Lévy Approximation Schemes

4.1. Random evolutionary systems with independent increments

4.1.1. *Poisson approximation scheme*

Consider the large deviations problem for random evolutionary systems in the Poisson approximation scheme (see section 1.1.10) with Markov switching (see section 1.1.3) from the point of view of the program proposed in section 1.2. Note that the second and third stages of this program are implemented in section 1.3, therefore, we will focus on the asymptotic analysis of the nonlinear process generator and the calculation of the rate functional by applying convex analysis methods.

V.S. Korolyuk (2010a, 2010b) proposed applying a solution of the singular perturbation problem in the study of large deviations for random evolutionary systems with independent increments in the scheme of asymptotically small diffusion.

In classical works, asymptotic analysis of the large deviations problem is usually performed using a large series parameter $n \to \infty$, sometimes several different parameters (see, for example, Mogulskii (1993)).

The normalization of random evolutionary systems by a small series parameter in order to solve the large deviations problem in the Poisson approximation scheme is performed using two small parameters (δ normalizes the intensity of jumps of the process, ε normalizes the time and the intensity of jumps):

$$\xi_\varepsilon^\delta(t) = \xi_\varepsilon^\delta(0) + \int_0^t \eta_\varepsilon^\delta(ds; x(s/\varepsilon^2)),\ t \geq 0,$$

$$\eta_\varepsilon^\delta(t) = \varepsilon\eta^\delta(t/\varepsilon^2),$$

$$\Gamma_\varepsilon^\delta(x)\varphi(u) = \varepsilon^{-2}\int_{\mathbf{R}}[\varphi(u+\varepsilon v) - \varphi(u)]\Gamma^\delta(dv;x), \; x \in E,$$

where $\varepsilon, \delta \to 0$, hence $\varepsilon^{-1}\delta \to 1$.

Let us rewrite the conditions of the Poisson approximation according to the new normalization:

C1: *Poisson approximation.* A family of processes with independent increments $\eta^\delta(t;x), x \in E, t \geq 0$ satisfies the conditions of the Poisson approximation:

PA1 Approximation of means:

$$b_\delta(x) = \int_{\mathbf{R}} v\Gamma^\delta(dv;x) = \delta[b(x) + \theta_a^\delta(x)],$$

and

$$c_\delta(x) = \int_{\mathbf{R}} v^2\Gamma^\delta(dv;x) = \delta[c(x) + \theta_c^\delta(x)],$$

where

$$\sup_{x\in E}|b(x)| \leq b < +\infty, \sup_{x\in E}|c(x)| \leq c < +\infty.$$

PA2 For the intensity kernel, the following asymptotic representation holds true:

$$\Gamma_g^\delta(x) = \int_{\mathbf{R}} g(v)\Gamma^\delta(dv;x) = \delta[\Gamma_g(x) + \theta_g^\delta(x)]$$

for all $g \in C_3(\mathbf{R})$ – a class of functions that determines a measure (see Jacod and Shiryaev (2003) and Korolyuk and Limnios (2005, Chapter 7)), $\Gamma_g(x)$ is a bounded kernel,

$$|\Gamma_g(x)| \leq \Gamma_g \quad \text{(a constant that depends on } g).$$

The kernel $\Gamma^0(dv;x)$ is set on a class of functions that defines the measure $C_3(\mathbf{R})$ by the relations

$$\Gamma_g(x) = \int_{\mathbf{R}} g(v)\Gamma^0(dv;x), \quad g \in C_3(\mathbf{R}).$$

The negligible terms $\theta_a^\delta, \theta_c^\delta, \theta_g^\delta$ satisfy the condition

$$\sup_{x\in E}|\theta_\cdot^\delta(x)| \to 0, \quad \delta \to 0.$$

PA3 There is a relationship

$$c(x) := \int_{\mathbf{R}} v^2 \Gamma^0(dv; x),$$

which causes the absence of a diffusion component in the exponential generator, which determines a solution of the large deviations problem.

C2: *Uniform quadratic integrability*:

$$\lim_{c \to \infty} \sup_{x \in E} \int_{|v|>c} v^2 \Gamma^0(dv; x) = 0.$$

C3: *Exponential limitations*:

$$\int_{\mathbf{R}} e^{p|v|} \Gamma^\delta(dv; x) < \infty, \forall p \in \mathbf{R}.$$

THEOREM 4.1.– *Solution of the large deviations problem for random evolutionary system*

$$\xi_\varepsilon^\delta(t) = \xi_\varepsilon^\delta(0) + \int_0^t \eta_\varepsilon^\delta(ds; x(s/\varepsilon^2)),\ t \geq 0,$$

which is set by the generator of the two-component Markov process $\xi_\varepsilon^\delta(t), x(t), t \geq 0$

$$L_\varepsilon^\delta \varphi(u, x) = \varepsilon^{-2} Q\varphi(\cdot, x) + \Gamma_\varepsilon^\delta(x)\varphi(u, \cdot), \qquad [4.1]$$

where

$$\Gamma_\varepsilon^\delta(x)\varphi(u) = \varepsilon^{-2} \int_{\mathbf{R}} [\varphi(u + \varepsilon v) - \varphi(u)] \Gamma^\delta(dv; x),\ x \in E \qquad [4.2]$$

is determined by the exponential generator

$$H^0\varphi(u) = \widehat{b}\varphi'(u) + \int_{\mathbf{R}} [e^{v\varphi'(u)} - 1 - v\varphi'(u)] \widehat{\Gamma}^0(dv), \qquad [4.3]$$

$$\widehat{b} = \Pi b(x) = \int_E \pi(dx) b(x), \widehat{\Gamma}^0(v) = \Pi\Gamma^0(v; x) = \int_E \pi(dx)\Gamma^0(v; x).$$

The averaging is performed by the stationary measure of the switching Markov process.

REMARK 4.1.– *In section 1.2.4 and Example 1.7, the large deviations problem for a process with independent increments in the Poisson approximation scheme is considered, and it is shown that this problem is solved by means of an exponential generator for a jump process with independent increments. The theorems presented in this section state that the large deviations for random evolutionary systems in the Poisson approximation scheme are also determined by an exponential generator for a jump process with independent increments. A comprehensive study of the problem of random deviations for a jump process with independent increments is presented in Freidlin and Ventzell (2012) (see Example 2.2).*

REMARK 4.2.– *The limit exponential generator in the space* $\mathbf{R}^d, d > 1$ *can be represented as follows:*

$$H^0\varphi(u) = \sum_{k=1}^{d}(\widehat{b}_k - \widehat{b}_k^0)\varphi_k' + \int_{\mathbf{R}^d}[e^{v\varphi'(u)} - 1]\widehat{\Gamma}^0(dv),$$

$$\varphi_k' := \partial\varphi(u)/\partial u_k, 1 \leq k \leq d,\ v\varphi'(u) := \sum_{k=1}^{d} v_k\varphi_k'.$$

The last exponential generator can be extended to the space of absolutely continuous functions (see Feng and Kurtz (2006))

$$C_b^1(\mathbf{R}^d) = \{\varphi : \exists \lim_{|u|\to\infty}\varphi(u) = \varphi(\infty),\ \lim_{|u|\to\infty}\varphi'(u) = 0\}.$$

To prove the theorem, we need the following lemma.

LEMMA 4.1.– *The exponential generator in a series scheme*

$$H_\Gamma^{\varepsilon,\delta}(x)\varphi(u) = e^{-\varphi/\varepsilon}\varepsilon\Gamma_\varepsilon^\delta(x)e^{\varphi/\varepsilon} \qquad [4.4]$$

has the following asymptotic representation:

$$H_\Gamma^{\varepsilon,\delta}(x)\varphi(u) = H_\Gamma(x)\varphi(u) + \theta_\Gamma^{\varepsilon,\delta}\varphi,$$

where $\sup_{x\in E}|\theta_\Gamma^{\varepsilon,\delta}\varphi| \to 0, \varepsilon, \delta \to 0,$

$$H_\Gamma(x)\varphi(u) = a(x)\varphi'(u) + \int_{\mathbf{R}}[e^{v\varphi'(u)} - 1 - v\varphi'(u)]\Gamma^0(dv;x).$$

PROOF.– Rewrite [4.4], taking into account the type of generator [4.2]. We have:

$$H_\Gamma^{\varepsilon,\delta}(x)\varphi(u) = \varepsilon^{-1}\int_{\mathbf{R}}[e^{\Delta_\varepsilon\varphi(u)} - 1]\Gamma^\delta(dv;x),$$

where

$$\Delta_\varepsilon\varphi(u) := \varepsilon^{-1}[\varphi(u + \varepsilon v) - \varphi(u)].$$

Rewrite the expression for the generator as follows:

$$H_\Gamma^{\varepsilon,\delta}(x)\varphi(u) = \varepsilon^{-1}\int_{\mathbf{R}}[e^{\Delta_\varepsilon\varphi(u)} - 1 - \Delta_\varepsilon\varphi(u) - \frac{1}{2}(\Delta_\varepsilon\varphi(u))^2]\Gamma^\delta(dv;x)$$

$$+\varepsilon^{-1}\int_{\mathbf{R}}[\Delta_\varepsilon\varphi(u) + \frac{1}{2}(\Delta_\varepsilon\varphi(u))^2]\Gamma^\delta(dv;x).$$

It is easy to see that the function $\psi_u^\varepsilon(v) = e^{\Delta_\varepsilon\varphi(u)} - 1 - \Delta_\varepsilon\varphi(u) - \frac{1}{2}(\Delta_\varepsilon\varphi(u))^2$ belongs to the class $C_3(\mathbf{R})$. In fact,

$$\psi_u^\varepsilon(v)/v^2 \to 0, v \to 0.$$

In addition, this function is continuous and bounded for any ε under the condition that the function $\varphi(u)$ is bounded. Moreover, the boundedness of the function $\psi_u^\varepsilon(v)$ is uniform over u under conditions **C2** and **C3** and the boundedness of the derivative $\varphi'(u)$.

Thus, we have:

$$H_\Gamma^{\varepsilon,\delta}(x)\varphi(u) = \varepsilon^{-1}\delta\int_{\mathbf{R}}[e^{\Delta_\varepsilon\varphi(u)} - 1 - \Delta_\varepsilon\varphi(u) - \frac{1}{2}(\Delta_\varepsilon\varphi(u))^2]\Gamma^0(dv;x)$$

$$+\varepsilon^{-1}\int_{\mathbf{R}}[\Delta_\varepsilon\varphi(u) - v\varphi'(u) - \varepsilon\frac{v^2}{2}\varphi''(u)]\Gamma^\delta(dv;x)$$

$$+\varepsilon^{-1}\delta a(x)\varphi'(u) + \delta c(x)\varphi''(u)$$

$$+\varepsilon^{-1}\int_{\mathbf{R}}[\frac{1}{2}(\Delta_\varepsilon\varphi(u))^2 - \frac{v^2}{2}(\varphi'(u))^2]\Gamma^\delta(dv;x) + \varepsilon^{-1}\delta\frac{1}{2}c(x)(\varphi'(u))^2.$$

Applying Taylor's formula to test functions $\varphi(u) \in C^3(\mathbf{R})$, and condition **PA2**, we obtain:

$$H_\Gamma^{\varepsilon,\delta}(x)\varphi(u) = \varepsilon^{-1}\delta\int_{\mathbf{R}}[e^{v\varphi'(u)} - 1 - v\varphi'(u) - \frac{v^2}{2}(\varphi'(u))^2]\Gamma^0(dv;x)$$

$$+\varepsilon^{-1}\delta\int_{\mathbf{R}}(e^{v\varphi'(u)}\varepsilon\frac{v^2}{2}\varphi''(\widehat{u}) - \varepsilon\frac{v^2}{2}\varphi''(\widehat{u}) - \varepsilon^2\frac{v^4}{8}(\varphi''(\widetilde{u}))^2)\Gamma^0(dv;x)$$

$$+\varepsilon^{-1}\delta\int_{\mathbf{R}}\varepsilon^2\frac{v^3}{3!}\varphi'''(\widetilde{u})\Gamma^0(dv;x) + \varepsilon^{-1}\delta a(x)\varphi'(u) + \delta c(x)\varphi''(u)$$

$$+\varepsilon^{-1}\delta\int_{\mathbf{R}}\varepsilon^2\frac{v^4}{4}(\varphi''(\widetilde{u}))^2\Gamma^0(dv;x) + \varepsilon^{-1}\delta\frac{1}{2}c(x)(\varphi'(u))^2.$$

Considering **PA3** and the limit condition $\varepsilon^{-1}\delta \to 1$, we finally have:

$$H_\Gamma^{\varepsilon,\delta}(x)\varphi(u) = H_\Gamma(x)\varphi(u) + \theta_\Gamma^{\varepsilon,\delta}\varphi,$$

where $\sup_{x\in E} |\theta_\Gamma^{\varepsilon,\delta}\varphi| \to 0, \varepsilon, \delta \to 0.$

Lemma 4.1 is proved.

PROOF OF THEOREM 4.1.– The limit transition for a nonlinear exponential generator of random evolutionary systems is performed on perturbed test functions

$$\varphi_\varepsilon^\delta(u,x) = \varphi(u) + \varepsilon \ln[1 + \delta\varphi_1(u,x)].$$

Thus, we have:

$$\mathbb{H}_\varepsilon^\delta \varphi^\varepsilon = e^{-\varphi^\varepsilon/\varepsilon}\varepsilon L_\varepsilon^\delta e^{\varphi^\varepsilon/\varepsilon} = e^{-\varphi/\varepsilon}[1+\delta\varphi_1]^{-1}\varepsilon L_\varepsilon^\delta e^{\varphi/\varepsilon}[1+\delta\varphi_1].$$

Calculating the asymptotic behavior of the last exponential generator gives the following result.

LEMMA 4.2.– *There is an asymptotic representation*

$$\mathbb{H}_\varepsilon^\delta \varphi^\varepsilon = Q\varphi_1 + H_\Gamma^{\varepsilon,\delta}(x)\varphi(u) + \theta^{\varepsilon,\delta}\varphi,$$

where $\sup_{x\in E} |\theta^{\varepsilon,\delta}\varphi| \to 0, \varepsilon, \delta \to 0.$

PROOF.– Considering [4.1], we have:

$$\mathbb{H}_\varepsilon^\delta \varphi^\varepsilon = e^{-\varphi/\varepsilon}[1 - \delta\varphi_1 + \frac{\delta^2\varphi_1^2}{1+\delta\varphi_1}]\left\{\varepsilon^{-1}Q + \varepsilon\Gamma_\varepsilon^\delta(x)\right\} e^{\varphi/\varepsilon}[1+\delta\varphi_1]$$

$$= e^{-\varphi/\varepsilon}[1-\delta\varphi_1+\frac{\delta^2\varphi_1^2}{1+\delta\varphi_1}]\left\{\varepsilon^{-1}\delta e^{\varphi/\varepsilon}Q\varphi_1+\varepsilon\Gamma_\varepsilon^\delta(x)e^{\varphi/\varepsilon} + \varepsilon\delta\Gamma_\varepsilon^\delta(x)e^{\varphi/\varepsilon}\varphi_1\right\}$$

$$= Q\varphi_1 + H_\Gamma^{\varepsilon,\delta}(x)\varphi(u) + \theta^{\varepsilon,\delta}\varphi,$$

where

$$\theta^{\varepsilon,\delta}\varphi = \delta\left[\frac{\varepsilon}{1+\delta\varphi_1}e^{-\varphi/\varepsilon}\Gamma^\delta(x)e^{\varphi/\varepsilon}\varphi_1 - \frac{\varepsilon^{-1}\delta\varphi_1}{1+\delta\varphi_1}Q\varphi_1\right.$$

$$\left. - \frac{\delta\varphi_1}{1+\delta\varphi_1}e^{-\varphi/\varepsilon}\Gamma^\delta(x)e^{\varphi/\varepsilon}\right].$$

Lemma 4.2 is proved.

Given Lemma 4.1, we have

$$\mathbb{H}_\varepsilon^\delta \varphi^\varepsilon = Q\varphi_1 + H_\Gamma(x)\varphi(u) + h^{\varepsilon,\delta}\varphi,$$

where $h^{\varepsilon,\delta}\varphi = \theta^{\varepsilon,\delta}\varphi + \theta_\Gamma^{\varepsilon,\delta}\varphi$.

Now we use the solution of the singular perturbation problem for a reducibly invertible operator Q (see Korolyuk and Limnios (2005, Chapter 1)). From the solvability condition, we have:

$$Q\varphi_1 + H_\Gamma(x)\varphi(u) = H^0\varphi(u),$$

where

$$H^0\varphi(u) = \Pi Q \Pi \varphi_1 + \Pi H_\Gamma(x)\Pi\varphi(u).$$

Thus, we finally obtain [4.3]:

$$H^0\varphi(u) = \widehat{b}\varphi'(u) + \int_{\mathbf{R}} [e^{v\varphi'(u)} - 1 - v\varphi'(u)]\widehat{\Gamma}^0(dv).$$

The negligible term $h^{\varepsilon,\delta}(x)$ can be calculated explicitly, using the solution of the Poisson equation (see section 1.1.2. For more details, see Korolyuk and Limnios (2005))

$$\varphi_1(u,x) = R_0\widehat{H}(x)\varphi(u), \widehat{H}(x) := H_\Gamma(x) - H^0.$$

Theorem 4.1 is proved.

To calculate the rate functional, we need to find the function $L(q)$ by the formula (see Example 1.7):

$$L(q) = \sup_{p \in \mathbf{R}}(qp - H(p)), \qquad [4.5]$$

where $H(p), p = \varphi'(u)$ is determined from the explicit form of the generator:

$$H(p) := H^0(\varphi') = (\widehat{b} - \widehat{b}_0)p + \int_{\mathbf{R}} [e^{vp} - 1]\widehat{\Gamma}^0(dv).$$

So, let us differentiate by p and equate the expression to 0:

$$(qp - H(p))'_p = (qp - (\widehat{b} - \widehat{b}_0)p - \int_{\mathbf{R}} [e^{vp} - 1]\widehat{\Gamma}^0(dv))'_p$$

$$= q - (\widehat{b} - \widehat{b}_0) - \int_{\mathbf{R}} ve^{vp}\widehat{\Gamma}^0(dv) = 0,$$

i.e. $p(q)$ is determined from the relation

$$\int_{\mathbf{R}} v e^{vp} \widehat{\Gamma}^0(dv) = q - (\widehat{b} - \widehat{b}_0).$$

Substituting the expression for $p(q)$ in formula [4.5], we have a function $L(q)$ and the rate functional is

$$I(x) = I_0(x(0)) + \int_0^{\infty} L(x'(s))ds.$$

4.1.2. *Lévy approximation scheme*

The normalization of random evolutionary systems by a small series parameter in order to solve the large deviations problem in the Lévy approximation scheme is carried out as follows:

$$\xi_\varepsilon^\delta(t) = \xi_\varepsilon^\delta(0) + \int_0^t \eta_\varepsilon^\delta(ds; x(s/\varepsilon^3)),\ t \geq 0,$$

$$\eta_\varepsilon^\delta(t) = \varepsilon \eta^\delta(t/\varepsilon^3),$$

$$\Gamma_\varepsilon^\delta(x)\varphi(u) = \varepsilon^{-3} \int_{\mathbf{R}} [\varphi(u + \varepsilon v) - \varphi(u)]\Gamma^\delta(dv; x),\ x \in E,$$

where $\varepsilon, \delta \to 0$, hence $\varepsilon^{-1}\delta \to 1$.

Let us recall the conditions of the Lévy approximation.

C1': *Lévy approximation.* A family of processes with independent increments $\eta^\delta(t; x), x \in E, t \geq 0$ satisfies the Lévy approximation conditions:

LA1 Approximation of means:

$$a_\delta(x) = \int_{\mathbf{R}} v\Gamma^\delta(dv; x) = \delta b_1(x) + \delta^2[b(x) + \theta_b^\delta(x)],$$

and

$$c_\delta(x) = \int_{\mathbf{R}} v^2\Gamma^\delta(dv; x) = \delta^2[c(x) + \theta_c^\delta(x)],$$

where

$$\sup_{x \in E} |b_1(x)| \leq b_1 < +\infty, \sup_{x \in E} |b(x)| \leq b < +\infty, \sup_{x \in E} |c(x)| \leq c < +\infty.$$

LA2 For the intensity kernel, the following asymptotic representation holds true:

$$\Gamma_g^\delta(x) = \int_{\mathbf{R}} g(v)\Gamma^\delta(dv;x) = \delta^2[\Gamma_g(x) + \theta_g^\delta(x)]$$

for all $g \in C_3(\mathbf{R})$ – a class of functions that determines a measure (see Jacod and Shiryaev (2003) and Korolyuk and Limnios (2005, Chapter 7)), $\Gamma_g(x)$ is a bounded kernel

$$|\Gamma_g(x)| \leq \Gamma_g \quad \text{(a constant that depends on } g\text{)}.$$

The kernel $\Gamma^0(dv;x)$ is set on the class of functions that define the measure $C_3(\mathbf{R})$ by the relation

$$\Gamma_g(x) = \int_{\mathbf{R}} g(v)\Gamma^0(dv;x), \quad g \in C_3(\mathbf{R}).$$

The negligible terms $\theta_b^\delta, \theta_c^\delta, \theta_g^\delta$ satisfy the condition

$$\sup_{x\in E} |\theta_\cdot^\delta(x)| \to 0, \quad \delta \to 0.$$

LA3 Balance condition:

$$\int_E \pi(dx)b_1(x) = 0.$$

C2: *Uniform quadratic integrability*:

$$\lim_{c\to\infty} \sup_{x\in E} \int_{|v|>c} v^2\Gamma^0(dv;x) = 0.$$

C3: *Exponential boundedness*:

$$\int_{\mathbf{R}} e^{p|v|}\Gamma^\delta(dv;x) < \infty, \forall p \in \mathbf{R}.$$

THEOREM 4.2.– *Solution of the large deviations problem for a random evolutionary system*

$$\xi_\varepsilon^\delta(t) = \xi_\varepsilon^\delta(0) + \int_0^t \eta_\varepsilon^\delta(ds; x(s/\varepsilon^3)),\ t \geq 0,$$

which is set by the generator of the two-component Markov process $\xi_\varepsilon^\delta(t), x(t), t \geq 0$

$$L_\varepsilon^\delta\varphi(u,x) = \varepsilon^{-3}Q\varphi(\cdot,x) + \Gamma_\varepsilon^\delta(x)\varphi(u,\cdot),$$

where

$$\Gamma_\varepsilon^\delta(x)\varphi(u) = \varepsilon^{-3}\int_{\mathbf{R}}[\varphi(u+\varepsilon v) - \varphi(u)]\Gamma^\delta(dv;x),\ x \in E \tag{4.6}$$

is determined by the exponential generator

$$H^0\varphi(u) = (\widehat{b} - \widehat{b}_0)\varphi'(u) + \frac{1}{2}\sigma^2(\varphi'(u))^2 + \int_{\mathbf{R}}[e^{v\varphi'(u)} - 1]\widehat{\Gamma}^0(dv), \tag{4.7}$$

$$\widehat{b} = \Pi b(x) = \int_E \pi(dx)b(x), \widehat{b}_0 = \Pi b_0(x) = \int_E \pi(dx)b_0(x),$$

$$b_0(x) = \int_{\mathbf{R}} v\Gamma^0(dv;x), \sigma^2 = (\widehat{c} - \widehat{c}_0) + 2\Pi b_1(x)R_0 b_1(x)\Pi,$$

$$\widehat{c} = \Pi c(x) = \int_E \pi(dx)c(x), \widehat{c}_0 = \Pi c_0(x) = \int_E \pi(dx)c_0(x),$$

$$c_0(x) = \int_{\mathbf{R}} v^2\Gamma^0(dv;x), \widehat{\Gamma}^0(v) = \Pi\Gamma^0(v;x) = \int_E \pi(dx)\Gamma^0(v;x).$$

The averaging is performed by a stationary measure of the switching Markov process.

REMARK 4.3.– *In section 1.2.4 and Example 1.7, we consider the large deviations problem for the process with independent increments in the Lévy approximation scheme and show that this problem is solved by the exponential generator for the jumping process with independent increments. The theorems presented in this section state that large deviations for random evolutionary systems in Lévy approximation schemes are also defined by the exponential generator for the jumping process with independent increments. A comprehensive study of the large deviations problem for the jump process with independent increments is presented in Freidlin and Ventzell (2012).*

REMARK 4.4.– *The limit exponential generator in the space* $\mathbf{R}^d, d > 1$ *can be represented as follows:*

$$H^0\varphi(u) = \sum_{k=1}^{d}(\widehat{a}_k - \widehat{a}_k^0)\varphi_k' + \frac{1}{2}\sum_{k,r=1}^{d}\sigma_{kr}\varphi_k'\varphi_r' + \int_{\mathbf{R}^d}[e^{v\varphi'(u)} - 1]\widehat{\Gamma}^0(dv),$$

$$\varphi_k' := \partial\varphi(u)/\partial u_k, 1 \le k \le d.$$

Here $\sigma^2 = [\sigma_{kr}; 1 \le k, r \le d]$ *is a variation matrix.*

The last exponential generator can be extended to the space of absolutely continuous functions (see Feng and Kurtz (2006)).

$$C_b^1(R^d) = \{\varphi : \exists \lim_{|u|\to\infty}\varphi(u) = \varphi(\infty), \lim_{|u|\to\infty}\varphi'(u) = 0\}.$$

PROOF.– The limit transition for a nonlinear exponential generator of random evolutionary systems is realized on perturbed test functions

$$\varphi_\varepsilon^\delta(u,x) = \varphi(u) + \varepsilon \ln[1 + \delta\varphi_1(u,x) + \delta^2\varphi_2(u,x)].$$

Thus, we have:

$$\begin{aligned}\mathbb{H}_\varepsilon^\delta \varphi_\varepsilon^\delta &= e^{-\varphi_\varepsilon^\delta/\varepsilon}\varepsilon L_\varepsilon^\delta e^{\varphi_\varepsilon^\delta/\varepsilon} \\ &= e^{-\varphi/\varepsilon}[1 + \delta\varphi_1 + \delta^2\varphi_2]^{-1}\varepsilon L_\varepsilon^\delta e^{\varphi/\varepsilon}[1 + \delta\varphi_1 + \delta^2\varphi_2].\end{aligned}$$

To calculate the asymptotic behavior of the last exponential generator, we need the following lemmas.

LEMMA 4.3.– *By conditions **C1'–C3**, the exponential generator in a series scheme*

$$H_\Gamma^{\varepsilon,\delta}(x)\varphi_\varepsilon^\delta(u,x) = e^{-\varphi_\varepsilon^\delta/\varepsilon}\varepsilon\Gamma_\varepsilon^\delta(x)e^{\varphi_\varepsilon^\delta/\varepsilon}$$

has the following asymptotic representation:

$$H_\Gamma^{\varepsilon,\delta}(x)\varphi(u) = H_\Gamma(x)\varphi(u) + \varepsilon^{-1}b_1(x)\varphi'(u) + \theta_\Gamma^{\varepsilon,\delta}\varphi,$$

where $\sup_{x\in E} |\theta_\Gamma^{\varepsilon,\delta}\varphi| \to 0, \varepsilon, \delta \to 0,$

$$\begin{aligned}H_\Gamma(x)\varphi(u) &= (b(x) - b_0(x))\varphi'(u) + \frac{1}{2}(c(x) - c_0(x))(\varphi'(u))^2 \\ &\quad + \int_{\mathbf{R}} [e^{v\varphi'(u)} - 1]\Gamma^0(dv;x).\end{aligned}$$

PROOF.– We have:

$$H_\Gamma^{\varepsilon,\delta}(x)\varphi_\varepsilon^\delta = e^{-\varphi/\varepsilon}[1 + \delta\varphi_1 + \delta^2\varphi_2]^{-1}\varepsilon\Gamma_\varepsilon^\delta(x)e^{\varphi/\varepsilon}[1 + \delta\varphi_1 + \delta^2\varphi_2]$$

$$= e^{-\varphi/\varepsilon}\left[1 - \delta\varphi_1 + \delta^2\frac{\varphi_1^2 + \delta\varphi_1\varphi_2 - \varphi_2}{1 + \delta\varphi_1 + \delta^2\varphi_2}\right][\varepsilon\Gamma_\varepsilon^\delta(x)e^{\varphi/\varepsilon} + \varepsilon\delta\Gamma_\varepsilon^\delta(x)e^{\varphi/\varepsilon}\varphi_1$$

$$+ \varepsilon\delta^2\Gamma_\varepsilon^\delta(x)e^{\varphi/\varepsilon}\varphi_2].$$

Thus,

$$\begin{aligned}H_\Gamma^{\varepsilon,\delta}(x)\varphi_\varepsilon^\delta = H_\Gamma^{\varepsilon,\delta}(x)\varphi(u) + e^{-\varphi/\varepsilon}\varepsilon\delta\{\Gamma_\varepsilon^\delta(x)e^{\varphi/\varepsilon}\varphi_1 - \varphi_1\Gamma_\varepsilon^\delta(x)e^{\varphi/\varepsilon}\} \quad [4.8] \\ + \widetilde{\theta}_\Gamma^{\varepsilon,\delta}\varphi,\end{aligned}$$

where

$$\widetilde{\theta}_{\Gamma}^{\varepsilon,\delta}\varphi=\varepsilon\delta^2[e^{-\varphi/\varepsilon}\Gamma_{\varepsilon}^{\delta}(x)e^{\varphi/\varepsilon}\varphi_2-e^{-\varphi/\varepsilon}\varphi_1\Gamma_{\varepsilon}^{\delta}(x)e^{\varphi/\varepsilon}\varphi_1]+\varepsilon\delta^2\frac{\varphi_1^2+\delta\varphi_1\varphi_2-\varphi_2}{1+\delta\varphi_1+\delta^2\varphi_2}$$

$$\times[e^{-\varphi/\varepsilon}\Gamma_{\varepsilon}^{\delta}(x)e^{\varphi/\varepsilon}+e^{-\varphi/\varepsilon}\delta\Gamma_{\varepsilon}^{\delta}(x)e^{\varphi/\varepsilon}\varphi_1+e^{-\varphi/\varepsilon}\delta^2\Gamma_{\varepsilon}^{\delta}(x)e^{\varphi/\varepsilon}\varphi_2]$$

$$-\varepsilon\delta^3e^{-\varphi/\varepsilon}\varphi_1\Gamma_{\varepsilon}^{\delta}(x)e^{\varphi/\varepsilon}\varphi_2.$$

Let us rewrite $H_{\Gamma}^{\varepsilon,\delta}(x)\varphi(u)$, taking into account the type of generator [4.6]. We have:

$$H_{\Gamma}^{\varepsilon,\delta}(x)\varphi(u)=\varepsilon^{-2}\int_{\mathbf{R}}[e^{\Delta_\varepsilon\varphi(u)}-1]\Gamma^{\delta}(dv;x),$$

where $\Delta_\varepsilon\varphi(u):=\varepsilon^{-1}[\varphi(u+\varepsilon v)-\varphi(u)]$.

Rewrite the expression for the generator as follows:

$$H_{\Gamma}^{\varepsilon,\delta}(x)\varphi(u)=\varepsilon^{-2}\int_{\mathbf{R}}[e^{\Delta_\varepsilon\varphi(u)}-1-\Delta_\varepsilon\varphi(u)-\frac{1}{2}(\Delta_\varepsilon\varphi(u))^2]\Gamma^{\delta}(dv;x)$$

$$+\varepsilon^{-2}\int_{\mathbf{R}}[\Delta_\varepsilon\varphi(u)+\frac{1}{2}(\Delta_\varepsilon\varphi(u))^2]\Gamma^{\delta}(dv;x).$$

It is easy to see that the function $\psi_u^{\varepsilon}(v)=e^{\Delta_\varepsilon\varphi(u)}-1-\Delta_\varepsilon\varphi(u)-\frac{1}{2}(\Delta_\varepsilon\varphi(u))^2$ belongs to the class $C_3(\mathbf{R})$. In fact,

$$\psi_u^{\varepsilon}(v)/v^2\to 0, v\to 0.$$

In addition, this function is continuous and bounded for each ε under the condition that the function $\varphi(u)$ is bounded. Moreover, the boundedness of the function $\psi_u^{\varepsilon}(v)$ is uniform over u under conditions **C3** and **C4** and the boundedness of the derivative $\varphi'(u)$.

Thus, we have:

$$H_{\Gamma}^{\varepsilon,\delta}(x)\varphi(u)=\varepsilon^{-2}\delta^2\int_{\mathbf{R}}[e^{\Delta_\varepsilon\varphi(u)}-1-\Delta_\varepsilon\varphi(u)-\frac{1}{2}(\Delta_\varepsilon\varphi(u))^2]\Gamma^{0}(dv;x)$$

$$+\varepsilon^{-2}\int_{\mathbf{R}}[\Delta_\varepsilon\varphi(u)-v\varphi'(u)-\varepsilon\frac{v^2}{2}\varphi''(u)]\Gamma^{\delta}(dv;x)+\varepsilon^{-2}\delta b_1(x)\varphi'(u)$$

$$+\varepsilon^{-2}\delta^2b(x)\varphi'(u)+\varepsilon^{-1}\delta^2c(x)\varphi''(u)$$

$$+\varepsilon^{-2}\int_{\mathbf{R}}[\frac{1}{2}(\Delta_\varepsilon\varphi(u))^2-\frac{v^2}{2}(\varphi'(u))^2]\Gamma^{\delta}(dv;x)+\varepsilon^{-2}\delta^2\frac{1}{2}c(x)(\varphi'(u))^2.$$

Obviously, the subintegral functions in the second and third integrals also belong to the class $C_3(\mathbf{R})$. Applying Taylor's formula to test functions $\varphi(u) \in C^3(\mathbf{R})$, and condition **PA2**, we obtain:

$$H_\Gamma^{\varepsilon,\delta}(x)\varphi(u) = \varepsilon^{-2}\delta^2 \int_{\mathbf{R}} [e^{v\varphi'(u)} - 1 - v\varphi'(u) - \frac{v^2}{2}(\varphi'(u))^2]\Gamma^0(dv;x)$$

$$+\varepsilon^{-2}\delta^2 \int_{\mathbf{R}} (e^{v\varphi'(u)}\varepsilon\frac{v^2}{2}\varphi''(\widetilde{u}) - \varepsilon\frac{v^2}{2}\varphi''(\widetilde{u}) - \varepsilon^2\frac{v^4}{8}(\varphi''(\widetilde{u}))^2)\Gamma^0(dv;x)$$

$$+\varepsilon^{-2}\delta^2 \int_{\mathbf{R}} \varepsilon^2\frac{v^3}{3!}\varphi'''(\widetilde{u})\Gamma^0(dv;x) + \varepsilon^{-2}\delta b_1(x)\varphi'(u) + \varepsilon^{-2}\delta^2 b(x)\varphi'(u)$$

$$+\varepsilon^{-1}\delta^2 c(x)\varphi''(u) + \varepsilon^{-2}\delta^2\int_{\mathbf{R}} \varepsilon^2\frac{v^4}{4}(\varphi''(\widetilde{u}))^2\Gamma^0(dv;x) + \varepsilon^{-2}\delta^2\frac{1}{2}c(x)(\varphi'(u))^2.$$

Taking into account the limit condition $\varepsilon^{-1}\delta \to 1$, we have:

$$H_\Gamma^{\varepsilon,\delta}(x)\varphi(u) = H_\Gamma(x)\varphi(u) + \varepsilon^{-1}b_1(x)\varphi'(u) + \theta_\Gamma^{\varepsilon,\delta}\varphi,$$

where $\sup_{x\in E} |\theta_\Gamma^{\varepsilon,\delta}\varphi| \to 0, \varepsilon, \delta \to 0.$

To complete the proof, we should show that the term in the parentheses in formula [4.8] is equal to 0. We use the following lemma.

LEMMA 4.4.–

$$\Gamma_\varepsilon^\delta(x)e^{\varphi(u)/\varepsilon}\varphi_1(u,x) = \varphi_1(u,x)\Gamma_\varepsilon^\delta(x)e^{\varphi(u)/\varepsilon} + (\varepsilon\delta)^{-1}\widehat{\theta}_\Gamma^{\varepsilon,\delta}\varphi,$$

where the negligible term

$$\sup_{x\in E} |\widehat{\theta}_\Gamma^{\varepsilon,\delta}\varphi| \to 0, \varepsilon, \delta \to 0.$$

PROOF.– Indeed, from [4.6], we have:

$$\Gamma_\varepsilon^\delta(x)e^{\varphi(u)/\varepsilon}\varphi_1(u,x) = \varepsilon^{-3}\int_{\mathbf{R}} [e^{\varphi(u+\varepsilon v)/\varepsilon}\varphi_1(u+\varepsilon v, x)$$

$$-e^{\varphi(u)/\varepsilon}\varphi_1(u,x)]\Gamma^\delta(dv;x) = \varphi_1(u,x)\Gamma_\varepsilon^\delta(x)e^{\varphi(u)/\varepsilon}$$

$$+(\varepsilon\delta)^{-1}\left[\varphi_1'(u,x)\varepsilon^{-1}\delta\int_{\mathbf{R}} e^{\varphi(u+\varepsilon v)/\varepsilon}v\Gamma^\delta(dv;x)\right].$$

Let us estimate the last integral. Since the function $\varphi(u)$ is bounded, we have for a fixed ε:

$$\int_{\mathbf{R}} e^{\varphi(u+\varepsilon v)/\varepsilon} v\Gamma^{\delta}(dv;x) < e^{C} \int_{\mathbf{R}} v\Gamma^{\delta}(dv;x) = \delta e^{C}[a_1(x) + \delta a(x) + \delta\theta_a^{\delta}\varphi].$$

Therefore, the last term is negligible by $\varepsilon, \delta \to 0$.

Lemma 4.4 is proved.

Finally, we have:

$$H_{\Gamma}^{\varepsilon,\delta}(x)\varphi_{\varepsilon}^{\delta} = H_{\Gamma}(x)\varphi(u) + \varepsilon^{-1}a_1(x)\varphi'(u) + \theta_{\Gamma}^{\varepsilon,\delta}\varphi,$$

where $\sup_{x\in E} |\theta_{\Gamma}^{\varepsilon,\delta}\varphi| \to 0, \varepsilon, \delta \to 0.$

Lemma 4.3 is proved.

LEMMA 4.5.– *The exponential generator*

$$H_Q^{\varepsilon}\varphi_{\varepsilon}^{\delta}(u,x) = e^{-\varphi_{\varepsilon}^{\delta}/\varepsilon}\varepsilon^{-2}Qe^{\varphi_{\varepsilon}^{\delta}/\varepsilon} \quad [4.9]$$

has the following asymptotic representation:

$$H_Q^{\varepsilon}\varphi_{\varepsilon}^{\delta} = \varepsilon^{-1}Q\varphi_1 + Q\varphi_2 - \varphi_1 Q\varphi_1 + \theta_Q^{\varepsilon,\delta}\varphi, \quad [4.10]$$

where $\sup_{x\in E} |\theta_Q^{\varepsilon,\delta}\varphi| \to 0, \varepsilon, \delta \to 0.$

PROOF.– We have:

$$H_Q^{\varepsilon}\varphi_{\varepsilon}^{\delta} = e^{-\varphi/\varepsilon}[1 + \delta\varphi_1 + \delta^2\varphi_2]^{-1}\varepsilon^{-2}Qe^{\varphi/\varepsilon}[1 + \delta\varphi_1 + \delta^2\varphi_2]$$

$$= \left[1 - \delta\varphi_1 + \delta^2\frac{\varphi_1^2 + \delta\varphi_1\varphi_2 - \varphi_2}{1 + \delta\varphi_1 + \delta^2\varphi_2}\right][\delta\varepsilon^{-2}Q\varphi_1 + \delta^2\varepsilon^{-2}Q\varphi_2]$$

$$= \delta\varepsilon^{-2}Q\varphi_1 + \delta^2\varepsilon^{-2}Q\varphi_2 - \delta^2\varepsilon^{-2}\varphi_1 Q\varphi_1 + \theta_Q^{\varepsilon,\delta}\varphi,$$

where

$$\theta_Q^{\varepsilon,\delta}\varphi = \delta^3\varepsilon^{-2}\frac{\varphi_1^2 + \delta\varphi_1\varphi_2 - \varphi_2}{1 + \delta\varphi_1 + \delta^2\varphi_2}[Q\varphi_1 + \delta Q\varphi_2] - \delta^3\varepsilon^{-2}\varphi_1 Q\varphi_2.$$

Taking into account the limit condition $\varepsilon^{-1}\delta \to 1, \varepsilon, \delta \to 0$, we finally have [4.10].

Lemma 4.5 is proved.

So, obviously,

$$\mathbb{H}_\varepsilon^\delta \varphi_\varepsilon^\delta = H_Q^\varepsilon \varphi_\varepsilon^\delta(u,x) + H_\Gamma^{\varepsilon,\delta}(x)\varphi_\varepsilon^\delta(u,x).$$

Using Lemma 4.3 and Lemma 4.5, we obtain the asymptotic representation

$$\mathbb{H}_\varepsilon^\delta \varphi_\varepsilon^\delta = \varepsilon^{-1}[Q\varphi_1 + b_1(x)\varphi'(u)] + [Q\varphi_2 - \varphi_1 Q\varphi_1 + H_\Gamma(x)\varphi(u)] + h^{\varepsilon,\delta}\varphi,$$

where $h^{\varepsilon,\delta}\varphi = \theta_Q^{\varepsilon,\delta}\varphi + \theta_\Gamma^{\varepsilon,\delta}\varphi$.

Now we use the solution of the singular perturbation problem (Korolyuk and Limnios 2005), which is determined by the equations:

$$\begin{aligned} Q\varphi_1 + b_1(x)\varphi'(u) &= 0, \\ Q\varphi_2 - \varphi_1 Q\varphi_1 + H_\Gamma(x)\varphi(u) &= H^0\varphi(u). \end{aligned}$$

From the balance condition **LA3**, we have, from the first equation:

$$\varphi_1(u,x) = R_0 b_1(x)\varphi'(u), \qquad Q\varphi_1(u,x) = -b_1(x)\varphi'(u).$$

After substitution into the second equation, we obtain:

$$Q\varphi_2 + b_1(x)R_0 b_1(x)(\varphi'(u))^2 + H_\Gamma(x)\varphi(u) = H^0\varphi(u)$$

From the solvability condition, we see:

$$H^0\varphi(u) = \Pi H_\Gamma(x)\Pi\varphi(u) + \Pi b_1(x)R_0 b_1(x)(\varphi'(u))^2.$$

Therefore, it follows from Lemma 4.3 that the generator H^0 has the form [4.7].

The negligible term $h^{\varepsilon,\delta}\varphi$ can be found explicitly from the solution of the Poisson equation (see section 1.1.2. For more details, see Korolyuk and Limnios (2005))

$$\varphi_2(u,x) = R_0\widetilde{H}(x)\varphi(u) - R_0 b_1(x)R_0 b_1(x)(\varphi'(u))^2, \widetilde{H}(x) := H^0 - H_\Gamma(x).$$

Theorem 4.2 is proved.

To calculate the rate functional, we need to find the function $L(q)$ by the formula (see Example 1.7):

$$L(q) = \sup_{p\in\mathbf{R}}(qp - H(p)), \qquad [4.11]$$

where $H(p), p = \varphi'(u)$ is determined from the explicit form of the generator:

$$H(p) := H^0(\varphi') = (\widehat{b} - \widehat{b}_0)p + \frac{1}{2}\sigma^2 p^2 + \int_{\mathbf{R}} [e^{vp} - 1]\widehat{\Gamma}^0(dv).$$

So, let us differentiate by p and equate the expression to 0:

$$(qp - H(p))'_p = (qp - (\widehat{b} - \widehat{b}_0)p - \frac{1}{2}\sigma^2 p^2 - \int_{\mathbf{R}} [e^{vp} - 1]\widehat{\Gamma}^0(dv))'_p$$
$$= q - (\widehat{b} - \widehat{b}_0) - \sigma^2 p - \int_{\mathbf{R}} v e^{vp}\widehat{\Gamma}^0(dv) = 0,$$

i.e. $p(q)$ is determined from the relation

$$\sigma^2 p + \int_{\mathbf{R}} v e^{vp}\widehat{\Gamma}^0(dv) = q - (\widehat{b} - \widehat{b}_0).$$

Substituting the expression for $p(q)$ in formula [4.11], we have a function $L(q)$ and the rate functional is

$$I(x) = I_0(x(0)) + \int_0^\infty L(x'(s))ds.$$

4.2. Impulsive processes

4.2.1. *Poisson approximation scheme*

We perform an asymptotic analysis of the exponential nonlinear generator of the impulsive process in the Poisson approximation scheme (see section 1.1.10) with Markov switching (see section 1.1.3).

The normalization of the impulsive process by small series parameters aims to solve the large deviations problem in the Poisson approximation scheme and is performed using two small parameters $\varepsilon, \delta \to 0$ such that $\varepsilon^{-1}\delta \to 1$:

$$\widetilde{\xi}^{\varepsilon,\delta}(t) = \widetilde{\xi}_0^{\varepsilon,\delta} + \varepsilon \sum_{k=1}^{\nu(t/\varepsilon^2)} \alpha_k^\delta(x_{k-1}),\ t \geq 0,$$

$$L^{\varepsilon,\delta}\varphi(u,x) = \varepsilon^{-2}q(x)\int_E P(x,dy)\int_{\mathbf{R}} G_y^\delta(dv)[\varphi(u+\varepsilon v, y) - \varphi(u,x)],$$

where the kernel $G_x^\delta(v)$ satisfies the Poisson approximation conditions.

We will recall these conditions.

C1: *Poisson approximation.* A family of processes $\alpha_k^\delta(x)$, $k \geq 1, x \in E, t \geq 0$ satisfies the conditions of the Poisson approximation if the following constrains are met:

PA1 Approximation of means:

$$b_\delta(x) = \int_{\mathbf{R}} vG_x^\delta(dv) = \delta[b(x) + \theta_b^\delta(x)],$$

and

$$c_\delta(x) = \int_{\mathbf{R}} v^2 G_x^\delta(dv) = \delta[c(x) + \theta_c^\delta(x)],$$

where $\sup_{x\in E} |b(x)| \leq b < +\infty$, $\sup_{x\in E} |c(x)| \leq c < +\infty$.

PA2 For the intensity kernel, the following asymptotic representation holds true:

$$G_{g,x}^\delta = \int_{\mathbf{R}} g(v) G_x^\delta(dv) = \delta[G_{g,x} + \theta_g^\delta(x)]$$

for all $g \in C_3(\mathbf{R})$ – a class of functions that determines a measure (see Jacod and Shiryaev (2003) and Korolyuk and Limnios (2005, Chapter 7)), $G_{g,x}$ is a bounded kernel

$$|G_{g,x}| \leq G_g (\text{a constant that depends on } g).$$

The kernel $G_x^0(dv)$ is defined on the class of functions that determines a measure $C_3(\mathbf{R})$ by the relation

$$G_{g,x} = \int_{\mathbf{R}} g(v) G_x^0(dv), \quad g \in C_3(\mathbf{R}).$$

The negligible terms $\theta_b^\delta, \theta_c^\delta, \theta_g^\delta$ satisfy the condition

$$\sup_{x\in E} |\theta_\cdot^\delta(x)| \to 0, \quad \delta \to 0.$$

PA3 The relationship

$$c(x) := \int_{\mathbf{R}} v^2 G_x^0(dv)$$

holds true, which causes the absence of a diffusion component in the exponential generator, which determines the solution of the large deviations problem.

C2: *Uniform quadratic integrability*:

$$\lim_{c\to\infty} \sup_{x\in E} \int_{|v|>c} v^2 G_x^0(dv) = 0.$$

C3: *Exponential boundedness*:

$$\int_{\mathbf{R}} e^{p|v|} G_x^\delta(dv) < \infty, \forall p \in \mathbf{R}.$$

THEOREM 4.3.– *The solution of the large deviations problem for the impulse process*

$$\widetilde{\xi}^{\varepsilon,\delta}(t) = \widetilde{\xi}_0^{\varepsilon,\delta} + \varepsilon \sum_{k=1}^{\nu(t/\varepsilon^2)} \alpha_k^\delta(x_{k-1}),\ t \geq 0,$$

given by the generator of the two-component Markov process $\widetilde{\xi}^{\varepsilon,\delta}(t), x(t/\varepsilon^2), t \geq 0$

$$L^{\varepsilon,\delta}\varphi(u,x) = \varepsilon^{-2}q(x)\int_E P(x,dy)\int_{\mathbf{R}} G_y^{\delta}(dv)[\varphi(u+\varepsilon v,y)-\varphi(u,x)],$$

or (here, $Q_0\varphi(x) := q(x)\int_E P(x,dy)\varphi(y)$*)*

$$L^{\varepsilon,\delta}\varphi(u,x) = [\varepsilon^{-2}Q + Q_0 G_x^{\varepsilon,\delta}]\varphi(u,x), \tag{4.12}$$

where

$$G_x^{\varepsilon,\delta}\varphi(u) := \varepsilon^{-2}\int_{\mathbf{R}} G_x^{\delta}(dv)[\varphi(u+\varepsilon v)-\varphi(u)] \tag{4.13}$$

determined by the exponential generator

$$\begin{aligned} H^0\varphi(u) &= \widehat{b}\varphi'(u) + \int_{\mathbf{R}} [e^{v\varphi'(u)} - 1 - v\varphi'(u)]\widehat{G}^0(dv) \\ &= (\widehat{b}-\widehat{b}_0)\varphi'(u) + \int_{\mathbf{R}} [e^{v\varphi'(u)} - 1]\widehat{G}^0(dv), \end{aligned} \tag{4.14}$$

where

$$\widehat{G}^0(dv) := q\int_E \rho(dx)G_x^0(dv), \widehat{b}_0 := \int_{\mathbf{R}} v\widehat{G}^0(dv),$$

is the parameter that determines the deterministic drift

$$\widehat{b} := q\int_E \rho(dx)b(x).$$

The averaging is performed by a stationary measure of the embedded Markov chain of the switching Markov process.

To prove the theorem, we need the following lemma.

LEMMA 4.6.– *The exponential generator in a series scheme*

$$H_G^{\varepsilon,\delta}(x)\varphi(u) = e^{-\varphi/\varepsilon}\varepsilon Q_0 G_x^{\varepsilon,\delta}e^{\varphi/\varepsilon} \tag{4.15}$$

has the following asymptotic representation:

$$H_G^{\varepsilon,\delta}(x)\varphi(u) = H_G(x)\varphi(u) + \theta_G^{\varepsilon,\delta}\varphi,$$

where $\sup_{x\in E}|\theta_G^{\varepsilon,\delta}\varphi| \to 0, \varepsilon,\delta \to 0,$

$$\begin{aligned} H_G(x)\varphi(u) = Q_0\mathcal{H}_G(x)\varphi(u) &:= Q_0[b(x)\varphi'(u) \\ &+ \int_{\mathbf{R}} [e^{v\varphi'(u)} - 1 - v\varphi'(u)]G_x^0(dv)]. \end{aligned} \tag{4.16}$$

PROOF.– Let us rewrite [4.15], taking into account the form of the generator [4.13]. We have:

$$H_G^{\varepsilon,\delta}(x)\varphi(u) = \varepsilon^{-1}Q_0 \int_{\mathbf{R}} [e^{\Delta_\varepsilon\varphi(u)} - 1]G_x^\delta(dv),$$

where

$$\Delta_\varepsilon\varphi(u) := \varepsilon^{-1}[\varphi(u+\varepsilon v) - \varphi(u)].$$

Rewrite the expression for the generator as follows:

$$H_G^{\varepsilon,\delta}(x)\varphi(u) = \varepsilon^{-1}Q_0 \int_{\mathbf{R}} [e^{\Delta_\varepsilon\varphi(u)} - 1 - \Delta_\varepsilon\varphi(u) - \frac{1}{2}(\Delta_\varepsilon\varphi(u))^2]G_x^\delta(dv)$$

$$+\varepsilon^{-1}Q_0 \int_{\mathbf{R}} [\Delta_\varepsilon\varphi(u) + \frac{1}{2}(\Delta_\varepsilon\varphi(u))^2]G_x^\delta(dv).$$

It is easy to see that the function $\psi_u^\varepsilon(v) = e^{\Delta_\varepsilon\varphi(u)} - 1 - \Delta_\varepsilon\varphi(u) - \frac{1}{2}(\Delta_\varepsilon\varphi(u))^2$ belongs to the class $C_3(\mathbf{R})$. In fact,

$$\psi_u^\varepsilon(v)/v^2 \to 0, v \to 0.$$

In addition, this function is continuous and bounded for any ε under the condition that the function $\varphi(u)$ is bounded. Moreover, the boundedness of the function $\psi_u^\varepsilon(v)$ is uniform by u under the conditions **C3** and **C4** and the boundedness of the derivative $\varphi'(u)$.

Thus, we have:

$$H_G^{\varepsilon,\delta}(x)\varphi(u) = \varepsilon^{-1}\delta Q_0 \int_{\mathbf{R}} [e^{\Delta_\varepsilon\varphi(u)} - 1 - \Delta_\varepsilon\varphi(u) - \frac{1}{2}(\Delta_\varepsilon\varphi(u))^2]G_x^0(dv)$$

$$+\varepsilon^{-1}Q_0 \int_{\mathbf{R}} [\Delta_\varepsilon\varphi(u) - v\varphi'(u) - \varepsilon\frac{v^2}{2}\varphi''(u)]G_x^\delta(dv) + \varepsilon^{-1}\delta Q_0 b(x)\varphi'(u)$$

$$+\delta Q_0 c(x)\varphi''(u) + \varepsilon^{-1}Q_0 \int_{\mathbf{R}} [\frac{1}{2}(\Delta_\varepsilon\varphi(u))^2 - \frac{v^2}{2}(\varphi'(u))^2]G_x^\delta(dv)$$

$$+\varepsilon^{-1}\delta\frac{1}{2}Q_0 c(x)(\varphi'(u))^2.$$

Applying Taylor's formula to test functions $\varphi(u) \in C^3(\mathbf{R})$, and condition **PA2**, we obtain:

$$H_G^{\varepsilon,\delta}(x)\varphi(u) = \varepsilon^{-1}\delta Q_0 \int_{\mathbf{R}} [e^{v\varphi'(u)} - 1 - v\varphi'(u) - \frac{v^2}{2}(\varphi'(u))^2]G_x^0(dv)$$

$$+\varepsilon^{-1}\delta Q_0 \int_{\mathbf{R}} (e^{v\varphi'(u)}\varepsilon\frac{v^2}{2}\varphi''(\widehat{u}) - \varepsilon\frac{v^2}{2}\varphi''(\widehat{u}) - \varepsilon^2\frac{v^4}{8}(\varphi''(\widehat{u}))^2)G_x^0(dv)$$

$$+\varepsilon^{-1}\delta Q_0 \int_{\mathbf{R}} \varepsilon^2 \frac{v^3}{3!} \varphi'''(\widehat{u}) G_x^0(dv) + \varepsilon^{-1}\delta Q_0 b(x)\varphi'(u) + \delta Q_0 c(x)\varphi''(u)$$

$$+\varepsilon^{-1}\delta Q_0 \int_{\mathbf{R}} \varepsilon^2 \frac{v^4}{4} (\varphi''(\widehat{u}))^2 G_x^0(dv) + \varepsilon^{-1}\delta \frac{1}{2} Q_0 c(x)(\varphi'(u))^2.$$

Taking into account **PA3** and the limit condition $\varepsilon^{-1}\delta \to 1$, we finally have:

$$H_G^{\varepsilon,\delta}(x)\varphi(u) = H_G(x)\varphi(u) + \theta_G^{\varepsilon,\delta}\varphi,$$

where $\sup_{x\in E} |\theta_G^{\varepsilon,\delta}\varphi| \to 0, \varepsilon, \delta \to 0.$

Lemma 4.6 is proved.

PROOF OF THEOREM 4.3.– The limit transition in the exponential nonlinear generator of the impulsive process is realized on perturbed test functions

$$\varphi^{\varepsilon,\delta}(u,x) = \varphi(u) + \varepsilon \ln[1 + \delta\varphi_1(u,x)].$$

Thus, we have:

$$\mathbb{H}_\varepsilon^\delta \varphi^{\varepsilon,\delta} = e^{-\varphi^\varepsilon/\varepsilon} \varepsilon L^{\varepsilon,\delta} e^{\varphi^\varepsilon/\varepsilon} = e^{-\varphi/\varepsilon}[1+\delta\varphi_1]^{-1} \varepsilon L^{\varepsilon,\delta} e^{\varphi/\varepsilon}[1+\delta\varphi_1].$$

The calculation of the asymptotic behavior of the last exponential generator gives the following result.

LEMMA 4.7.– *There is an asymptotic representation*

$$\mathbb{H}_\varepsilon^\delta \varphi^{\varepsilon,\delta} = Q\varphi_1 + H_G^{\varepsilon,\delta}(x)\varphi(u) + \theta^{\varepsilon,\delta}\varphi,$$

where $\sup_{x\in E} |\theta^{\varepsilon,\delta}\varphi| \to 0, \varepsilon, \delta \to 0.$

PROOF.– Considering [4.12], we have:

$$\mathbb{H}^{\varepsilon,\delta}\varphi^{\varepsilon,\delta} = e^{-\varphi/\varepsilon}[1 - \delta\varphi_1 + \frac{\delta^2\varphi_1^2}{1+\delta\varphi_1}] \left\{\varepsilon^{-1}Q + \varepsilon Q_0 G_x^{\varepsilon,\delta}\right\} e^{\varphi/\varepsilon}[1+\delta\varphi_1]$$

$$= e^{-\varphi/\varepsilon}[1-\delta\varphi_1+\frac{\delta^2\varphi_1^2}{1+\delta\varphi_1}]\left\{\varepsilon^{-1}\delta e^{\varphi/\varepsilon}Q\varphi_1 + \varepsilon Q_0 G_x^{\varepsilon,\delta} e^{\varphi/\varepsilon} + \varepsilon\delta Q_0 G_x^{\varepsilon,\delta} e^{\varphi/\varepsilon}\varphi_1\right\}$$

$$= Q\varphi_1 + H_G^{\varepsilon,\delta}(x)\varphi(u) + \theta^{\varepsilon,\delta}\varphi,$$

where

$$\theta^{\varepsilon,\delta}\varphi = \delta\left[\frac{\varepsilon}{1+\delta\varphi_1}e^{-\varphi/\varepsilon}Q_0G_x^{\varepsilon,\delta}e^{\varphi/\varepsilon}\varphi_1 - \frac{\varepsilon^{-1}\delta\varphi_1}{1+\delta\varphi_1}Q\varphi_1\right.$$

$$\left. - \frac{\delta\varphi_1}{1+\delta\varphi_1}e^{-\varphi/\varepsilon}Q_0G_x^{\varepsilon,\delta}e^{\varphi/\varepsilon}\right].$$

Lemma 4.7 is proved.

Taking into account Lemma 4.6, we have

$$\mathbb{H}_\varepsilon^\delta\varphi^{\varepsilon,\delta} = Q\varphi_1 + H_G(x)\varphi(u) + h^{\varepsilon,\delta}\varphi,$$

where $h^{\varepsilon,\delta}\varphi = \theta^{\varepsilon,\delta}\varphi + \theta_G^{\varepsilon,\delta}\varphi$.

We now use the solution of the singular perturbation problem for the reducibly invertible operator Q (see Korolyuk and Limnios (2005, Chapter 1)). From the solvability condition, we have:

$$Q\varphi_1 + H_G(x)\varphi(u) = H^0\varphi(u),$$

where

$$H^0\varphi(u) = \Pi Q\Pi\varphi_1 + \Pi H_G(x)\Pi\varphi(u).$$

Given the relationship [4.16] and the properties of the Markov process from section 1.1.3, we can rewrite

$$H^0\varphi(u) = \int_E \pi(dx)q(x)\int_E P(x,dy)\mathcal{H}_G(y)\varphi(u) = q\int_E \rho(dx)\mathcal{H}_G(x)\varphi(u),$$

therefore, we finally obtain [4.14]:

$$H^0\varphi(u) = \widehat{b}\varphi'(u) + \int_{\mathbf{R}}[e^{v\varphi'(u)} - 1 - v\varphi'(u)]\widehat{\Gamma}^0(dv).$$

The negligible term $h^{\varepsilon,\delta}\varphi$ can be calculated explicitly, using the solution of the Poisson equation (see section 1.1.2. For more details, see Korolyuk and Limnios (2005))

$$\varphi_1(u,x) = R_0\widehat{H}(x)\varphi(u), \widehat{H}(x) := H_G(x) - H^0.$$

Theorem 4.3 is proved.

4.2.2. *Lévy approximation scheme*

The normalization of the impulse process by small series parameters, aimed to solve the large deviations problem in the Lévy approximation scheme, is also performed using two small parameters $\varepsilon, \delta \to 0$ such that $\varepsilon^{-1}\delta \to 1$:

$$\widetilde{\xi}^{\varepsilon,\delta}(t) = \widetilde{\xi}_0^{\varepsilon,\delta} + \varepsilon \sum_{k=1}^{\nu(t/\varepsilon^3)} \alpha_k^\delta(x_{k-1}),\ t \geq 0,$$

$$L^{\varepsilon,\delta}\varphi(u,x) = \varepsilon^{-3}q(x)\int_E P(x,dy)\int_{\mathbf{R}} G_y^\delta(dv)[\varphi(u+\varepsilon v, y) - \varphi(u,x)],$$

where the kernel $G_x^\delta(v)$ satisfies the conditions of the Lévy approximation.

We will recall the main conditions of the Lévy approximation.

C1': *Lévy approximation*. The family of processes $\alpha_k^\delta(x)$, $k \geq 1, x \in E, t \geq 0$ satisfies the Lévy approximation conditions:

LA1 Approximation of means:

$$b_\delta(x) = \int_{\mathbf{R}} vG_x^\delta(dv) = \delta b_1(x) + \delta^2[b(x) + \theta_b^\delta(x)],$$

and

$$c_\delta(x) = \int_{\mathbf{R}} v^2 G_x^\delta(dv) = \delta^2[c(x) + \theta_c^\delta(x)],$$

where

$$\sup_{x\in E}|b_1(x)| \leq a < +\infty, \sup_{x\in E}|b(x)| \leq a < +\infty,$$

$$\sup_{x\in E}|c(x)| \leq c < +\infty.$$

LA2 For the intensity kernel, there is an asymptotic representation

$$G_{g,x}^\delta = \int_{\mathbf{R}} g(v)G_x^\delta(dv) = \delta^2[G_{g,x} + \theta_g^\delta(x)]$$

for all $g \in C_3(\mathbf{R})$ – a class of functions that determines a measure, $G_{g,x}$ is a bounded kernel

$$|G_{g,x}| \leq G_g \quad \text{(a constant that depends on } g).$$

The kernel $G_x^0(dv)$ is set on the class of function that defines the measure $C_3(\mathbf{R})$ by the relation

$$G_{g,x} = \int_{\mathbf{R}} g(v) G_x^0(dv), \quad g \in C_3(\mathbf{R}).$$

The negligible terms $\theta_a^\delta, \theta_c^\delta, \theta_g^\delta$ satisfy the condition

$$\sup_{x \in E} |\theta_\cdot^\delta(x)| \to 0, \quad \delta \to 0.$$

LA3 Balance condition:

$$q \int_E \rho(dx) b_1(x) = 0.$$

C2: *Uniform quadratic integrability*:

$$\lim_{c \to \infty} \sup_{x \in E} \int_{|v|>c} v^2 G_x^0(dv) = 0.$$

C3: *Exponential boundedness*:

$$\int_{\mathbf{R}} e^{p|v|} G_x^\delta(dv) < \infty, \forall p \in \mathbf{R}.$$

THEOREM 4.4.– *Solution of the large deviations problem for the impulsive process*

$$\widetilde{\xi}^{\varepsilon,\delta}(t) = \widetilde{\xi}_0^{\varepsilon,\delta} + \varepsilon \sum_{k=1}^{\nu(t/\varepsilon^3)} \alpha_k^\delta(x_{k-1}), \ t \geq 0,$$

given by the generator of the two-component Markov process $\widetilde{\xi}^{\varepsilon,\delta}(t), x(t/\varepsilon^3), t \geq 0$

$$L^{\varepsilon,\delta}\varphi(u,x) = \varepsilon^{-3} q(x) \int_E P(x,dy) \int_{\mathbf{R}} G_y^\delta(dv)[\varphi(u+\varepsilon v, y) - \varphi(u,x)],$$

or

$$L^{\varepsilon,\delta}\varphi(u,x) = [\varepsilon^{-3} Q + Q_0 G_x^{\varepsilon,\delta}]\varphi(u,x), \quad [4.17]$$

where

$$G_x^{\varepsilon,\delta}\varphi(u) := \varepsilon^{-3} \int_{\mathbf{R}} G_x^\delta(dv)[\varphi(u+\varepsilon v) - \varphi(u)] \quad [4.18]$$

is determined by the exponential generator

$$H^0\varphi(u) = (\widehat{b} - \widehat{b}_0)\varphi'(u) + \frac{1}{2}\sigma^2(x)(\varphi'(u))^2 + \int_{\mathbf{R}} [e^{v\varphi'(u)} - 1]\widehat{G}^0(dv),$$

where $\widehat{b} = \Pi b(x) := q \int_E \rho(dx)b(x), \widehat{a}_0 = \Pi b_0(x) := q \int_E \rho(dx)b_0(x),$

$$b_0(x) = \int_{\mathbf{R}} vG^0(dv), \widehat{c} = \Pi c(x) := q \int_E \rho(dx)c(x),$$

$$\widehat{c}_0 = \Pi c_0(x) := q \int_E \rho(dx)c_0(x), c_0(x) = \int_{\mathbf{R}} v^2 G^0(dv),$$

$$\sigma^2 = (\widehat{c} - \widehat{c}_0) + 2q \int_E \rho(dx)b_1(x)R_0 b_1(x), \widehat{G}^0(v)$$

$$= \Pi G^0(v) := q \int_E \rho(dx)G^0(v).$$

The averaging is performed by a stationary measure of the embedded Markov switching process.

PROOF.– The limit transition for a nonlinear exponential generator is implemented on the test functions

$$\varphi_\varepsilon^\delta(u,x) = \varphi(u) + \varepsilon \ln[1 + \delta\varphi_1(u,x) + \delta^2\varphi_2(u,x)],$$

where $\varphi(u) \in C^3(\mathbf{R})$ (the space of continuous bounded functions with continuous bounded derivatives up to the third order). Taking into account the type of generator [4.17], we have

$$\mathbb{H}_\varepsilon^\delta \varphi_\varepsilon^\delta = e^{-\varphi_\varepsilon^\delta/\varepsilon} \varepsilon L^{\varepsilon,\delta} e^{\varphi_\varepsilon^\delta/\varepsilon} = e^{-\varphi_\varepsilon^\delta/\varepsilon}[\varepsilon^{-2}Q + \varepsilon Q_0 G_x^{\varepsilon,\delta}]e^{\varphi_\varepsilon^\delta/\varepsilon}$$

$$= e^{-\varphi/\varepsilon}[1 + \delta\varphi_1 + \delta^2\varphi_2]^{-1}[\varepsilon^{-2}Q + \varepsilon Q_0 G_x^{\varepsilon,\delta}]e^{\varphi/\varepsilon}[1 + \delta\varphi_1 + \delta^2\varphi_2].$$

For a further proof of the theorem we need the following lemmas.

LEMMA 4.8.– *The exponential generator*

$$H_Q^\varepsilon \varphi_\varepsilon^\delta(u,x) = e^{-\varphi_\varepsilon^\delta/\varepsilon}\varepsilon^{-2}Q e^{\varphi_\varepsilon^\delta/\varepsilon} \quad [4.19]$$

has the following asymptotic representation:

$$H_Q^\varepsilon \varphi_\varepsilon^\delta = \varepsilon^{-1}Q\varphi_1 + Q\varphi_2 - \varphi_1 Q\varphi_1 + \theta_Q^{\varepsilon,\delta}\varphi, \quad [4.20]$$

where $\sup_{x\in E} |\theta_Q^{\varepsilon,\delta}\varphi| \to 0, \varepsilon, \delta \to 0.$

PROOF.– We have:

$$H_Q^\varepsilon \varphi_\varepsilon^\delta = e^{-\varphi/\varepsilon}[1+\delta\varphi_1+\delta^2\varphi_2]^{-1}\varepsilon^{-2}Qe^{\varphi/\varepsilon}[1+\delta\varphi_1+\delta^2\varphi_2]$$

$$= \left[1-\delta\varphi_1+\delta^2\frac{\varphi_1^2+\delta\varphi_1\varphi_2-\varphi_2}{1+\delta\varphi_1+\delta^2\varphi_2}\right][\delta\varepsilon^{-2}Q\varphi_1+\delta^2\varepsilon^{-2}Q\varphi_2]$$

$$= \delta\varepsilon^{-2}Q\varphi_1+\delta^2\varepsilon^{-2}Q\varphi_2-\delta^2\varepsilon^{-2}\varphi_1 Q\varphi_1+\theta_Q^{\varepsilon,\delta}(x),$$

where

$$\theta_Q^{\varepsilon,\delta}\varphi = \delta^3\varepsilon^{-2}\frac{\varphi_1^2+\delta\varphi_1\varphi_2-\varphi_2}{1+\delta\varphi_1+\delta^2\varphi_2}[Q\varphi_1+\delta Q\varphi_2]-\delta^3\varepsilon^{-2}\varphi_1 Q\varphi_2.$$

Due to the limit condition $\varepsilon^{-1}\delta\to 1, \varepsilon,\delta\to 0$, we finally obtain [4.20].

Lemma 4.8 is proved.

LEMMA 4.9.– *The exponential generator*

$$H_G^{\varepsilon,\delta}(x)\varphi_\varepsilon^\delta(u,x) = e^{-\varphi_\varepsilon^\delta/\varepsilon}\varepsilon Q_0 G_x^{\varepsilon,\delta}e^{\varphi_\varepsilon^\delta/\varepsilon} \quad [4.21]$$

has the following asymptotic representation:

$$H_G^{\varepsilon,\delta}(x)\varphi_\varepsilon^\delta = H_G(x)\varphi(u)+\varepsilon^{-1}b_1(x)\varphi'(u)+\theta_G^{\varepsilon,\delta}\varphi,$$

where

$$H_G(x)\varphi(u) = Q_0\mathcal{H}_G(x) := Q_0[(b(x)-b_0(x))\varphi'(u)$$
$$+\frac{1}{2}(c(x)--c_0(x))(\varphi'(u))^2+\int_{\mathbf{R}}[e^{v\varphi'(u)}-1]G^0(dv)], \quad [4.22]$$

and $\sup_{x\in iE}|\theta_G^{\varepsilon,\delta}\varphi|\to 0, \varepsilon,\delta\to 0.$

PROOF.– We have:

$$H_G^{\varepsilon,\delta}(x)\varphi_\varepsilon^\delta = e^{-\varphi/\varepsilon}[1+\delta\varphi_1+\delta^2\varphi_2]^{-1}\varepsilon Q_0 G_x^{\varepsilon,\delta}e^{\varphi/\varepsilon}[1+\delta\varphi_1+\delta^2\varphi_2]$$

$$= e^{-\varphi/\varepsilon}\left[1-\delta\varphi_1+\delta^2\frac{\varphi_1^2+\delta\varphi_1\varphi_2-\varphi_2}{1+\delta\varphi_1+\delta^2\varphi_2}\right][\varepsilon Q_0 G_x^{\varepsilon,\delta}e^{\varphi/\varepsilon}+\varepsilon\delta Q_0 G_x^{\varepsilon,\delta}e^{\varphi/\varepsilon}\varphi_1$$

$$+\varepsilon\delta^2 Q_0 G_x^{\varepsilon,\delta}e^{\varphi/\varepsilon}\varphi_2] = H_G^{\varepsilon,\delta}(x)\varphi(u)$$

$$+e^{-\varphi/\varepsilon}\varepsilon\delta[Q_0 G_x^{\varepsilon,\delta}e^{\varphi/\varepsilon}\varphi_1-\varphi_1 Q_0 G_x^{\varepsilon,\delta}e^{\varphi/\varepsilon}]+\widehat{\theta}_\Gamma^{\varepsilon,\delta}\varphi, \quad [4.23]$$

where

$$\widehat{\theta}_{\Gamma}^{\varepsilon,\delta}\varphi = \varepsilon\delta^2[e^{-\varphi/\varepsilon}Q_0G_x^{\varepsilon,\delta}e^{\varphi/\varepsilon}\varphi_2 - e^{-\varphi/\varepsilon}\varphi_1 Q_0 G_x^{\varepsilon,\delta}e^{\varphi/\varepsilon}\varphi_1]$$

$$+\varepsilon\delta^2\frac{\varphi_1^2+\delta\varphi_1\varphi_2-\varphi_2}{1+\delta\varphi_1+\delta^2\varphi_2}[e^{-\varphi/\varepsilon}Q_0G_x^{\varepsilon,\delta}e^{\varphi/\varepsilon} + e^{-\varphi/\varepsilon}\delta Q_0G_x^{\varepsilon,\delta}e^{\varphi/\varepsilon}\varphi_1$$

$$+e^{-\varphi/\varepsilon}\delta^2 Q_0 G_x^{\varepsilon,\delta}e^{\varphi/\varepsilon}\varphi_2] - \varepsilon\delta^3 e^{-\varphi/\varepsilon}\varphi_1 Q_0 G_x^{\varepsilon,\delta}e^{\varphi/\varepsilon}\varphi_2$$

To complete the proof, we use the following two statements:

LEMMA 4.10.–

$$Q_0G_x^{\varepsilon,\delta}e^{\varphi(u)/\varepsilon}\varphi_1(u,x) = \varphi_1(u,x)Q_0G_x^{\varepsilon,\delta}e^{\varphi(u)/\varepsilon} + (\varepsilon\delta)^{-1}\widehat{\theta}_{\Gamma}^{\varepsilon,\delta}\varphi,$$

where the negligible term

$$\sup_{x\in E}|\widehat{\theta}_{\Gamma}^{\varepsilon,\delta}\varphi| \to 0, \varepsilon,\delta\to 0.$$

PROOF.– Indeed, according to [4.17], we have:

$$Q_0G_x^{\varepsilon,\delta}e^{\varphi(u)/\varepsilon}\varphi_1(u,x) = \varepsilon^{-3}Q_0\int_{\mathbf{R}}[e^{\varphi(u+\varepsilon v)/\varepsilon}\varphi_1(u+\varepsilon v,x)$$

$$-e^{\varphi(u)/\varepsilon}\varphi_1(u,x)]G_x^{\delta}(dv) = \varphi_1(u,x)Q_0G_x^{\varepsilon,\delta}e^{\varphi(u)/\varepsilon}$$

$$+(\varepsilon\delta)^{-1}\left[\varphi_1'(u,x)\varepsilon^{-1}\delta\int_{\mathbf{R}}e^{\varphi(u+\varepsilon v)/\varepsilon}vG_x^{\delta}(dv)\right].$$

Let us estimate the last integral. As soon as the function $\varphi(u)$ is bounded, we have, for each fixed ε:

$$\int_{\mathbf{R}}e^{\varphi(u+\varepsilon v)/\varepsilon}vG_x^{\delta}(dv) < e^C\int_{\mathbf{R}}vG_x^{\delta}(dv) = \delta e^C[b_1(x)+\delta b(x)+\delta\theta_b^{\delta}\varphi].$$

Thus, we see that the last term is negligible by the condition $\varepsilon,\delta\to 0$.

Lemma 4.10 is proved.

LEMMA 4.11.– *The exponential generator*

$$H_G^{\varepsilon,\delta}(x)\varphi(u) = e^{-\varphi/\varepsilon}\varepsilon Q_0G_x^{\varepsilon,\delta}e^{\varphi/\varepsilon} \quad [4.24]$$

has the following asymptotic representation:

$$H_G^{\varepsilon,\delta}(x)\varphi(u) = H_G(x)\varphi(u) + \varepsilon^{-1}a_1(x)\varphi'(u) + \theta^{\varepsilon,\delta}\varphi,$$

where $\sup_{x\in E}|\theta^{\varepsilon,\delta}\varphi| \to 0, \varepsilon,\delta\to 0.$

PROOF.– Rewrite [4.24], taking into account the form of the generator [4.17]. We have:

$$H_G^{\varepsilon,\delta}(x)\varphi(u) = \varepsilon^{-2}Q_0 \int_{\mathbf{R}} [e^{\Delta_\varepsilon \varphi(u)} - 1] G_x^\delta(dv),$$

where

$$\Delta_\varepsilon \varphi(u) := \varepsilon^{-1}[\varphi(u+\varepsilon v) - \varphi(u)].$$

Thus, the expression for the generator is as follows:

$$H_G^{\varepsilon,\delta}(x)\varphi(u) = \varepsilon^{-2}Q_0 \int_{\mathbf{R}} [e^{\Delta_\varepsilon \varphi(u)} - 1 - \Delta_\varepsilon\varphi(u) - \frac{1}{2}(\Delta_\varepsilon\varphi(u))^2] G_x^\delta(dv)$$

$$+\varepsilon^{-2}Q_0 \int_{\mathbf{R}} [\Delta_\varepsilon\varphi(u) + \frac{1}{2}(\Delta_\varepsilon\varphi(u))^2] G_x^\delta(dv).$$

It is easy to see that the function $\psi_u^\varepsilon(v) = e^{\Delta_\varepsilon \varphi(u)} - 1 - \Delta_\varepsilon\varphi(u) - \frac{1}{2}(\Delta_\varepsilon\varphi(u))^2$ belongs to the class $C_3(\mathbf{R})$. In fact,

$$\psi_u^\varepsilon(v)/v^2 \to 0, v \to 0.$$

In addition, this function is continuous and bounded for any ε under the condition of the boundedness of the function $\varphi(u)$. Moreover, the boundedness of the function $\psi_u^\varepsilon(v)$ is uniform over u under conditions **C2** and **C3** and the boundedness of the derivative $\varphi'(u)$.

Thus, we have:

$$H_G^{\varepsilon,\delta}(x)\varphi(u) = \varepsilon^{-2}\delta^2 Q_0 \int_{\mathbf{R}} [e^{\Delta_\varepsilon \varphi(u)} - 1 - \Delta_\varepsilon\varphi(u) - \frac{1}{2}(\Delta_\varepsilon\varphi(u))^2] G_x^0(dv)$$

$$+\varepsilon^{-2}Q_0 \int_{\mathbf{R}} [\Delta_\varepsilon\varphi(u) - v\varphi'(u) - \varepsilon\frac{v^2}{2}\varphi''(u)] G_x^\delta(dv) + \varepsilon^{-2}\delta Q_0 b_1(x)\varphi'(u)$$

$$+\varepsilon^{-2}\delta^2 Q_0 b(x)\varphi'(u) + \varepsilon^{-1}\delta^2 Q_0 c(x)\varphi''(u)$$

$$+\varepsilon^{-2}Q_0 \int_{\mathbf{R}} [\frac{1}{2}(\Delta_\varepsilon\varphi(u))^2 - \frac{v^2}{2}(\varphi'(u))^2] G_x^\delta(dv) + \varepsilon^{-2}\delta^2 \frac{1}{2} Q_0 c(x)(\varphi'(u))^2.$$

Applying Taylor's formula to test functions $\varphi(u) \in C^3(\mathbf{R})$, and condition **LA2**, we obtain:

$$H_G^{\varepsilon,\delta}(x)\varphi(u) = \varepsilon^{-2}\delta^2 Q_0 \int_{\mathbf{R}} [e^{v\varphi'(u)} - 1 - v\varphi'(u) - \frac{v^2}{2}(\varphi'(u))^2] G_x^0(dv)$$

$$+\varepsilon^{-2}\delta^2 Q_0 \int_{\mathbf{R}} (e^{v\varphi'(u)}\varepsilon\frac{v^2}{2}\varphi''(\widehat{u}) - \varepsilon\frac{v^2}{2}\varphi''(\widehat{u}) - \varepsilon^2\frac{v^4}{8}(\varphi''(\widehat{u}))^2) G_x^0(dv)$$

$$+\varepsilon^{-2}\delta^2 Q_0 \int_{\mathbf{R}} \varepsilon^2 \frac{v^3}{3!}\varphi'''(\widehat{u})G_x^0(dv)+\varepsilon^{-2}\delta Q_0 b_1(x)\varphi'(u)+\varepsilon^{-2}\delta^2 Q_0 b(x)\varphi'(u)$$

$$+\,\varepsilon^{-1}\delta^2 Q_0 c(x)\varphi''(u) + \varepsilon^{-2}\delta^2 Q_0 \int_{\mathbf{R}} \varepsilon^2 \frac{v^4}{4}(\varphi''(\widehat{u}))^2 G_x^0(dv)$$

$$+\,\varepsilon^{-2}\delta^2 \frac{1}{2} Q_0 c(x)(\varphi'(u))^2.$$

Given the limit condition $\varepsilon^{-1}\delta \to 1$, we finally have:

$$H_G^{\varepsilon,\delta}(x)\varphi(u) = H_G(x)\varphi(u) + \theta_G^{\varepsilon,\delta}\varphi,$$

where $\sup_{x\in E} |\theta_G^{\varepsilon,\delta}\varphi| \to 0, \varepsilon, \delta \to 0.$

Lemma 4.11 is proved.

The application of the relations obtained in Lemma 4.10 and Lemma 4.11 to formula [4.23] finally completes the proof of Lemma 4.9.

Lemma 4.9 is proved.

From formulas [4.19] and [4.21], we see

$$\mathbb{H}_\varepsilon^\delta \varphi_\varepsilon^\delta = H_Q^\varepsilon \varphi_\varepsilon^\delta(u,x) + H_G^{\varepsilon,\delta}(x)\varphi_\varepsilon^\delta(u,x)$$

Thus, with Lemma 4.8 and Lemma 4.9, we obtain an asymptotic representation

$$\mathbb{H}_\varepsilon^\delta \varphi_\varepsilon^\delta = \varepsilon^{-1}[Q\varphi_1 + b_1(x)\varphi'(u)] + Q\varphi_2 - \varphi_1 Q\varphi_1 + H_G(x)\varphi(u) + h^{\varepsilon,\delta}\varphi,$$

where $h^{\varepsilon,\delta}\varphi = \theta_Q^{\varepsilon,\delta}\varphi + \theta_G^{\varepsilon,\delta}\varphi$.

We can now apply the solution of the singular perturbation problem to the reducibly invertible operator Q (see Korolyuk and Limnios (2005, Chapter 1)):

$$\begin{aligned} Q\varphi_1 + b_1(x)\varphi'(u) &= 0, \\ Q\varphi_2 - \varphi_1 Q\varphi_1 + H_G(x)\varphi(u) &= H^0\varphi(u). \end{aligned}$$

Due to the balance condition **LA3**, we have, from the first equation:

$$Q\varphi_1(u,x) = -b_1(x)\varphi'(u), \qquad \varphi_1(u,x) = R_0 b_1(x)\varphi'(u).$$

After substitution into the second equation, we obtain

$$Q\varphi_2 + b_1(x)R_0 b_1(x)(\varphi'(u))^2 + H_G(x)\varphi(u) = H^0\varphi(u)$$

From the solvability condition:

$$H^0\varphi(u) = \Pi H_G(x)\Pi\varphi(u) + \Pi b_1(x)R_0 b_1(x)(\varphi'(u))^2.$$

Given the relationship [4.22] and the properties of the Markov process described in section 1.1.3, we can rewrite

$$H^0\varphi(u) = \int_E \pi(dx)q(x)\int_E P(x,dy)\mathcal{H}_G(y)\varphi(u) = q\int_E \rho(dx)\mathcal{H}_G(x)\varphi(u),$$

therefore, we finally obtain:

$$H^0\varphi(u) = (\widehat{b} - \widehat{b}_0)\varphi'(u) + \frac{1}{2}\sigma^2(x)(\varphi'(u))^2 + \int_{\mathbf{R}} [e^{v\varphi'(u)} - 1]\widehat{G}^0(dv).$$

The negligible term $h^{\varepsilon,\delta}\varphi$ can be calculated explicitly, using the solution of the Poisson equation (see section 1.1.2. For more details, see Korolyuk and Limnios (2005))

$$\varphi_1(u,x) = R_0\widehat{H}(x)\varphi(u), \widehat{H}(x) := H_G(x) - H^0.$$

Theorem 4.4 is proved.

5

Large Deviations of Systems in the Scheme of Splitting and Double Merging

As we have already mentioned, the classical formulations of the problem in the study of large deviations relate to the small diffusion scheme. A problem that is similar to the Poisson approximation scheme in the case of processes with independent increments was investigated in Mogulskii (1993).

In this section, we describe how the approach proposed in Chapters 3 and 4 can be applied in cases that could not be studied using classical methods. That is, we describe the situation when the switching Markov process has a complicated structure. We introduce the switching Markov process $x^\varepsilon(t), t \geq 0$ on the standard phase (state) space $(E, \mathcal{E})$ in the series scheme with a small series parameter $\varepsilon \to 0, \varepsilon > 0$ on the split phase space

$$E = \bigcup_{k=1}^{N} E_k, E_k \cap E_{k'} = \emptyset, k \neq k'.$$

The Markov kernel is

$$Q^\varepsilon(x, B, t) = P^\varepsilon(x, B)[1 - e^{-q(x)t}], \ x \in E, B \in \mathcal{E}, t \geq 0.$$

The problem of studying large deviations in the phase merging scheme (see section 1.1.5 and conditions **ME1–ME4**), both in the small diffusion scheme and the Poisson approximation scheme, is new and still unexplored. It should be noted that an attempt to set such a task was made in Mogulskii (1993), but its formulation and method of solution do not meet the stated goal.

5.1. Small diffusion scheme

5.1.1. *Large deviations under the local balance (LB) condition*

Random evolutionary systems are considered under the local balance condition **LB** $b(u;x) := \int_{\mathbf{R}} v\Gamma(u,dv;x) \equiv 0$

with the following normalization:

$$\xi^{\varepsilon}(t) = \xi_0 + \varepsilon^2 \int_0^t \eta_{\varepsilon}(ds; x(s/\varepsilon^3)), x_t^{\varepsilon} := x^{\varepsilon}(t/\varepsilon^3). \qquad [5.1]$$

The random evolutionary system generator [5.1] has the form

$$\mathbf{L}_{\Lambda}^{\varepsilon}\varphi(u,x) = [\varepsilon^{-3}Q + \varepsilon^{-2}Q_1 + \Gamma^{\varepsilon}(x)]\varphi(u,x), \qquad [5.2]$$

$$\Gamma^{\varepsilon}(x)\varphi(u) = \varepsilon^{-3}\int_{\mathbf{R}} [\varphi(u+\varepsilon^2 v) - \varphi(u)]\Gamma(u,dv;x).$$

The generator [5.2] has the following asymptotic representation:

$$\mathbf{L}_{\Lambda}^{\varepsilon}\varphi(u,x) = [\varepsilon^{-3}Q + \varepsilon^{-2}Q_1 + \varepsilon\mathbf{B}(x)]\varphi(u,x) + \varepsilon\delta^{\varepsilon}(u,x)\varphi(u,x).$$

Here,

$$\mathbf{B}(x)\varphi(u) = \frac{1}{2}B(u;x)\varphi''(u), B(u;x) = \int_{\mathbf{R}} v^2\Gamma(u,dv;x).$$

THEOREM 5.1.– *The exponential generator of large deviations for random evolutionary systems* [5.2] *under the* ***ME1–ME4*** *and* **LB** *conditions is determined by the ratio*

$$\widehat{\widehat{H}}\varphi(u) = \frac{1}{2}\widehat{\widehat{B}}(u)[\varphi'(u)]^2,$$

$$\widehat{\widehat{B}}(u) = \sum_{k=1}^{N}\widehat{\pi}_k \int_{E_k} \pi_k(dx)B(u;x), \; B(u;x) = \int_{\mathbf{R}} v^2\Gamma(u,dv;x).$$

PROOF.– The proof is based on the following lemma.

LEMMA 5.1.– *The exponential generator on perturbed test functions*

$$\varphi^{\varepsilon}(u,x) = \varphi(u) + \varepsilon\ln[1 + \varepsilon\varphi_1(u,x) + \varepsilon^2\varphi_2(u,x)]$$

has the following asymptotic representation:

$$\mathbf{H}^{\varepsilon}\varphi^{\varepsilon}(u,x) = \varepsilon^{-1}Q\varphi_1 + Q\varphi_2 + Q_1\varphi_1 - \varphi_1 Q\varphi_1 + \widetilde{\mathbf{B}}(x)\varphi(u) + \delta_H^{\varepsilon}(u,x)\varphi,$$

and the negligible term converges uniformly by u, x on functions $\varphi(u) \in C^3(\mathbf{R})$:

$$|\delta_H^\varepsilon(u,x)\varphi| \to 0, \ \varepsilon \to 0.$$

The operator

$$\widetilde{\mathbf{B}}(x)\varphi(u) = \frac{1}{2}B(u;x)[\varphi'(u)]^2. \quad [5.3]$$

PROOF.– We will perform an asymptotic analysis of terms

$$H_Q^\varepsilon\varphi^\varepsilon(u,x) = e^{-\varphi(u)/\varepsilon}[1+\varepsilon\varphi_1+\varepsilon^2\varphi_2]^{-1}$$

$$[\varepsilon^{-2}Q+\varepsilon^{-1}Q_1][1+\varepsilon\varphi_1+\varepsilon^2\varphi_2]e^{\varphi(u)/\varepsilon}$$

$$= e^{-\varphi(u)/\varepsilon}[1-\varepsilon\varphi_1][\varepsilon^{-2}Q+\varepsilon^{-1}Q_1][1+\varepsilon\varphi_1+\varepsilon^2\varphi_2]e^{\varphi(u)/\varepsilon} + \delta^\varepsilon(x)\varphi$$

$$= \varepsilon^{-1}Q\varphi_1 + Q\varphi_2 + Q_1\varphi_1 - \varphi_1 Q\varphi_1 + \delta^\varepsilon(x)\varphi$$

and

$$H_\Gamma^\varepsilon\varphi^\varepsilon(u,x) = e^{-\varphi(u)/\varepsilon}[1+\varepsilon\varphi_1+\varepsilon^2\varphi_2]^{-1}\varepsilon\Gamma^\varepsilon(x)[1+\varepsilon\varphi_1+\varepsilon^2\varphi_2]e^{\varphi(u)/\varepsilon}$$

$$= e^{-\varphi(u)/\varepsilon}[1-\varepsilon\varphi_1]\varepsilon\Gamma^\varepsilon(x)[1+\varepsilon\varphi_1+\varepsilon^2\varphi_2]e^{\varphi(u)/\varepsilon} + \delta^\varepsilon(u,x)\varphi$$

$$= \varepsilon^{-2}\int_{\mathbf{R}}[e^{\Delta_v^\varepsilon\varphi(u)}-1]\Gamma(u,dv;x) + \delta^\varepsilon(u,x)\varphi.$$

Here,

$$\Delta_v^\varepsilon\varphi(u) = \varepsilon^{-1}[\varphi(u+\varepsilon^2 v)-\varphi(u)] = \varepsilon v\varphi'(u) + \varepsilon^3\widehat{\varphi}_v''(u)$$

and due to the **LB** condition, we have

$$\varepsilon^{-2}\int_{\mathbf{R}}[e^{\Delta_v^\varepsilon\varphi(u)}-1]\Gamma(u,dv;x) = \varepsilon^{-2}\int_{\mathbf{R}}\left[\varepsilon v\varphi'(u)+\frac{1}{2}(\varepsilon v)^2[\varphi'(u)]^2\right]\Gamma(u,dv;x)$$

$$+\delta^\varepsilon(u,x)\varphi(u) = \widetilde{\mathbf{B}}(x)\varphi(u) + \delta^\varepsilon(u,x)\varphi.$$

So,

$$\mathbf{H}_\Gamma^\varepsilon\varphi^\varepsilon(u,x) = \widetilde{\mathbf{B}}(x)\varphi(u) + \delta^\varepsilon(u,x)\varphi$$

with the main term [5.3].

Lemma 5.1 is proved.

We complete the proof of the theorem by applying the solution of the singular perturbation problem to the equations:

$$Q\varphi_1(u,x) = 0$$

$$Q\varphi_2 + Q_1\varphi_1 + \widetilde{\mathbf{B}}(x)\varphi(u) = \widehat{\widehat{H}}\varphi(u).$$

From the first equation, it follows that $\varphi_1(u,x) = \varphi_1(u,\widehat{x}) \in \mathcal{N}_Q$, and therefore, from the solvability condition for the second equation, we have a new relation

$$\Pi Q_1 \Pi \varphi_1 + \Pi \widetilde{\mathbf{B}}(x) \Pi \varphi(u) = \widehat{\widehat{H}}\varphi(u),$$

or

$$\widehat{Q}_1\widehat{\varphi}_1 + \widehat{\widetilde{\mathbf{B}}(x)}\widehat{\varphi}(u) = \widehat{\widehat{H}}\varphi(u).$$

Finally, the solvability condition for the averaged equation gives:

$$\widehat{\Pi}\widehat{\widetilde{\mathbf{B}}(x)}\widehat{\Pi}\widehat{\varphi}(u) = \widehat{\widehat{H}}\varphi(u).$$

The equality

$$\mathbf{H}^{\varepsilon}\varphi^{\varepsilon}(u,x) = \widehat{\widehat{H}}\varphi(u) + \delta_H^{\varepsilon}(u,x)\varphi(u)$$

completes the proof of the theorem.

Theorem 5.1 is proved.

5.1.2. *Large deviations under the total balance (TB) condition*

Under the total balance condition:

TB $\quad b(u;x) = \int_{\mathbf{R}} v\Gamma(u,dv;x) \not\equiv 0,$
$\sum_{k=1}^{N} \widehat{\pi}_k \widehat{b}_k(u) = 0, \widehat{b}_k(u) = \int_{E_k} \pi_k(dx) b(u;x), 1 \le k \le N$

we use the following normalization:

$$\xi^{\varepsilon}(t) = \xi_0 + \varepsilon^2 \int_0^t \eta_{\varepsilon}(ds; x(s/\varepsilon^4)),\ x_t^{\varepsilon} := x^{\varepsilon}(t/\varepsilon^4). \qquad [5.4]$$

The random evolutionary system generator is set as follows:

$$\mathbf{L}_T^{\varepsilon}\varphi(u,x) = [\varepsilon^{-4}Q + \varepsilon^{-3}Q_1 + \Gamma^{\varepsilon}(x)]\varphi(u,x), \qquad [5.5]$$

where

$$\Gamma^\varepsilon(x)\varphi(u) = \varepsilon^{-3}\int_{\mathbf{R}}[\varphi(u+\varepsilon^2 v) - \varphi(u)]\Gamma(u, dv; x).$$

The generator [5.5] has the following asymptotic representation:

$$\mathbf{L}_T^\varepsilon\varphi(u,x) = [\varepsilon^{-4}Q + \varepsilon^{-3}Q_1 + \varepsilon^{-1}\Gamma(x) + \varepsilon\mathbf{B}(x)]\varphi(u,x) + \varepsilon\delta^\varepsilon(u,x)\varphi(u,x).$$

Here,

$$\Gamma(x)\varphi(u) := b(u;x)\varphi'(u).$$

REMARK 5.1.– *The statement of the large deviations problem is possible in the absence of total balance*

$$b(u;x) = \int_{\mathbf{R}} v\Gamma(u, dv; x) \not\equiv 0, \sum_{k=1}^N \widehat{\pi}_k \widehat{b}_k(u) \not\equiv 0.$$

In this case, it is necessary to consider evolution under equilibrium conditions similar to those in Korolyuk and Limnios (2005).

THEOREM 5.2.– *The exponential generator of large deviations for random evolutionary systems* [5.4] *under the **ME1–ME4** and **TB** conditions is determined by the ratio*

$$\widehat{\widehat{H}}\varphi(u) = \frac{1}{2}\widehat{\widehat{B}}_T(u)[\varphi'(u)]^2, \ \widehat{\widehat{B}}_T(u) = \widehat{\widehat{B}}(u) + \widehat{\widehat{B}}_0(u).$$

Here,

$$\widehat{\widehat{B}}(u) := \sum_{k=1}^N \widehat{\pi}_k \int_{E_k} \pi_k(dx)B(u;x), \ B(u;x) = \int_{\mathbf{R}} v^2\Gamma(u, dv; x),$$

$$\widehat{\widehat{B}}_0(u) := \widehat{\Pi}\widehat{b}(u,\widehat{x})\widehat{R}_0\widehat{b}(u,\widehat{x})\widehat{\Pi} = \sum_{k,l=1}^N \widehat{\pi}_k\widehat{b}_k\widehat{R}_{kl}^0\widehat{b}_l.$$

REMARK 5.2.– *The limit exponential generator consists of two terms: the first is the average diffusion coefficient and the second is determined by the merged first moments of jumps, which are averaged by the potential operators of the limit merged Markov switching process.*

PROOF.– The proof is based on the following lemma.

LEMMA 5.2.– *The exponential generator on perturbed test functions*

$$\varphi^\varepsilon(u,x) = \varphi(u) + \varepsilon\ln[1 + \varepsilon\varphi_1(u,x) + \varepsilon^2\varphi_2(u,x) + \varepsilon^3\varphi_3(u,x)]$$

has the following asymptotic representation:

$$\mathbf{H}^\varepsilon\varphi^\varepsilon(u,x) = \varepsilon^{-2}Q\varphi_1 + \varepsilon^{-1}[Q\varphi_2 + Q_1\varphi_1 - \varphi_1 Q\varphi_1 + \Gamma(x)\varphi(u)]$$

$$+[Q\varphi_3 + Q_1\varphi_2 - \varphi_1 Q\varphi_2 - \varphi_2 Q\varphi_1 - \varphi_1 Q_1\varphi_1 + \widetilde{\mathbf{B}}(x)\varphi(u)] + \delta^\varepsilon(u,x)\varphi$$

and the negligible terms converge $|\delta^\varepsilon(u,x)\varphi| \to 0,\ \varepsilon \to 0$ *uniformly by* u, x *on functions* $\varphi(u) \in C^3(\mathbf{R})$. *The operators*

$$\Gamma(x)\varphi(u) := b(u;x)\varphi'(u),\ \widetilde{\mathbf{B}}(x)\varphi(u) := \frac{1}{2}B(u;x)[\varphi'(u)]^2. \qquad [5.6]$$

PROOF.– We will perform an asymptotic analysis of terms

$$H_Q^\varepsilon\varphi^\varepsilon(u,x) = e^{-\varphi(u)/\varepsilon}[1 + \varepsilon\varphi_1 + \varepsilon^2\varphi_2 + \varepsilon^3\varphi_3]^{-1}[\varepsilon^{-3}Q + \varepsilon^{-2}Q_1]$$

$$\times[1+\varepsilon\varphi_1+\varepsilon^2\varphi_2+\varepsilon^3\varphi_3]e^{\varphi(u)/\varepsilon} = e^{-\varphi(u)/\varepsilon}[1 - \varepsilon\varphi_1 - \varepsilon^2\varphi_2 + \varepsilon^2\varphi_1^2 - \varepsilon^3\varphi_3]$$

$$\times[\varepsilon^{-3}Q + \varepsilon^{-2}Q_1][1 + \varepsilon\varphi_1 + \varepsilon^2\varphi_2 + \varepsilon^3\varphi_3]e^{\varphi(u)/\varepsilon} + \delta^\varepsilon(x)\varphi$$

$$= \varepsilon^{-2}Q\varphi_1 + \varepsilon^{-1}[Q\varphi_2 + Q_1\varphi_1 - \varphi_1 Q\varphi_1] + [Q\varphi_3 + Q_1\varphi_2 - \varphi_1 Q\varphi_2$$

$$-\varphi_2 Q\varphi_1 + \varphi_1^2 Q\varphi_1 - \varphi_1 Q_1\varphi_1] + \delta^\varepsilon(x)\varphi$$

and

$$H_\Gamma^\varepsilon\varphi^\varepsilon(u,x) = e^{-\varphi(u)/\varepsilon}[1 + \varepsilon\varphi_1 + \varepsilon^2\varphi_2 + \varepsilon^3\varphi_3]^{-1}\varepsilon\Gamma^\varepsilon(x)[1 + \varepsilon\varphi_1 + \varepsilon^2\varphi_2$$

$$+\varepsilon^3\varphi_3]e^{\varphi(u)/\varepsilon} = e^{-\varphi(u)/\varepsilon}[1 - \varepsilon\varphi_1 - \varepsilon^2\varphi_2 + \varepsilon^2\varphi_1^2]\varepsilon\Gamma^\varepsilon(x)[1 + \varepsilon\varphi_1 + \varepsilon^2\varphi_2$$

$$+\varepsilon^3\varphi_3]e^{\varphi(u)/\varepsilon} + \delta^\varepsilon(u,x)\varphi = \varepsilon^{-2}\int_{\mathbf{R}}[e^{\Delta_v^\varepsilon\varphi(u)} - 1]\Gamma(u,dv;x)$$

$$+\varepsilon^2 e^{-\varphi(u)/\varepsilon}[\Gamma^\varepsilon(x)\varphi_1 e^{\varphi(u)/\varepsilon} - \varphi_1\Gamma^\varepsilon(x)e^{\varphi(u)/\varepsilon}] + \delta_\Gamma^\varepsilon(u,x)\varphi.$$

Here, $\Delta_v^\varepsilon\varphi(u) = \varepsilon^{-1}[\varphi(u+\varepsilon^2 v) - \varphi(u)] = \varepsilon v\varphi'(u) + \varepsilon^3\widetilde{\varphi}_v''(u)$ and, due to the **TB** condition, we obtain

$$\varepsilon^{-2}\int_{\mathbf{R}}[e^{\Delta_v^\varepsilon\varphi(u)} - 1]\Gamma(u,dv;x) = \varepsilon^{-2}\int_{\mathbf{R}}\Big[\varepsilon v\varphi'(u)$$

$$+\frac{1}{2}(\varepsilon v)^2[\varphi'(u)]^2\Big]\Gamma(u,dv;x) + \delta^\varepsilon(u,x)\varphi,$$

i.e. taking into account [5.6], we can write

$$\varepsilon^{-2} \int_{\mathbf{R}} [e^{\Delta_v^\varepsilon \varphi(u)} - 1]\Gamma(u, dv; x) = \varepsilon^{-1}\Gamma(x)\varphi(u) + \widetilde{\mathbf{B}}(x)\varphi(u) + \delta^\varepsilon(u, x)\varphi.$$

Each of the terms in square brackets is not negligible, but their difference is equal to 0. In fact:

$$\varepsilon^2 \varphi_1 e^{-\varphi(u)/\varepsilon} \Gamma^\varepsilon(x) e^{\varphi(u)/\varepsilon} = \varepsilon \varphi_1 e^{-\varphi(u)/\varepsilon} \varepsilon \Gamma^\varepsilon(x) e^{\varphi(u)/\varepsilon}$$
$$= \varphi_1 \Gamma(x)\varphi(u) + \delta^\varepsilon(u, x)\varphi,$$

with

$$\Gamma^\varepsilon(x) e^{\varphi(u)/\varepsilon} \varphi_1 = \varepsilon^{-3} \int_{\mathbf{R}} [e^{\varphi(u+\varepsilon^2 v)/\varepsilon} \varphi_1(u+\varepsilon v, x) - e^{\varphi(u)/\varepsilon} \varphi_1(u, x)]\Gamma(u, dv; x)$$

$$= \varphi_1(u, x)\Gamma^\varepsilon(x) e^{\varphi(u)/\varepsilon} + o(\varepsilon^2).$$

So,

$$H_\Gamma^\varepsilon \varphi^\varepsilon(u, x) = \varepsilon^{-1}\Gamma(x)\varphi(u) + \widetilde{\mathbf{B}}(x)\varphi(u) + \delta^\varepsilon(u, x)\varphi(u),$$

with the main terms [5.6].

Lemma 5.2 is proved.

To complete the proof of the theorem, we should apply the solution of the singular perturbation problem to the equations:

$$Q\varphi_1 = 0,$$

$$Q\varphi_2 + Q_1\varphi_1 + b(u; x)\varphi'(u) = 0,$$

$$Q\varphi_3 + Q_1\varphi_2 - \varphi_1 Q\varphi_2 - \varphi_1 Q_1 \varphi_1 + \widetilde{\mathbf{B}}(x)\varphi(u) = \widehat{\widehat{H}}\varphi(u).$$

From the first equation, we have $\varphi_1(u, x) = \varphi_1(u, \widehat{x}) \in \mathcal{N}_Q$, and hence, from the solvability condition for the second equation

$$\widehat{Q\varphi_2} + \widehat{Q}_1 \widehat{\varphi}_1 + \widehat{b}(u; \widehat{x})\varphi'(u) = 0, \; \widehat{Q\varphi_2} = 0, \tag{5.7}$$

we obtain a new ratio

$$\widehat{Q}_1 \widehat{\varphi}_1 + \widehat{b}(u, \widehat{x})\varphi'(u) = 0, \; \widehat{\Pi}\widehat{b}(u; \widehat{x}) \equiv 0,$$

from which it follows that

$$\widehat{\varphi}_1(u, \widehat{x}) = \widehat{R}_0 \widehat{b}(u; \widehat{x})\varphi'(u), \widehat{Q}_1 \widehat{\varphi}_1 = -\widehat{b}(u, \widehat{x})\varphi'(u). \tag{5.8}$$

The solvability condition for the equation

$$Q\varphi_3 + Q_1\varphi_2 - \varphi_1 Q\varphi_2 - \varphi_1 Q_1\varphi_1 + \widetilde{\mathbf{B}}(x)\varphi(u) = \widehat{\widehat{B}}\varphi(u)$$

gives

$$\widehat{Q}_1\widehat{\varphi}_2 - \widehat{\varphi}_1\widehat{Q\varphi_2} - \widehat{\varphi}_1\widehat{Q}_1\widehat{\varphi}_1 + \widehat{\widetilde{\mathbf{B}}}(x)\widehat{\varphi}(u) = \widehat{\widehat{H}}\varphi(u),$$

but, from [5.7],

$$\widehat{Q\varphi_2} = -[\widehat{Q_1\varphi_1 + \widehat{b}(u,\widehat{x})\varphi'(u)}] = 0,$$

and, using relation [5.8], we have

$$\widehat{Q}_1\widehat{\varphi}_2 + \widehat{B}_T(x)\widehat{\varphi}(u) = \widehat{\widehat{H}}\varphi(u).$$

The application of the solvability condition to the last equation completes the proof of the theorem.

Theorem 5.2 is proved.

5.2. Poisson approximation scheme

Consider the asymptotic analysis of a nonlinear exponential generator that solves the large deviations problem for random evolutionary systems with independent increments in the Poisson approximation scheme (section 1.1.10) with split and double merging (section 1.1.5).

The normalization of random evolutionary systems by a small series parameter in order to solve the large deviations problem in the Poisson approximation scheme is implemented as follows:

$$\xi_\varepsilon^\delta(t) = \xi_\varepsilon^\delta(0) + \int_0^t \eta_\varepsilon^\delta(ds; x^\varepsilon(s/\varepsilon^3)),\ t \geq 0,$$

$$\eta_\varepsilon^\delta(t) = \varepsilon\eta^\delta(t/\varepsilon^2),$$

$$\Gamma_\varepsilon^\delta(x)\varphi(u) = \varepsilon^{-2}\int_{\mathbf{R}}[\varphi(u+\varepsilon v) - \varphi(u)]\Gamma^\delta(dv;x),\ x \in E,$$

where $\varepsilon, \delta \to 0$, therefore $\varepsilon^{-1}\delta \to 1$.

We recall that the conditions of the Poisson approximation **C1–C3** were formulated in section 4.1.1, and the phase merging conditions **ME1–ME4** in section 1.1.5.

THEOREM 5.3.– *Under the conditions* ***ME1–ME4*** *and* ***C1–C3****, the solution of the large deviations problem for random evolutionary systems*

$$\xi_\varepsilon^\delta(t) = \xi_\varepsilon^\delta(0) + \int_0^t \eta_\varepsilon^\delta(ds; x^\varepsilon(s/\varepsilon^3)),\ t \geq 0,$$

which is set by the generator of the two-component Markov process $\xi_\varepsilon^\delta(t), x^\varepsilon(t), t \geq 0$

$$\mathbf{L}_\varepsilon^\delta \varphi(u,x) = \varepsilon^{-3} Q\varphi(\cdot,x) + \varepsilon^{-2} Q_1 \varphi(\cdot,x) + \Gamma_\varepsilon^\delta(x)\varphi(u,\cdot), \qquad [5.9]$$

where

$$\Gamma_\varepsilon^\delta(x)\varphi(u) = \varepsilon^{-2}\int_{\mathbf{R}} [\varphi(u+\varepsilon v) - \varphi(u)]\Gamma^\delta(dv;x),\ x \in E \qquad [5.10]$$

is determined by the exponential generator

$$\widehat{\widehat{H}}\varphi(u) = \widehat{\widehat{b}}\varphi'(u) + \int_{\mathbb{R}} [e^{v\varphi'(u)} - 1 - v\varphi'(u)]\widehat{\widehat{\Gamma^0}}(dv), \varphi(u) \in C^3(\mathbf{R}), \quad [5.11]$$

$$\widehat{\widehat{b}} = \sum_{k=1}^N \widehat{\pi}_k \int_{E_k} \pi_k(dx) b(x), \widehat{\widehat{\Gamma^0}}(v) = \sum_{k=1}^N \widehat{\pi}_k \int_{E_k} \pi_k(dx)\Gamma^0(v;x).$$

To prove the theorem, we use the following lemma.

LEMMA 5.3.– *The exponential generator in a series scheme*

$$H_\Gamma^{\varepsilon,\delta}(x)\varphi(u) = e^{-\varphi/\varepsilon}\varepsilon\Gamma_\varepsilon^\delta(x)e^{\varphi/\varepsilon} \qquad [5.12]$$

has the following asymptotic representation:

$$H_\Gamma^{\varepsilon,\delta}(x)\varphi(u) = H_\Gamma(x)\varphi(u) + \theta_\Gamma^{\varepsilon,\delta}\varphi,$$

where $\sup_{x\in E} |\theta_\Gamma^{\varepsilon,\delta}\varphi| \to 0, \varepsilon, \delta \to 0,$

$$H_\Gamma(x)\varphi(u) = b(x)\varphi'(u) + \int_{\mathbf{R}} [e^{v\varphi'(u)} - 1 - v\varphi'(u)]\Gamma^0(dv;x).$$

PROOF.– Let us rewrite [5.12], taking into account the form of the generator [5.10]. We have:

$$H_\Gamma^{\varepsilon,\delta}(x)\varphi(u) = \varepsilon^{-1}\int_{\mathbf{R}} [e^{\Delta_\varepsilon\varphi(u)} - 1]\Gamma^\delta(dv;x),$$

where $\Delta_\varepsilon\varphi(u) := \varepsilon^{-1}[\varphi(u+\varepsilon v) - \varphi(u)]$.

Rewrite the expression for the generator as follows:

$$H_{\Gamma}^{\varepsilon,\delta}(x)\varphi(u) = \varepsilon^{-1}\int_{\mathbf{R}}[e^{\Delta_{\varepsilon}\varphi(u)} - 1 - \Delta_{\varepsilon}\varphi(u) - \frac{1}{2}(\Delta_{\varepsilon}\varphi(u))^2]\Gamma^{\delta}(dv;x)$$

$$+\varepsilon^{-1}\int_{\mathbf{R}}[\Delta_{\varepsilon}\varphi(u) + \frac{1}{2}(\Delta_{\varepsilon}\varphi(u))^2]\Gamma^{\delta}(dv;x).$$

It is easy to see that the function $\psi_u^{\varepsilon}(v) = e^{\Delta_{\varepsilon}\varphi(u)} - 1 - \Delta_{\varepsilon}\varphi(u) - \frac{1}{2}(\Delta_{\varepsilon}\varphi(u))^2$ belongs to the class $C_3(\mathbf{R})$. In fact,

$$\psi_u^{\varepsilon}(v)/v^2 \to 0, v \to 0$$

uniformly by u, as soon as $\varphi(u) \in C^3(\mathbf{R})$. In addition, the function $\psi_u^{\varepsilon}(v)$ is continuous and bounded for any ε due to the boundedness of the function $\varphi(u)$. Moreover, the boundedness of the function $\psi_u^{\varepsilon}(v)$ is uniform by u, due to the conditions **C2** and **C3** and the boundedness of the derivative $\varphi'(u)$.

Thus, we have:

$$H_{\Gamma}^{\varepsilon,\delta}(x)\varphi(u) = \varepsilon^{-1}\delta\int_{\mathbf{R}}[e^{\Delta_{\varepsilon}\varphi(u)} - 1 - \Delta_{\varepsilon}\varphi(u) - \frac{1}{2}(\Delta_{\varepsilon}\varphi(u))^2]\Gamma^{0}(dv;x)$$

$$+\varepsilon^{-1}\int_{\mathbf{R}}[\Delta_{\varepsilon}\varphi(u) - v\varphi'(u) - \varepsilon\frac{v^2}{2}\varphi''(u)]\Gamma^{\delta}(dv;x)$$

$$+\varepsilon^{-1}\delta b(x)\varphi'(u) + \delta c(x)\varphi''(u)$$

$$+\varepsilon^{-1}\int_{\mathbf{R}}[\frac{1}{2}(\Delta_{\varepsilon}\varphi(u))^2 - \frac{v^2}{2}(\varphi'(u))^2]\Gamma^{\delta}(dv;x) + \varepsilon^{-1}\delta\frac{1}{2}c(x)(\varphi'(u))^2.$$

Applying Taylor's formula to the test functions $\varphi(u) \in C^3(\mathbf{R})$, and condition **PA2**, we obtain:

$$H_{\Gamma}^{\varepsilon,\delta}(x)\varphi(u) = \varepsilon^{-1}\delta\int_{\mathbf{R}}[e^{v\varphi'(u)} - 1 - v\varphi'(u) - \frac{v^2}{2}(\varphi'(u))^2]\Gamma^{0}(dv;x)$$

$$+\varepsilon^{-1}\delta\int_{\mathbf{R}}(e^{v\varphi'(u)}\varepsilon\frac{v^2}{2}\varphi''(\widetilde{u}) - \varepsilon\frac{v^2}{2}\varphi''(\widetilde{u}) - \varepsilon^2\frac{v^4}{8}(\varphi''(\widetilde{u}))^2)\Gamma^{0}(dv;x)$$

$$+\varepsilon^{-1}\delta\int_{\mathbf{R}}\varepsilon^2\frac{v^3}{3!}\varphi'''(\widetilde{u})\Gamma^{0}(dv;x) + \varepsilon^{-1}\delta b(x)\varphi'(u) + \delta c(x)\varphi''(u)$$

$$+\varepsilon^{-1}\delta\int_{\mathbf{R}}\varepsilon^2\frac{v^4}{4}(\varphi''(\widetilde{u}))^2\Gamma^{0}(dv;x) + \varepsilon^{-1}\delta\frac{1}{2}c(x)(\varphi'(u))^2.$$

Given **PA3** and the limit condition $\varepsilon^{-1}\delta \to 1$, we finally have:

$$H_\Gamma^{\varepsilon,\delta}(x)\varphi(u) = H_\Gamma(x)\varphi(u) + \theta_\Gamma^{\varepsilon,\delta}\varphi,$$

where $\sup_{x\in E} |\theta_\Gamma^{\varepsilon,\delta}\varphi| \to 0, \varepsilon, \delta \to 0.$

Lemma 5.3 is proved.

PROOF OF THEOREM 5.3.– The limit transition in the exponential nonlinear generator of random evolutionary systems is realized on perturbed test functions

$$\varphi_\varepsilon^\delta(u,x) = \varphi(u) + \varepsilon \ln[1 + \delta\varphi_1(u,x) + \delta^2\varphi_2(u,x)].$$

Thus, we have:

$$\mathbf{H}_\varepsilon^\delta\varphi^\varepsilon = e^{-\varphi^\varepsilon/\varepsilon}\varepsilon\mathbf{L}_\varepsilon^\delta e^{\varphi^\varepsilon/\varepsilon} = e^{-\varphi/\varepsilon}[1+\delta\varphi_1+\delta^2\varphi_2]^{-1}\varepsilon\mathbf{L}_\varepsilon^\delta e^{\varphi/\varepsilon}[1+\delta\varphi_1+\delta^2\varphi_2].$$

The analysis of the asymptotic behavior of the last exponential generator gives the following result.

LEMMA 5.4.– *The following asymptotic representation holds true*

$$\mathbf{H}_\varepsilon^\delta\varphi^\varepsilon = \varepsilon^{-1}Q\varphi_1 + Q\varphi_2 + Q_1\varphi_1 - \varphi_1 Q\varphi_1 + H_\Gamma^{\varepsilon,\delta}(x)\varphi(u) + \theta^{\varepsilon,\delta}\varphi,$$

where $\sup_{x\in E} |\theta^{\varepsilon,\delta}\varphi| \to 0, \varepsilon, \delta \to 0.$

PROOF.– Given [5.9] and the limit ratio $\varepsilon^{-1}\delta \to 1, \varepsilon, \delta \to 0$, we have:

$$\mathbf{H}_\varepsilon^\delta\varphi^\varepsilon = e^{-\varphi/\varepsilon}[1 - \delta\varphi_1 + \frac{\delta^2\varphi_1^2}{1+\delta\varphi_1}]\left\{\varepsilon^{-2}Q + \varepsilon^{-1}Q_1 + \varepsilon\Gamma_\varepsilon^\delta(x)\right\}e^{\varphi/\varepsilon}[1 + \delta\varphi_1$$

$$+\delta^2\varphi_2] = e^{-\varphi/\varepsilon}[1 - \delta\varphi_1 + \delta^2\frac{\varphi_1^2 - \varphi_2 + \delta\varphi_1\varphi_2}{1 + \delta\varphi_1 + \delta^2\varphi_2}]\left\{\varepsilon^{-2}\delta e^{\varphi/\varepsilon}Q\varphi_1 + \varepsilon^{-2}\delta^2 e^{\varphi/\varepsilon}Q\varphi_2\right.$$

$$+\varepsilon^{-1}\delta e^{\varphi/\varepsilon}Q_1\varphi_1 + \varepsilon^{-1}\delta^2 e^{\varphi/\varepsilon}Q_1\varphi_2 + \varepsilon\Gamma_\varepsilon^\delta(x)e^{\varphi/\varepsilon} + \varepsilon\delta\Gamma_\varepsilon^\delta(x)e^{\varphi/\varepsilon}\varphi_1$$

$$\left.+\varepsilon\delta^2\Gamma_\varepsilon^\delta(x)e^{\varphi/\varepsilon}\varphi_2\right\} = \varepsilon^{-1}Q\varphi_1 + Q\varphi_2 + Q_1\varphi_1 - \varphi_1 Q\varphi_1 + H_\Gamma^{\varepsilon,\delta}(x)\varphi(u) + \theta^{\varepsilon,\delta}\varphi,$$

where

$$\theta^{\varepsilon,\delta}\varphi = \delta\left[Q_1\varphi_2 - \varphi_1 Q\varphi_2 - \varphi_1 Q_1\varphi_1 - \delta\varphi_1 Q_1\varphi_2 + e^{-\varphi/\varepsilon}\varepsilon\Gamma_\varepsilon^\delta(x)e^{\varphi/\varepsilon}\varphi_1\right.$$

$$+\delta e^{-\varphi/\varepsilon}\varepsilon\Gamma_\varepsilon^\delta(x)e^{\varphi/\varepsilon}\varphi_2 - \varphi_1 e^{-\varphi/\varepsilon}\varepsilon\Gamma_\varepsilon^\delta(x)e^{\varphi/\varepsilon} - \delta\varphi_1 e^{-\varphi/\varepsilon}\varepsilon\Gamma_\varepsilon^\delta(x)e^{\varphi/\varepsilon}\varphi_1$$

$$-\delta^2\varphi_1 e^{-\varphi/\varepsilon}\varepsilon\Gamma^\delta_\varepsilon(x)e^{\varphi/\varepsilon}\varphi_2 + \frac{\varphi_1^2 - \varphi_2 + \delta\varphi_1\varphi_2}{1 + \delta\varphi_1 + \delta^2\varphi_2}\{\varepsilon^{-2}\delta^2 Q\varphi_1$$

$$+\varepsilon^{-2}\delta^3 Q\varphi_2 + \varepsilon^{-1}\delta^2 Q_1\varphi_1 + \varepsilon^{-1}\delta^3 Q_1\varphi_2 + \delta e^{-\varphi/\varepsilon}\varepsilon\Gamma^\delta_\varepsilon(x)e^{\varphi/\varepsilon}$$

$$+\delta^2 e^{-\varphi/\varepsilon}\varepsilon\Gamma^\delta_\varepsilon(x)e^{\varphi/\varepsilon}\varphi_1 + \delta^3 e^{-\varphi/\varepsilon}\varepsilon\Gamma^\delta_\varepsilon(x)e^{\varphi/\varepsilon}\varphi_2\}\Big].$$

Lemma 5.4 is proved.

Using Lemma 5.3, we obtain

$$\mathbf{H}^\delta_\varepsilon\varphi^\varepsilon = \varepsilon^{-1}Q\varphi_1 + Q\varphi_2 + Q_1\varphi_1 - \varphi_1 Q\varphi_1 + H_\Gamma(x)\varphi(u) + h^{\varepsilon,\delta}\varphi,$$

where $h^{\varepsilon,\delta}\varphi = \theta^{\varepsilon,\delta}\varphi + \theta^{\varepsilon,\delta}_\Gamma\varphi$.

We now use the solution of the singular perturbation problem for the system of equations

$$Q\varphi_1 = 0,$$
$$Q\varphi_2 + Q_1\varphi_1 - \varphi_1 Q\varphi_1 + H_\Gamma(x)\varphi(u) = H\varphi(u).$$

From the first equation, we have $\varphi_1(u,x) = \varphi_1(u,\widehat{x}) \in \mathcal{N}_Q$, so, from the solvability condition for the second equation:

$$\widehat{Q}_1\widehat{\varphi}_1 + \widehat{H}_\Gamma(x)\varphi(u) = H\varphi(u),$$

where

$$H\varphi(u) = \Pi Q_1\Pi\varphi_1 + \Pi H_\Gamma(x)\Pi\varphi(u).$$

By applying the solvability condition for the last equation, we finally obtain [5.11]:

$$\widehat{\widehat{H}}\varphi(u) = \widehat{\widehat{b}}\varphi'(u) + \int_{\mathbf{R}}[e^{v\varphi'(u)} - 1 - v\varphi'(u)]\widehat{\widehat{\Gamma^0}}(dv).$$

Theorem 5.3 is proved.

5.3. Lévy approximation scheme

Consider the asymptotic analysis of a nonlinear exponential generator that solves the large deviations problem for random evolutionary systems with independent increments in the Lévy approximation scheme (section 1.1.10) with split and double merging (section 1.1.5).

The normalization of random evolutionary systems by a small series parameter in order to solve the large deviations problem in the Lévy approximation scheme is implemented as follows (see sections 4.1.2 and 5.2):

$$\xi^\delta_\varepsilon(t) = \xi^\delta_\varepsilon(0) + \int_0^t \eta^\delta_\varepsilon(ds; x^\varepsilon(s/\varepsilon^4)),\ t \geq 0,$$

$$\eta^\delta_\varepsilon(t) = \varepsilon\eta^\delta(t/\varepsilon^3).$$

$$\Gamma^\delta_\varepsilon(x)\varphi(u) = \varepsilon^{-3}\int_{\mathbf{R}}[\varphi(u+\varepsilon v) - \varphi(u)]\Gamma^\delta(dv; x),\ x \in E,$$

where $\varepsilon, \delta \to 0$, therefore $\varepsilon^{-1}\delta \to 1$.

Recall that the Lévy approximation conditions **C1'–C3** were formulated in section 4.1.2, and the phase merging conditions **ME1–ME4** in section 1.1.5.

THEOREM 5.4.– *Under the conditions **ME1–ME4** and **C1'–C3**, the solution of the large deviations problem for random evolutionary systems*

$$\xi^\delta_\varepsilon(t) = \xi^\delta_\varepsilon(0) + \int_0^t \eta^\delta_\varepsilon(ds; x^\varepsilon(s/\varepsilon^4)),\ t \geq 0,$$

which is set by the generator of the two-component Markov process $\xi^\delta_\varepsilon(t), x^\varepsilon(t), t \geq 0$

$$\mathbf{L}^\delta_\varepsilon\varphi(u,x) = \varepsilon^{-4}Q\varphi(u,\cdot)(x) + \varepsilon^{-3}Q_1\varphi(u,\cdot)(x) + \Gamma^\delta_\varepsilon(x)\varphi(\cdot,x)(u), \quad [5.13]$$

where

$$\Gamma^\delta_\varepsilon(x)\varphi(u) = \varepsilon^{-3}\int_{\mathbf{R}}[\varphi(u+\varepsilon v) - \varphi(u)]\Gamma^\delta(dv; x),\ x \in E \quad [5.14]$$

is determined by the exponential generator

$$\widehat{\widehat{H}}\varphi(u) = (\widehat{\widehat{b}} - \widehat{\widehat{b}}_0)\varphi'(u) + \frac{1}{2}\sigma^2(\varphi'(u))^2 + \int_{\mathbf{R}}[e^{v\varphi'(u)} - 1]\widehat{\widehat{\Gamma^0}}(dv),$$

$$\widehat{\widehat{b}} = \sum_{k=1}^N \widehat{\pi}_k \int_{E_k} \pi_k(dx)b(x), \widehat{\widehat{b}}_0 = \sum_{k=1}^N \widehat{\pi}_k \int_{E_k} \pi_k(dx)b_0(x), b_0(x) = \int_{\mathbf{R}} v\Gamma^0(dv; x),$$

$$\widehat{\widehat{c}} = \sum_{k=1}^N \widehat{\pi}_k \int_{E_k} \pi_k(dx)c(x), \widehat{\widehat{c}}_0 = \sum_{k=1}^N \widehat{\pi}_k \int_{E_k} \pi_k(dx)c(x)_0, c_0(x) = \int_{\mathbf{R}} v^2\Gamma^0(dv; x),$$

$$\sigma^2 = (\widehat{\widehat{c}} - \widehat{\widehat{c}}_0) + 2\sum_{k,l=1}^N \widehat{\pi}_k\widehat{b}^k_1\widehat{R}^0_{kl}\widehat{b}^l_1, \widehat{b}^r_1 := \int_{E_r} \pi_r(dx)b_1(x),$$

$$\widehat{\widehat{\Gamma^0}}(v) = \sum_{k=1}^N \widehat{\pi}_k \int_{E_k} \pi_k(dx)\Gamma^0(v; x).$$

PROOF.– The limit transition for a nonlinear exponential generator will be performed on perturbed test functions

$$\varphi_\varepsilon^\delta(u,x) = \varphi(u) + \varepsilon \ln[1 + \delta\varphi_1(u,x) + \delta^2\varphi_2(u,x) + \delta^3\varphi_3(u,x)],$$

where $\varphi(u) \in C^3(\mathbf{R})$. Thus, from [5.13], we have:

$$\mathbf{H}_\varepsilon^\delta \varphi_\varepsilon^\delta = e^{-\varphi_\varepsilon^\delta/\varepsilon}\varepsilon\mathbf{L}_\varepsilon^\delta e^{\varphi_\varepsilon^\delta/\varepsilon} = e^{-\varphi_\varepsilon^\delta/\varepsilon}[\varepsilon^{-3}Q + \varepsilon^{-2}Q_1 + \varepsilon\Gamma_\varepsilon^\delta(x)]e^{\varphi_\varepsilon^\delta/\varepsilon}$$

$$= e^{-\varphi/\varepsilon}[1 + \delta\varphi_1 + \delta^2\varphi_2 + \delta^3\varphi_3]^{-1}[\varepsilon^{-3}Q + \varepsilon^{-2}Q_1 + \varepsilon\Gamma_\varepsilon^\delta(x)]$$
$$e^{\varphi/\varepsilon}[1 + \delta\varphi_1 + \delta^2\varphi_2 + \delta^3\varphi_3].$$

The asymptotic behavior of the last generator is demonstrated by the following lemmas.

LEMMA 5.5.– *The exponential generator*

$$H_Q^\varepsilon \varphi_\varepsilon^\delta(u,x) = e^{-\varphi_\varepsilon^\delta/\varepsilon}[\varepsilon^{-3}Q + \varepsilon^{-2}Q_1]e^{\varphi_\varepsilon^\delta/\varepsilon} \qquad [5.15]$$

has the following asymptotic representation:

$$H_Q^\varepsilon \varphi_\varepsilon^\delta = \varepsilon^{-2}Q\varphi_1 + \varepsilon^{-1}[Q\varphi_2 + Q_1\varphi_1 - \varphi_1 Q\varphi_1]$$
$$+ [Q\varphi_3 + Q_1\varphi_2 - \varphi_1 Q\varphi_2 - \varphi_1 Q_1\varphi_1 - \varphi_2 Q\varphi_1 + \varphi_1^2 Q\varphi_1] + \theta_Q^{\varepsilon,\delta}\varphi, \quad [5.16]$$

where $\sup_{x\in E} |\theta_Q^{\varepsilon,\delta}\varphi| \to 0, \varepsilon, \delta \to 0.$

PROOF.– We have:

$$H_Q^\varepsilon \varphi_\varepsilon^\delta = e^{-\varphi/\varepsilon}[1 + \delta\varphi_1 + \delta^2\varphi_2 + \delta^3\varphi_3]^{-1}[\varepsilon^{-3}Q + \varepsilon^{-2}Q_1]e^{\varphi/\varepsilon}[1 + \delta\varphi_1$$

$$+\delta^2\varphi_2 + \delta^3\varphi_3] = \Big[1 - \delta\varphi_1 - \delta^2\varphi_2 + \delta^2\varphi_1^2$$

$$+ \delta^3 \frac{2\varphi_1\varphi_2 + \delta\varphi_1\varphi_3 + \delta\varphi_2^2 + \delta^2\varphi_2\varphi_3 - \varphi_3 - \varphi_1^3 - \delta\varphi_1^2\varphi_2 - \delta^2\varphi_1^2\varphi_3}{1 + \delta\varphi_1 + \delta^2\varphi_2 + \delta^3\varphi_3}\Big]$$

$$\times[\delta\varepsilon^{-3}Q\varphi_1 + \delta^2\varepsilon^{-3}Q\varphi_2 + \delta^3\varepsilon^{-3}Q\varphi_3 + \delta\varepsilon^{-2}Q_1\varphi_1 + \delta^2\varepsilon^{-2}Q_1\varphi_2] = \delta\varepsilon^{-3}Q\varphi_1$$

$$+[\delta^2\varepsilon^{-3}Q\varphi_2 + \delta\varepsilon^{-2}Q_1\varphi_1 - \delta^2\varepsilon^{-3}\varphi_1 Q\varphi_1] + [\delta^3\varepsilon^{-3}Q\varphi_3 + \delta^2\varepsilon^{-2}Q_1\varphi_2$$

$$-\delta^3\varepsilon^{-3}\varphi_1 Q\varphi_2 - \delta^2\varepsilon^{-2}\varphi_1 Q_1\varphi_1 - \delta^3\varepsilon^{-3}\varphi_2 Q\varphi_1 + \delta^3\varepsilon^{-3}\varphi_1^2 Q\varphi_1] + \theta_Q^{\varepsilon,\delta}\varphi,$$

where

$$\theta_Q^{\varepsilon,\delta}\varphi = \delta^4\varepsilon^{-2}\frac{2\varphi_1\varphi_2 + \delta\varphi_1\varphi_3 + \delta\varphi_2^2 + \delta^2\varphi_2\varphi_3 - \varphi_3 - \varphi_1^3 - \delta\varphi_1^2\varphi_2 - \delta^2\varphi_1^2\varphi_3}{1 + \delta\varphi_1 + \delta^2\varphi_2 + \delta^3\varphi_3}$$

$$\times[\varepsilon^{-1}Q\varphi_1 + \delta\varepsilon^{-1}Q\varphi_2 + \delta^2\varepsilon^{-1}Q\varphi_3 + Q_1\varphi_1 + \delta Q_1\varphi_2] - \delta^4\varepsilon^{-3}\varphi_1 Q\varphi_3$$

$$-\delta^3\varepsilon^{-2}\varphi_1 Q_1\varphi_2 - \delta^4\varepsilon^{-3}\varphi_2 Q\varphi_2 - \delta^5\varepsilon^{-3}\varphi_2 Q\varphi_3 - \delta^3\varepsilon^{-2}\varphi_2 Q_1\varphi_1 - \delta^4\varepsilon^{-2}\varphi_2 Q_1\varphi_2$$

$$+\delta^4\varepsilon^{-3}\varphi_1^2 Q\varphi_2 + \delta^5\varepsilon^{-3}\varphi_1^2 Q\varphi_3 + \delta^3\varepsilon^{-2}\varphi_1^2 Q_1\varphi_1 + \delta^2\varepsilon^{-2}\varphi_1^2 Q_1\varphi_2.$$

Given the limit condition $\varepsilon^{-1}\delta \to 1, \varepsilon, \delta \to 0$, we finally obtain [5.16].

Lemma 5.5 is proved.

LEMMA 5.6.– *Under the conditions **C1'–C3**, the exponential generator*

$$H_\Gamma^{\varepsilon,\delta}(x)\varphi_\varepsilon^\delta(u,x) = e^{-\varphi_\varepsilon^\delta/\varepsilon}\varepsilon\Gamma_\varepsilon^\delta(x)e^{\varphi_\varepsilon^\delta/\varepsilon} \quad [5.17]$$

has the following asymptotic representation:

$$H_\Gamma^{\varepsilon,\delta}(x)\varphi_\varepsilon^\delta = H_\Gamma(x)\varphi(u) + \varepsilon^{-1}b_1(x)\varphi'(u) + \theta_\Gamma^{\varepsilon,\delta}\varphi,$$

where

$$H_\Gamma(x)\varphi(u) = (b(x) - b_0(x))\varphi'(u) + \frac{1}{2}(c(x) - c_0(x))(\varphi'(u))^2$$
$$+ \int_{\mathbf{R}}[e^{v\varphi'(u)} - 1]\Gamma^0(dv;x), \quad [5.18]$$

and $\sup_{x\in E}|\theta_\Gamma^{\varepsilon,\delta}\varphi| \to 0, \varepsilon, \delta \to 0.$

PROOF.– We have:

$$H_\Gamma^{\varepsilon,\delta}(x)\varphi_\varepsilon^\delta = e^{-\varphi/\varepsilon}[1 + \delta\varphi_1 + \delta^2\varphi_2 + \delta^3\varphi_3]^{-1}\varepsilon\Gamma_\varepsilon^\delta(x)e^{\varphi/\varepsilon}[1+\delta\varphi_1 + \delta^2\varphi_2+\delta^3\varphi_3]$$

$$= e^{-\varphi/\varepsilon}\left[1 - \delta\varphi_1 + \delta^2\frac{\varphi_1^2 + \delta\varphi_1\varphi_2 + \delta^2\varphi_1\varphi_3 - \varphi_2 - \delta\varphi_3}{1 + \delta\varphi_1 + \delta^2\varphi_2 + \delta^3\varphi_3}\right][\varepsilon\Gamma_\varepsilon^\delta(x)e^{\varphi/\varepsilon}$$

$$+\varepsilon\delta\Gamma_\varepsilon^\delta(x)e^{\varphi/\varepsilon}\varphi_1 + \varepsilon\delta^2\Gamma_\varepsilon^\delta(x)e^{\varphi/\varepsilon}\varphi_2 + \varepsilon\delta^3\Gamma_\varepsilon^\delta(x)e^{\varphi/\varepsilon}\varphi_3].$$

So,

$$H_\Gamma^{\varepsilon,\delta}(x)\varphi_\varepsilon^\delta = H_\Gamma^{\varepsilon,\delta}(x)\varphi(u) + e^{-\varphi/\varepsilon}\varepsilon\delta\{\Gamma_\varepsilon^\delta(x)e^{\varphi/\varepsilon}\varphi_1 - \varphi_1\Gamma_\varepsilon^\delta(x)e^{\varphi/\varepsilon}\} + \widetilde{\theta}_\Gamma^{\varepsilon,\delta}\varphi, \quad [5.19]$$

where

$$\widetilde{\theta}_\Gamma^{\varepsilon,\delta}\varphi = \varepsilon\delta^3\Gamma_\varepsilon^\delta(x)e^{\varphi/\varepsilon}\varphi_3 + \varepsilon\delta^2[e^{-\varphi/\varepsilon}\Gamma_\varepsilon^\delta(x)e^{\varphi/\varepsilon}\varphi_2 - e^{-\varphi/\varepsilon}\varphi_1\Gamma_\varepsilon^\delta(x)e^{\varphi/\varepsilon}\varphi_1]$$

$$+\varepsilon\delta^2\frac{\varphi_1^2 + \delta\varphi_1\varphi_2 + \delta^2\varphi_1\varphi_3 - \varphi_2 - \delta\varphi_3}{1 + \delta\varphi_1 + \delta^2\varphi_2 + \delta^3\varphi_3}[e^{-\varphi/\varepsilon}\Gamma_\varepsilon^\delta(x)e^{\varphi/\varepsilon}$$

$$+e^{-\varphi/\varepsilon}\delta\Gamma_\varepsilon^\delta(x)e^{\varphi/\varepsilon}\varphi_1 + e^{-\varphi/\varepsilon}\delta^2\Gamma_\varepsilon^\delta(x)e^{\varphi/\varepsilon}\varphi_2 + \delta^3\Gamma_\varepsilon^\delta(x)e^{\varphi/\varepsilon}\varphi_3]$$

$$-\varepsilon\delta^3 e^{-\varphi/\varepsilon}\varphi_1\Gamma_\varepsilon^\delta(x)e^{\varphi/\varepsilon}\varphi_2 - \varepsilon\delta^4\varphi_1\Gamma_\varepsilon^\delta(x)e^{\varphi/\varepsilon}\varphi_3.$$

Rewrite $H_\Gamma^{\varepsilon,\delta}(x)\varphi(u)$, taking into account the form of the generator [5.14]. We get:

$$H_\Gamma^{\varepsilon,\delta}(x)\varphi(u) = \varepsilon^{-2}\int_{\mathbf{R}}[e^{\Delta_\varepsilon\varphi(u)} - 1]\Gamma^\delta(dv;x),$$

where

$$\Delta_\varepsilon\varphi(u) := \varepsilon^{-1}[\varphi(u+\varepsilon v) - \varphi(u)].$$

Now we can rewrite as follows:

$$H_\Gamma^{\varepsilon,\delta}(x)\varphi(u) = \varepsilon^{-2}\int_{\mathbf{R}}[e^{\Delta_\varepsilon\varphi(u)} - 1 - \Delta_\varepsilon\varphi(u) - \frac{1}{2}(\Delta_\varepsilon\varphi(u))^2]\Gamma^\delta(dv;x)$$

$$+\varepsilon^{-2}\int_{\mathbf{R}}[\Delta_\varepsilon\varphi(u) + \frac{1}{2}(\Delta_\varepsilon\varphi(u))^2]\Gamma^\delta(dv;x).$$

It is easy to see that the function $\psi_u^\varepsilon(v) = e^{\Delta_\varepsilon\varphi(u)} - 1 - \Delta_\varepsilon\varphi(u) - \frac{1}{2}(\Delta_\varepsilon\varphi(u))^2$ belongs to the class $C_3(\mathbf{R})$. In fact,

$$\psi_u^\varepsilon(v)/v^2 \to 0, v \to 0.$$

In addition, this function is continuous and bounded for each ε, provided that $\varphi(u)$ is bounded. Moreover, the function $\psi_u^\varepsilon(v)$ is uniformly bounded by u under conditions **C2** and **C3** and if $\varphi'(u)$ is bounded.

So, under condition **C1**, we have:

$$H_\Gamma^{\varepsilon,\delta}(x)\varphi(u) = \varepsilon^{-2}\delta^2\int_{\mathbf{R}}[e^{\Delta_\varepsilon\varphi(u)} - 1 - \Delta_\varepsilon\varphi(u) - \frac{1}{2}(\Delta_\varepsilon\varphi(u))^2]\Gamma^0(dv;x)$$

$$+\varepsilon^{-2}\int_{\mathbf{R}}[\Delta_\varepsilon\varphi(u) - v\varphi'(u) - \varepsilon\frac{v^2}{2}\varphi''(u)]\Gamma^\delta(dv;x) + \varepsilon^{-2}\delta b_1(x)\varphi'(u)$$

$$+\varepsilon^{-2}\delta^2 b(x)\varphi'(u) + \varepsilon^{-1}\delta^2 c(x)\varphi''(u) + \varepsilon^{-2}\int_{\mathbf{R}}[\frac{1}{2}(\Delta_\varepsilon\varphi(u))^2 - \frac{v^2}{2}(\varphi'(u))^2]\Gamma^\delta(dv;x)$$

$$+\varepsilon^{-2}\delta^2\frac{1}{2}c(x)(\varphi'(u))^2.$$

The functions in the second and third integrals obviously belong to $C_3(\mathbf{R})$. Using Taylor's formula for test functions $\varphi(u) \in C^3(\mathbf{R})$ and condition **LA2**, we obtain:

$$H_\Gamma^{\varepsilon,\delta}(x)\varphi(u) = \varepsilon^{-2}\delta^2 \int_{\mathbf{R}} [e^{v\varphi'(u)} - 1 - v\varphi'(u) - \frac{v^2}{2}(\varphi'(u))^2]\Gamma^0(dv;x)$$

$$+\varepsilon^{-2}\delta^2 \int_{\mathbf{R}} (e^{v\varphi'(u)}\varepsilon\frac{v^2}{2}\varphi''(\widetilde{u}) - \varepsilon\frac{v^2}{2}\varphi''(\widetilde{u}) - \varepsilon^2\frac{v^4}{8}(\varphi''(\widetilde{u}))^2)\Gamma^0(dv;x)$$

$$+\varepsilon^{-2}\delta^2 \int_{\mathbf{R}} \varepsilon^2\frac{v^3}{3!}\varphi'''(\widetilde{u})\Gamma^0(dv;x) + \varepsilon^{-2}\delta b_1(x)\varphi'(u) + \varepsilon^{-2}\delta^2 b(x)\varphi'(u)$$

$$+\varepsilon^{-1}\delta^2 c(x)\varphi''(u) + \varepsilon^{-2}\delta^2\int_{\mathbf{R}} \varepsilon^2\frac{v^4}{4}(\varphi''(\widetilde{u}))^2\Gamma^0(dv;x) + \varepsilon^{-2}\delta^2\frac{1}{2}c(x)(\varphi'(u))^2.$$

According to the limit condition $\varepsilon^{-1}\delta \to 1$, we finally have:

$$H_\Gamma^{\varepsilon,\delta}(x)\varphi(u) = H_\Gamma(x)\varphi(u) + \varepsilon^{-1}b_1(x)\varphi'(u) + \theta^{\varepsilon,\delta}\varphi, \qquad [5.20]$$

where

$$H_\Gamma(x)\varphi(u) = (b(x) - b_0(x))\varphi'(u) + \frac{1}{2}(c(x) - c_0(x))(\varphi'(u))^2$$

$$+ \int_{\mathbf{R}} [e^{v\varphi'(u)} - 1]\Gamma^0(dv;x),$$

and $\sup_{x\in E} |\theta^{\varepsilon,\delta}\varphi| \to 0, \varepsilon, \delta \to 0.$

To complete the proof, it is necessary to show that the term in the parentheses in [5.19] is equal to 0. We use the following lemma:

LEMMA 5.7.–

$$\Gamma_\varepsilon^\delta(x)e^{\varphi(u)/\varepsilon}\varphi_1(u,x) = \varphi_1(u,x)\Gamma_\varepsilon^\delta(x)e^{\varphi(u)/\varepsilon} + (\varepsilon\delta)^{-1}\widehat{\theta}_\Gamma^{\varepsilon,\delta}\varphi,$$

where the negligible term

$$\sup_{x\in E} |\widehat{\theta}_\Gamma^{\varepsilon,\delta}\varphi| \to 0, \varepsilon, \delta \to 0.$$

PROOF.– Indeed, from [5.14], we have:

$$\Gamma_\varepsilon^\delta(x)e^{\varphi(u)/\varepsilon}\varphi_1(u,x) = \varepsilon^{-3}\int_{\mathbf{R}} [e^{\varphi(u+\varepsilon v)/\varepsilon}\varphi_1(u+\varepsilon v,x) - e^{\varphi(u)/\varepsilon}\varphi_1(u,x)]\Gamma^\delta(dv;x)$$

$$= \varphi_1(u,x)\Gamma_\varepsilon^\delta(x)e^{\varphi(u)/\varepsilon} + (\varepsilon\delta)^{-1}\left[\varphi_1'(u,x)\varepsilon^{-1}\delta\int_{\mathbf{R}} e^{\varphi(u+\varepsilon v)/\varepsilon} v\Gamma^\delta(dv;x)\right].$$

Let us estimate the last integral. Because the function $\varphi(u)$ is bounded, we have, for a fixed ε:

$$\int_{\mathbf{R}} e^{\varphi(u+\varepsilon v)/\varepsilon} v\Gamma^\delta(dv;x) < e^C \int_{\mathbf{R}} v\Gamma^\delta(dv;x) = \delta e^C[b_1(x) + \delta b(x) + \delta\theta_b^\delta\varphi].$$

Thus, we see that the last term is negligible when $\varepsilon, \delta \to 0$.

Lemma 5.7 is proved.

Applying equation [5.20] and Lemma 5.7 to formula [5.19], we finally have

$$H_\Gamma^{\varepsilon,\delta}(x)\varphi_\varepsilon^\delta = H_\Gamma(x)\varphi(u) + \varepsilon^{-1}b_1(x)\varphi'(u) + \theta_\Gamma^{\varepsilon,\delta}\varphi,$$

where $\sup_{x\in E} |\theta_\Gamma^{\varepsilon,\delta}\varphi| \to 0, \varepsilon, \delta \to 0$.

Lemma 5.6 is proved.

We can now complete the proof of the theorem.

From [5.15] and [5.17], we see that

$$\mathbf{H}_\varepsilon^\delta\varphi_\varepsilon^\delta = H_Q^\varepsilon\varphi_\varepsilon^\delta(u,x) + H_\Gamma^{\varepsilon,\delta}(x)\varphi_\varepsilon^\delta(u,x).$$

Using Lemmas 5.5 and 5.6, we have an asymptotic representation

$$\begin{aligned}\mathbf{H}_\varepsilon^\delta\varphi_\varepsilon^\delta &= \varepsilon^{-2}Q\varphi_1 + \varepsilon^{-1}[Q\varphi_2 + Q_1\varphi_1 - \varphi_1 Q\varphi_1 + b_1(x)\varphi'(u)]\\ &+ [Q\varphi_3 + Q_1\varphi_2 - \varphi_1 Q\varphi_2 - \varphi_1 Q_1\varphi_1 - \varphi_2 Q\varphi_1\\ &+ \varphi_1^2 Q\varphi_1 + H_\Gamma(x)\varphi(u)] + h^{\varepsilon,\delta}\varphi,\end{aligned}$$

where $h^{\varepsilon,\delta}\varphi = \theta_Q^{\varepsilon,\delta}\varphi + \theta_\Gamma^{\varepsilon,\delta}\varphi$.

To complete the proof, the following singular perturbation problem should be solved:

$$Q\varphi_1 = 0,$$

$$Q\varphi_2 + Q_1\varphi_1 + b_1(x)\varphi'(u) = 0,$$

$$Q\varphi_3 + Q_1\varphi_2 - \varphi_1 Q\varphi_2 - \varphi_1 Q_1\varphi_1 + H_\Gamma(x)\varphi(u) = \widehat{\widehat{H}}\varphi(u).$$

From the first equation, it follows that $\varphi_1(u,x) = \varphi_1(u,\widehat{x}) \in \mathcal{N}_Q$, so, from the solvability condition for the second equation,

$$\widehat{Q\varphi_2} + \widehat{Q}_1\widehat{\varphi}_1 + \widehat{b}_1(\widehat{x})\varphi'(u) = 0,\ \widehat{Q\varphi_2} = 0,$$

and therefore we have new correlations

$$\widehat{Q}_1\widehat{\varphi}_1 + \widehat{b}_1(\widehat{x})\varphi'(u) = 0,\ \widehat{\Pi}\widehat{b}_1(\widehat{x}) \equiv 0,$$

which give us

$$\widehat{\varphi}_1(u,\widehat{x}) = \widehat{R}_0\widehat{b}_1(\widehat{x})\varphi'(u), \widehat{Q}_1\widehat{\varphi}_1 = -\widehat{b}_1(\widehat{x})\varphi'(u). \quad [5.21]$$

The solvability condition for the third equation

$$Q\varphi_3 + Q_1\varphi_2 - \varphi_1 Q\varphi_2 - \varphi_1 Q_1\varphi_1 + H_\Gamma(x)\varphi(u) = \widehat{\widehat{H}}\varphi(u)$$

now gives

$$\widehat{Q}_1\widehat{\varphi}_2 - \widehat{\varphi}_1\widehat{Q\varphi_2} - \widehat{\varphi}_1\widehat{Q}_1\widehat{\varphi}_1 + \widehat{H}_\Gamma(x)\varphi(u) = \widehat{\widehat{H}}\varphi(u),$$

and from [5.21]

$$\widehat{Q\varphi_2} = -[\widehat{Q_1\varphi_1} + \widehat{b}_1(\widehat{x})\varphi'(u)] = 0,$$

and using formula [5.18] and the second equality with [5.21], we have

$$\widehat{Q}_1\widehat{\varphi}_2 + \widehat{H}^0(x)\varphi(u) = \widehat{\widehat{H}}\varphi(u),$$

where

$$\widehat{H}^0(x)\varphi(u) = \Pi H_\Gamma(x)\Pi\varphi(u) + \widehat{b}_1(\widehat{x})\widehat{R}_0\widehat{b}_1(\widehat{x})(\varphi'(u))^2.$$

The application of the solvability condition to this equation completes the proof of the theorem.

Theorem 5.4 is proved.

6

Difference Diffusion Models with Equilibrium

6.1. Statistical experiments with linear persistent regression

In physical processes with equilibrium, the dynamics of concentration, or frequencies, of a predefined characteristic, can be described by a mathematical model of binary statistical experiments, based on the statistical data of the elementary hypothesis validation, about the presence or absence of a predefined attribute A in the set of elements that make up a complex system.

It is assumed that:

– all of the elements that make up the system can gain or lose the attribute A over time, i.e. the frequency of attribute A is a dynamic variable;

– the basic objects of our study are the statistical experiments, characterized by the relative frequencies of the presence or absence of the attribute A, in a sample of fixed volume at each time instant;

– it is assumed that the (average) results of the next experiment (at time instant $k+1$) depend on average result of the present experiment at time instant k, and do not depend on all previous time instants $k-1, k-2$, etc. This relationship is called the feature of persistent regression and is used as a fundamental condition for the subsequent analysis of the model.

The method of constructing and exploring the proposed mathematical model is based on the analysis of the following basic properties of statistical experiments:

– persistent regression;

– equilibrium value and fluctuations, as well as their asymptotic behavior;

– approximation by a normal process of autoregression.

Such a mathematical model can be formalized as binary statistical experiments, generated by a sample of random variables that only assume two values. The values of these random variables correspond to the presence or absence of a certain property, respectively. The main characteristic used throughout this paper to analyze the dynamics of statistical experiments is the relative frequencies of the presence and absence of this property.

The dynamics of the statistical experiments are defined in terms of a persistent regression, which is based on an assumption that the relative frequencies for the next experiment (at instant $k+1$) depend (in the mean sense) on the mean values of the frequencies for the current experiment, at instant k. The regression function of the increments of relative frequencies is expressed in terms of a linear action, which directs the frequencies toward an equilibrium state. The representation of binary statistical experiments in terms of relative frequencies provides the possibility of determining the equilibrium state.

6.1.1. *Persistent regression*

Statistical experiments are generated by sample $(\delta_r(k),\ 1 \le r \le N)$ variables that, for every fixed $k \ge 0$, are jointly independent and identically distributed and only assume two values, say ± 1.

The binary statistical experiments are defined by the normalized sums

$$S_N(k) := \frac{1}{N}\sum_{r=1}^{N}\delta_r(k),\ k \ge 0. \qquad [6.1]$$

Denote the conditional probabilities

$$P\{\delta_r(k+1) = \pm 1 \mid S_N(k) = s\} = P_\pm(s),\ |s| \le 1,\ k \ge 0. \qquad [6.2]$$

Then

$$P_+(s) + P_-(s) = 1,\ \forall s \in [-1, +1].$$

Therefore, we can only restrict the consideration to one of the frequencies, either $S_N^+(k) := \frac{1}{N}\sum_{r=1}^{N} I\{\delta_r(k) = +1\}$ or $S_N^-(k) := \frac{1}{N}\sum_{r=1}^{N} I\{\delta_r(k) = -1\}$, since a result for $S_N^+(k)$ can easily be transformed into a result for $S_N^-(k)$ and vice versa.

For statistical experiments, we have the representation

$$S_N(k) = S_N^+(k) - S_N^-(k). \qquad [6.3]$$

We study two-component experiments:

$$\overline{S_N}(k) := (\, S_N^+(k)\,,\, S_N^-(k)\,),\ k \geq 0, \tag{6.4}$$

Let

$$\bar{p} := (p_+, p_-),\ 0 \leq p_\pm \leq 1,\ p_+ + p_- = 1,$$

be the vector of probabilities of relative frequencies S_N^+ and S_N^- for the current experiment.

Let

$$\Delta S_N^\pm(k) := S_N^\pm(k+1) - S_N^\pm(k),\ k \geq 0.$$

be the increments of the statistical experiments.

The following is the main assumption.

Assumption. The increments of the statistical experiments possess the property of the persistent regression if the regression function is expressed as a linear directing action:

$$E[\Delta S_N^\pm(k) \,|\, \overline{S_N}(k) = \bar{p}] =: C_0^\pm(\bar{p}) = \mp[V_+ p_+ - V_- p_-], \tag{6.5}$$

where $0 < V_\pm < 1$.

The regression function of a two-component statistical experiment

$$C^\pm(\overline{p}) := E[S_N^\pm(k+1) \,|\, \overline{S_N}(k) = \overline{p}],\ k \geq 0, \tag{6.6}$$

is determined by the relation

$$C^\pm(\overline{p}) = p_\pm + C_0^\pm(\overline{p}). \tag{6.7}$$

According to [6.5], the balance condition holds:

$$C_0^+(\overline{p}) + C_0^-(\overline{p}) = 0. \tag{6.8}$$

It is also clear that

$$C^+(\overline{p}) + C^-(\overline{p}) = 1. \tag{6.9}$$

In particular, the persistent linear regression for the frequencies of the statistical experiments is given by

$$E[\, S_N^+(k+1) \,|\, \overline{S_N}(k) = \bar{p}\,] = p_+ - V_+ p_+ + V_- p_-, \tag{6.10}$$

$$E[\, S_N^-(k+1) \,|\, \overline{S_N}(k) = \bar{p}\,] = p_- - V_- p_- + V_+ p_+. \tag{6.11}$$

REMARK 6.1.– *Equalities* [6.10] *make the role of the directing parameters* V_+ *and* V_- *for determining the equilibrium state clear. It can be seen from* [6.10] *that the frequency of any property* $(\pm)$ *decreases proportionally to its frequencies with the directing coefficient* $V_\pm$, *and increases proportionally to the frequencies of the opposite sign with the directing coefficient* $V_\mp$.

The regression function

$$C(s) := E[S_N(k+1) \mid S_N(k) = s]$$

is expressed in terms of the regression function of the increments of the statistical experiments and possesses the property of the persistent linear regression as well, namely

$$C(s) = s + C_0(s), \tag{6.12}$$

$$C_0(s) := C_0^+(\bar{p}) - C_0^-(\bar{p}) = -V_+(1+s) + V_-(1-s),$$

where $\bar{p} = (p_+, p_-)$, $p_\pm := \frac{1}{2}(1 \pm s)$, $|s| \le 1$.

LEMMA 6.1.– *The stochastic experiments* [6.1], *defined by the regression function* [6.5]–[6.7], *are characterized by the conditional variance given by*

$$Var[\, S_N^\pm(k+1) \mid \overline{S_N}(k) = \overline{p}] = \frac{1}{N} C^\pm(\overline{p})[1 - C^\pm(\overline{p})], \; k \ge 0. \tag{6.13}$$

PROOF.– Using [6.6], we get

$$Var[\, S_N^\pm(k+1) \mid \overline{S_N}(k) = \overline{p}]$$
$$= E[\,(S_N^\pm(k+1))^2 \mid \overline{S_N}(k) = \overline{p}] - [C^\pm(\overline{p})]^2.$$

Denote

$$a_r^\pm := I\{\delta_r(k+1) = \pm 1\}.$$

We apply equality [6.2] to evaluate

$$(\, S_N^\pm(k+1)\,)^2 = \frac{1}{N^2}\left[\sum_{r=1}^{N}[a_r^\pm]^2 + \sum_{r,r'=1}^{N,\ r\neq r'} [a_r^\pm][a_{r'}^\pm]\right].$$

Taking into account $[a_r^\pm]^2 = a_r^\pm$, and that the latter sum contains $N(N-1)$ terms, we obtain

$$E[\,(S_N^\pm(k+1))^2 \mid \overline{S_N}(k) = \overline{p}] = \frac{1}{N}\left[C^\pm(\overline{p}) - (C^\pm)^2(\overline{p})\right] + (C^\pm)^2(\overline{p}),$$

whence [6.13] follows. Lemma [6.1] is proved. □

6.1.2. *Equilibrium state*

The equilibrium state for the probability frequencies ρ_+, ρ_- and ρ is characterized by the invariant values of the corresponding regression functions

$$C^+(\rho_+) = \rho_+ \,, \quad C^-(\rho_-) = \rho_- \,, \quad C(\rho) = \rho.$$

For the regression functions [6.5] and [6.12], this means that the directing actions vanish at points of the equilibrium state. In other words,

$$C_0^{\pm}(\bar{\rho}) = 0, \text{ that is } \quad V_+\rho_+ = V_-\rho_-, \qquad [6.14]$$

where $\bar{\rho} := (\rho_+, \rho_-)$, $\rho_+ + \rho_- = 1$. According to assumption [6.5], the mean value of the statistical experiment at the point of the equilibrium state coincides with the current value of the statistical experiment, i.e.

$$E[S_N^{\pm}(k+1) \,|\, \overline{S_N}(k) = \overline{\rho}] = \rho_\pm, \text{ for all } \; k \geq 0,$$

and

$$E[S_N(k+1) \,|\, \overline{S_N}(k) = \overline{\rho}] = \rho := \rho_+ - \rho_-, \text{ for all } \; k \geq 0.$$

The equilibrium state $\bar{\rho}$ can be evaluated by solving equation [6.14], i.e.

$$\begin{cases} \rho_+ V_+ = \rho_- V_- = C = \text{const} \\ \rho_+ + \rho_- = 1 \end{cases} \qquad [6.15]$$

In other words,

$$\begin{cases} \rho_\pm = C/V_\pm \\ C = [1/V_+ + 1/V_-]^{-1}. \end{cases} \qquad [6.16]$$

In particular,

$$\rho_\pm = V_\mp / V, \; V = V_+ + V_-. \qquad [6.17]$$

REMARK 6.2.– *The existence of the equilibrium state $\bar{\rho}$ implies the following properties of the directing action:*

$$C_0^{\pm}(\bar{p}) = -V(p_+ - \rho_+) = -V(p_- - \rho_-), \qquad [6.18]$$

$$C_0(s) = C_0^+(\bar{p}) - C_0^-(\bar{p}) = -V(s - \rho), \qquad [6.19]$$

where

$$s = p_+ - p_-, \; \rho := \rho_+ - \rho_-.$$

REMARK 6.3.– *As seen from* [6.18]–[6.19], *the directing action is a fluctuation function, in the sense that it depends on the deviation of the current value of the statistical experiment from the equilibrium state. Note that the directing action for every frequency only depends on the fluctuations of the corresponding frequency.*

Consider the fluctuation processes of the frequencies of the statistical experiments defined by

$$\widehat{S}_N^{\pm}(k) := S_N^{\pm}(k) - \rho_{\pm},\ k \geq 0, \qquad [6.20]$$

$$\widehat{p}_{\pm} := p_{\pm} - \rho_{\pm}$$

and their regression functions

$$\widehat{C}^{\pm}(\widehat{p}_{\pm}) = E[\,\widehat{S}_N^{\pm}(k+1)\,|\,\widehat{S}_N^{\pm}(k) = \widehat{p}_{\pm}\,] = (1-V)\widehat{p}_{\pm},\ k \geq 0. \qquad [6.21]$$

The regression function of the fluctuation of statistical experiment [6.1]

$$\widehat{S}_N(k) := S_N(k) - \rho,\ \ \rho := \rho_+ - \rho_-,\ k \geq 0, \qquad [6.22]$$

is given by

$$\widehat{C}(\widehat{s}) := E[\widehat{S}_N(k+1)\,|\,\widehat{S}_N(k) = \widehat{s}] = (1-V)\widehat{s},\ k \geq 0, \qquad [6.23]$$

$$\widehat{s} := s - \rho.$$

REMARK 6.4.– *There are two possible ways to analyze statistical experiments* [6.1]–[6.2]*:*

– *explicitly, with the help of* [6.18]–[6.19]*;*

– *implicitly, by analyzing the two-component statistical experiments* [6.2], *generated by the frequencies* $S_N^{\pm}(k)$, $k \geq 0$ *and then by using equality* [6.3] *to obtain a result for statistical experiments* [6.1].

REMARK 6.5.– *In the paper, Korolyouk (2015a), the binary statistical experiments* $S_N(k)$, $k \geq 0$ *are studied in the framework of regression models with the linear regression function* $C(s) = (1-a)s + a\rho$ *or, which is the same, with* $\widehat{C}(\widehat{s}) = (1-a)\widehat{s}$.

In the problem considered above, $a = V$, *and the values* $\rho = \widehat{\rho} := \rho_+ - \rho_-$ *of the equilibrium state are obvious.*

6.1.3. Convergence of statistical experiments to the equilibrium state

The directing actions [6.10] and [6.12] yield the existence of the equilibrium states for the statistical experiments [6.1].

THEOREM 6.1.– *Assume that the initial state converges with probability 1,*

$$S_N^{\pm}(0) \xrightarrow{P1} \rho_{\pm}, \; N \to \infty. \tag{6.24}$$

Then

$$S_N^{\pm}(k) \xrightarrow{P1} \rho_{\pm}, \; N \to \infty, \; \forall k > 0. \tag{6.25}$$

$$S_N(k) \xrightarrow{P1} \rho, \; N \to \infty, \; \forall k > 0. \tag{6.26}$$

PROOF OF THEOREM 6.1.– Clearly, it is sufficient to prove the almost sure convergence [6.25] for positive frequencies only,

$$S_N^{+}(k) \xrightarrow{P1} \rho_{+}, \; N \to \infty, \; \forall k > 0. \tag{6.27}$$

Then, [6.3] implies [6.25], as well as [6.26], for negative frequencies in view of representation [6.3].

Consider the martingale

$$\widehat{M}_N^{+}(n) := \sum_{k=0}^{n-1} \widehat{\mu}_N^{+}(k+1) \tag{6.28}$$

where

$$\widehat{\mu}_N^{+}(k+1) := \widehat{S}_N^{+}(k+1) - (1-V)\widehat{S}_N^{+}(k) \tag{6.29}$$

is a martingale difference. It is clear that

$$\widehat{\mu}_N^{+}(k+1) = \mu_N^{+}(k+1),$$

where

$$\mu_N^{+}(k+1) := S_N^{+}(k+1) - C^{+}(S_N^{+}(k)). \tag{6.30}$$

Thus, the quadratic characteristics of the martingale differences [6.30] converge and, in view of [6.13], are given by

$$\langle \widehat{\mu}_N^{+} \rangle_k = \langle \mu_N^{+} \rangle_k = C^{+}(S_N^{+}(k))(1 - C^{+}(S_N^{+}(k)))/N. \tag{6.31}$$

Thus, for every fixed $n \geq 0$

$$\langle M_N^{+} \rangle_n \to 0, \; N \to \infty. \tag{6.32}$$

This implies (see Korolyouk 2020) the almost sure convergence of martingales [6.28]

$$M_N^{+}(n) \xrightarrow{P1} 0, \; \text{as } N \to \infty, \; n \geq 0. \tag{6.33}$$

In particular, if $n = 1$, then

$$M_N^+(0) = \widehat{S}_N^+(1) - (1 - V)\widehat{S}_N^+(0).$$

Since $M_N^+(0)$ and $\widehat{S}_N^+(0)$ converge to zero as $N \to \infty$, we obtain

$$\widehat{S}_N^+(1) \xrightarrow{P1} 0, \quad \text{as} \quad N \to \infty.$$

Using the induction, we prove the convergence

$$\widehat{S}_N^+(k) \xrightarrow{P1} 0, \quad \text{as} \quad N \to \infty.$$

for every fixed $k \geq 1$. Theorem 6.1 is proved. □

6.1.4. *Approximation of statistical experiments by a normal process of autoregression*

The existence of an equilibrium state for the statistical experiments [6.1] and [6.2], stated in Theorem 6.1, is an important but simple property of statistical experiments. The main problem of the description of statistical experiments [6.1] and [6.2], as the size N of a sample increases, is to analyze the asymptotic behavior of the stochastic part corresponding to the martingale difference [6.29].

The asymptotic behavior of the martingale difference [6.29] is given below.

THEOREM 6.2.– *Let all of the assumptions of Theorem 6.2 hold. Then*

$$\sqrt{N}[S_N^{\pm}(k+1) - C^{\pm}(S_N^{\pm}(k))] \xrightarrow{d} \breve{\sigma} W^{\pm}(k+1), \; N \to \infty, \qquad [6.34]$$

$$\sqrt{N}[S_N(k+1) - C(S_N(k))] \xrightarrow{d} \sigma W(k+1), \; N \to \infty, \qquad [6.35]$$

where

$$\sigma^2 = 1 - \rho^2, \; \breve{\sigma}^2 = \rho_+\rho_-, \quad \sigma^2 = 4\breve{\sigma}^2.$$

Here, $W(k)^{\pm}$ and $W(k)$ are the standard normal random variables.

The limit relations [6.34] and [6.35] allow us to use the stochastic approximation for the three statistical experiments in the following form.

PROPOSITION 6.1.– *The normal process of autoregression that approximates the sequence of statistical experiments* [6.1], [6.2], [6.5]–[6.7] *(see also* [6.10]*) is given by*

$$\widetilde{S}_N^{\pm}(k+1) = C^{\pm}(\widetilde{S}_N^{\pm}(k)) + \frac{\breve{\sigma}}{\sqrt{N}} W^{\pm}(k+1) \qquad [6.36]$$

$$\widetilde{S}_N(k+1) = C(\widetilde{S}_N(k)) + \frac{\sigma}{\sqrt{N}} W(k+1) \qquad [6.37]$$

The regression functions $C^{\pm}(\overline{p})$ *and* $C(s)$ *are defined by* [6.5] *and* [6.12], *correspondingly.*

PROOF OF THEOREM 6.2.– As in the proof in Theorem 6.1, we can restrict the consideration to positive frequencies only:

$$\sqrt{N}[S_N^+(k+1) - C^{\pm}(S_N^+(k))] \xrightarrow{d} \breve{\sigma} W^+(k+1) \text{ as } N \to \infty. \quad [6.38]$$

This implies the other relations [6.34]–[6.35] in Theorem 6.2.

According to denotation [6.20], consider the normalized fluctuations of positive frequencies

$$\zeta_N^+(k) := \sqrt{N}\widehat{S}_N^+(k)$$

In view of equalities [6.21], the conditional expectations are given by

$$E[\zeta_N^+(k+1)|\zeta_N^+(k)] = (1-V)\zeta_N^+(k). \quad [6.39]$$

Relation [6.13] implies that

$$D[\zeta_N^+(k+1)|\zeta_N^+(k)] = C^+(S_N^+(k))[1 - C^+(S_N^+(k))] \quad [6.40]$$

Consider the martingale

$$M_N^+(n) := \sum_{k=0}^{n-1} \{\zeta_N^+(k+1) - E[\zeta_N^+(k+1)|\zeta_N^+(k)]\} \quad [6.41]$$

$$= \sum_{k=0}^{n-1} [\zeta_N^+(k+1) - (1-V)\zeta_N^+(k)]. \quad [6.42]$$

The quadratic characteristic of martingale [6.41] (see also [6.40]) can be written as follows:

$$\langle M_N^+ \rangle_n = \sum_{k=0}^{n-1} C^+(S_N^+(k))[1 - C^+(S_N^+(k))], \; n \geq 0 \quad [6.43]$$

By using Theorem 4.1, the quadratic characteristic converges with probability 1,

$$\langle M_N^+ \rangle_n \xrightarrow{P1} n\rho_+\rho_- = n\breve{\sigma} \text{ as } N \to \infty, \; n \geq 0. \quad [6.44]$$

The central limit theorem for martingale differences (see, for example, Korolyouk (2020)) implies that the martingale [6.41] converges in distribution as $N \to \infty$, to a sum of normal random variables

$$\sum_{k=0}^{n-1} [\zeta_N^+(k+1) - (1-V)\zeta_N^+(k)] \xrightarrow{d} \breve{\sigma} \sum_{k=0}^{n-1} W(k+1), n \geq 1. \quad [6.45]$$

The limit normal random variables $W(k), \quad k \geq 1$ are jointly independent, such that

$$EW(k) = 0, \quad EW^2(k) = 1, \quad k \geq 1$$

since, by [6.44], the limit variance of the martingale [6.41] equals the sum of variances of the martingale differences.

The convergence of martingale [6.41] implies the convergence in distribution [6.38] for every finite $k \geq 0$. Theorem 6.2 is proved. □

PROPOSITION 6.2.– *For the normalized fluctuations of statistical experiments, the corresponding normal process of autoregression is a solution of the following difference stochastic equation:*

$$\Delta\zeta_N(k) = -V\zeta_N(k) + \sigma\Delta W(k), \quad [6.46]$$

where

$$\Delta W(k) := W(k+1) - W(k), \ \sigma^2 = 1 - \rho^2.$$

REMARK 6.6.– *The normal process of autoregression can be used in practice when processing the results of a real statistical experiment. An approximation of the initial statistical experiments by the normal process of autoregression* [6.36], [6.37] *and* [6.46] *is determined by directing the parameters $V_\pm$, equilibrium values $\rho_\pm$, ρ, and variances $\breve{\sigma}$, σ of the normally distributed random variables $W^\pm$ and W.*

6.2. Exponential statistical experiments

Theorems 6.1 and 6.2 can be used for statistical experiments [6.1]–[6.2] to study the asymptotic behavior of exponential statistical experiments generated by a symmetric exponential statistic

$$\Pi_N(\lambda, k) := \prod_{r=1}^{N} [1 + \lambda\delta_r(k)], \ k \geq 0. \quad [6.47]$$

The parameter λ $(0 < \lambda < 1)$ plays an important role as a control parameter in various models. For example, it can be viewed as a discount parameter in some models of economy (see, for example, Shiryaev (1999)).

The sampling values $(\delta_r(k), 1 \leq r \leq N)$, $k \geq 0$ in representation [6.47] of an exponential statistical experiment are considered for the following three special cases:

$$\delta_r(k) = \pm 1, \quad \delta_r^\pm(k) := I(\delta_r(k) = \pm 1). \quad [6.48]$$

Thus, every exponential statistical experiment is generated by one of the above three statistical experiments given by [6.1] or [6.2].

The asymptotic behavior of the exponential statistical experiments as $N \to \infty$ can be studied with the help of Theorems 6.1 and 6.2. Thus, we can restrict the consideration to one of the exponential statistical experiments [6.47].

We consider the exponential statistical experiments [6.47], generated by statistical experiments [6.1]. Similar results can be proved for other exponential statistical experiments.

6.2.1. *Steady regime for exponential statistical experiments*

In the averaging scheme, the exponential statistical experiments [6.47] are considered under the normalization $\lambda_N = \lambda/N$.

Consider the following exponential martingale:

$$\mu_N^e(\lambda/N, k+1) = \Pi_N(\lambda/N, k+1)/\overline{\Pi_N}(\lambda/N, k), \ k \geq 0, \tag{6.49}$$

where

$$\overline{\Pi_N}(\lambda, k) := E[\Pi_N(\lambda, k+1)|S_N(k)] = [1 + \lambda C(S_N(k))]^N. \tag{6.50}$$

Its martingale property is obvious,

$$E[\mu_N^e(\lambda/N, k+1)|S_N(k)] = 1, \ k \geq 0. \tag{6.51}$$

The existence of a stable regime of an exponential statistical experiment is stated in the following result.

THEOREM 6.3.– *If the initial values of a statistical experiment converge with probability 1,*

$$S_N(0) \xrightarrow{P1} \rho, \ N \to \infty, \tag{6.52}$$

then the exponential statistical experiment [6.49] *converges in probability*

$$\Pi_N(\lambda/N, k) \xrightarrow{P} \exp(\lambda\rho), \ \ N \to \infty, \ k \geq 0, \tag{6.53}$$

and its mean value [6.50] *converges in probability*

$$\overline{\Pi_N}(\lambda/N, k) \xrightarrow{P} \exp(\lambda\rho), \ \ N \to \infty, \ k \geq 0. \tag{6.54}$$

COROLLARY 6.1.– *If condition* [6.52] *holds, then*

$$\mu_N^e(\lambda/N, k+1) \xrightarrow{P} 1, \quad N \to \infty. \tag{6.55}$$

PROOF OF THEOREM 6.3.– We use the approximation based on the Le Cam formula in the following form:

LEMMA 6.2.– *(see Borovskikh and Korolyuk (1997, Lemma 6.3.1)) Let the following convergence by probability hold:*

$$\max_{1 \le r \le N} |\delta_r(k+1)/N| \xrightarrow{P} 0, N \to \infty, \ \forall k \ge 0. \tag{6.56}$$

Then, there is a convergence in probability

$$\Big\{\sum_{r=1}^{N} \ln[1 + \lambda\delta_r(k+1)/N] - \lambda S_N(k+1)\Big\} \xrightarrow{P} 0, \ N \to \infty. \tag{6.57}$$

Here

$$S_N(k+1) := \frac{1}{N}\sum_{r=1}^{N} \delta_r(k+1).$$

Condition [6.56] holds in the case under consideration in view of Lemma 6.2, then by [6.1]

$$S_N(k) \xrightarrow{P1} \rho, \ N \to \infty. \tag{6.58}$$

Taking into account an obvious equality $\Pi = \exp\ln\Pi$, we rewrite [6.57] as follows:

$$\exp\Big\{\sum_{r=1}^{N} \ln[1 + \lambda\delta_r(k+1)/N] - \lambda S_N(k+1)\Big\} \xrightarrow{P} 1, \ N \to \infty.$$

Using [6.58], we obtain

$$P \cdot \lim_{N\to\infty} \Pi_N(\lambda/N, k+1)$$

$$= P \cdot \lim_{N\to\infty} \exp\sum_{r=1}^{N} \ln[1 + \lambda\delta_r(k+1)/N] = \exp(\lambda\rho). \tag{6.59}$$

The latter result coincides with [6.53]. Convergence [6.54] is proved similarly. Convergence [6.55] follows from [6.53] and [6.54]. The proof of Theorem 6.3 is complete. □

6.2.2. Approximation of an exponential statistical experiment by a normal process of autoregression

We consider an exponential statistical experiment in a series scheme with the parameter $\lambda_N = \lambda/\sqrt{N}$, i.e.

$$\Pi_N(\lambda/\sqrt{N}, k+1) = \prod_{r=1}^{N} [1 + \lambda\delta_r(k+1)/\sqrt{N}], \ k \geq 0. \qquad [6.60]$$

The mean value of the exponential statistical experiment [6.60] is given by

$$\overline{\Pi_N}(\lambda/\sqrt{N}, k) = [1 + \lambda C(S_N(k))/\sqrt{N}]^N, \ k \geq 0. \qquad [6.61]$$

Consider the following exponential martingale:

$$\mu_N^e(\lambda_N, k+1) := \Pi_N(\lambda/\sqrt{N}, k+1)/\overline{\Pi_N}(\lambda/\sqrt{N}, k), \ k \geq 0. \qquad [6.62]$$

Its martingale property is obvious:

$$E[\mu_N^e(\lambda_N, k+1)|S_N(k)] = 1, \ k \geq 0.$$

THEOREM 6.4.– *(Approximation of an exponential statistical experiment) Let all of the assumptions of Theorem 6.3 hold. Then*

$$\mu_N^e(\lambda/\sqrt{N}, k+1) \xrightarrow{d} \exp[\lambda\sigma W(k+1) - \lambda^2\sigma^2/2], \qquad [6.63]$$
$$N \to \infty, \ k \geq 0,$$

where $\sigma^2 = 1 - \rho^2$.

REMARK 6.7.– *The limit result* [6.63] *is an analogue of a property of exponential martingale [6.59].*

It is obvious that

$$E \exp[\lambda\sigma W(k+1) - \lambda^2\sigma^2/2] = 1, \ k \geq 0. \qquad [6.64]$$

PROOF OF THEOREM 6.4.– As in the proof of Theorem 6.3, we use the equality $\Pi = \exp \ln \Pi$, and the following result concerning the Le Cam approximation.

LEMMA 6.3.– *(Le Cam approximation (Borovskikh and Korolyuk 1997, Lemma 6.3.1)) Assume that*

$$\max_{1 \leq r \leq N} |\delta_r(k+1)/\sqrt{N}| \xrightarrow{P} 0, N \to \infty.$$

Let the sums

$$V_N(k+1) := \frac{1}{N}\sum_{r=1}^{N}(\delta_r(k+1))^2,$$

be bounded in probability.

Then

$$\Big\{\sum_{r=1}^{N}\ln[1+\lambda\delta_r(k+1)/\sqrt{N}] - \lambda\sqrt{N}S_N(k+1) + \lambda^2 V_N(k+1)/2\Big\} \xrightarrow{P} 0,$$

as $N \to \infty$.

Therefore, we get the convergence in probability as $N \to \infty$ with an extra factor,

$$\Pi_N(\lambda/\sqrt{N}, k+1)\exp[-\lambda\sqrt{N}S_N(k+1) + \lambda^2 V_N(k+1)/2] \xrightarrow{P} 1, \quad N \to \infty,\ k \geq 0. \qquad [6.65]$$

Also note that

$$V_N(k+1) = \frac{1}{N}\sum_{r=1}^{N}(\delta_r(k+1))^2 = 1.$$

Hence, convergence [6.65] is equivalent to

$$\Big\{\Pi_N(\lambda/\sqrt{N}, k+1)\exp[-\lambda\sqrt{N}S_N(k+1)]\Big\} \xrightarrow{P} \exp[-\lambda^2/2], \qquad [6.66]$$

for $N \to \infty,\ \ k \geq 0$. The assumptions of Lemma 6.3 imply the convergence in probability for the mean value [6.61], i.e.

$$\Big\{\overline{\Pi_N}(\lambda/\sqrt{N}, k)\exp[-\lambda\sqrt{N}C(S_N(k))]\Big\} \xrightarrow{P} \exp[-\lambda^2\rho^2/2], \qquad [6.67]$$

for $N \to \infty,\ k \geq 0$. Here, we have used the convergence proved in Theorem 6.2:

$$C(S_N(k)) \xrightarrow{P1} C(\rho) = \rho,\ N \to \infty,\ k \geq 0.$$

Thus,

$$C^2(S_N(k)) \xrightarrow{P1} \rho^2,\ N \to \infty,\ k \geq 0.$$

The exponential martingale [6.62] can be represented as

$$\mu_N^e(\lambda/\sqrt{N}, k+1) = \left[\Pi_N^0(\lambda/\sqrt{N}, k+1)\,\overline{\Pi_N^0}(\lambda/\sqrt{N}, k)\right] \times \exp\{\lambda\sqrt{N}[S_N(k+1) - C(S_N(k))]\},\ k \geq 0, \qquad [6.68]$$

where, for $k \geq 0$

$$\Pi_N^0(\lambda/\sqrt{N}, k+1) := \Pi_N(\lambda/\sqrt{N}, k+1) \exp[-\lambda\sqrt{N} S_N(k+1)], \quad [6.69]$$

$$\overline{\Pi_N^0}(\lambda/\sqrt{N}, k) := \overline{\Pi_N}(\lambda/\sqrt{N}, k) \exp[-\lambda\sqrt{N} C(S_N(k))]. \quad [6.70]$$

Theorem 6.2 implies that

$$\sqrt{N}[S_N(k+1) - C(S_N(k))] \xrightarrow{d} \sigma W(k+1), \ N \to \infty, \ k \geq 0. \quad [6.71]$$

Now, [6.66], [6.67] and [6.71] imply the convergence of the exponential martingale [6.68], whence relation [6.63] follows.

The proof of Theorem 6.4 is complete. □

Theorem 6.4, together with [6.62], allows us to express an exponential statistical experiment [6.60] considered in a series scheme as follows:

$$\Pi_N(\lambda/\sqrt{N}, k+1) = \overline{\Pi_N}(\lambda/\sqrt{N}, k) \mu_N^e(\lambda/\sqrt{N}, k+1), \quad [6.72]$$

according to [6.69]. Using [6.72], we obtain

$$\Pi_N(\lambda/\sqrt{N}, k+1) \quad [6.73]$$
$$= \overline{\Pi_N^0}(\lambda/\sqrt{N}, k) \exp[\lambda C(S_N(k))] \mu_N^e(\lambda/\sqrt{N}, k+1).$$

Considering [6.67], we have

$$\overline{\Pi_N^0}(\lambda/\sqrt{N}, k) = \exp[-\lambda^2\rho^2/2] e^{R_N} \quad [6.74]$$

Here and in what follows, the residual term R_N tends to zero in probability as $N \to \infty$.

Representation [6.73], Theorem 6.4 and asymptotic formula [6.74] prove the approximation [6.49] of the exponential statistical experiment by the normal process of autoregression.

PROPOSITION 6.3.– *Exponential statistical experiment* [6.49] *is approximated by a normal process of autoregression,*

$$\widetilde{\Pi}_N(\lambda/\sqrt{N}, k+1) := \prod_{r=1}^{N} [1 + \lambda\widetilde{\delta}_r(k+1)/\sqrt{N}]$$
$$= \exp[\lambda\sqrt{N} C(\widetilde{S_N}(k)) - \lambda^2\rho^2/2] \cdot \exp[\lambda\sigma W(k+1) - \lambda^2\sigma^2/2], \quad [6.75]$$

or, equivalently,

$$\widetilde{\Pi}_N(\lambda/\sqrt{N}, k+1) = \exp\left[[\sqrt{N}\lambda C(\widetilde{S}_N(k) + \sigma W(k+1)] - \lambda^2/2\right]. \quad [6.76]$$

REMARK 6.8.– *It is worth mentioning that the approximations* [6.75] *or* [6.76] *follow from the fact that the conditional expectation of the normal process of autoregression* [6.75] *(or* [6.76]*) asymptotically coincides with the regression function (conditional expectation) of the initial exponential statistical experiment* [6.47], *namely*

$$E\left[\widetilde{\Pi}_N(\lambda/\sqrt{N}, k+1)\right] = \exp[\sqrt{N}\lambda C(\widetilde{S_N}(k)) - \lambda^2\rho^2/2] \qquad [6.77]$$

$$= \overline{\Pi}_N(\lambda/\sqrt{N}, k)e^{R_N}.$$

6.3. Statistical experiments with nonlinear persistent regression

In this section, the regression function of increments is given by the product of two components, namely a linear directing action $[V_-p_- - V_+p_+]$ (defined in Korolyouk (2015a)) and the factor p_+p_-, which changes the directed action in a neighborhood of boundary values $-1, +1$. Such a nonlinear factor appears, for example, in the Wright–Fisher model (see, for example, Ethier and Kurtz (1986)).

PROPOSITION 6.4.– *(Main assumption.) The regression function of the increments of the stochastic experiments possesses the property of a persistent nonlinear regression determined by the following nonlinear directed action:*

$$C_0^{\pm}(\overline{p}) := E[\Delta S_N^{\pm}(k) \mid \overline{S_N}(k) = \overline{p}] = \pm p_+p_-[V_-p_- - V_+p_+], \qquad [6.78]$$

where the parameters of $V_{\pm}$ *are such that*

$$0 < V_{\pm} < 1. \qquad [6.79]$$

According to relation [6.78], *the regression function of the increments of binary stochastic experiments*

$$C_0(s) := E[\,\Delta S_N(k) \mid S_N(k) = s\,], \qquad [6.80]$$

admits the following representation:

$$C_0(s) = C_0^+(\overline{p}) - C_0^-(\overline{p}) = -\frac{1}{4}V(1-s^2)(s-\rho), \ |s| \le 1. \qquad [6.81]$$

It is clear that the directing parameters and their frequency components are such that

$$V_0 := V_- - V_+, \ V := V_- + V_+, \qquad [6.82]$$

and

$$s = p_+ - p_-, \ p_{\pm} = \frac{1}{2}(1 \pm s), \ \rho = V_0/V. \qquad [6.83]$$

The balance condition for the directing action is obvious, namely

$$C_0^+(\overline{p}) + C_0^-(\overline{p}) = 0. \qquad [6.84]$$

REMARK 6.9.– *The main assumption* [6.78] *means that the increment of frequencies of the stochastic experiments evaluated at the moment* $(k+1)$ *increases or decreases proportionally to the additional frequency with coefficients* V_+ *or* V_-*, respectively.*

REMARK 6.10.– *Another consequence of the main assumption is that the regression functions of the stochastic experiments* [6.78]–[6.81] *are given by*

$$C^{\pm}(\overline{p}) := E[\,S_N^{\pm}(k+1) \,|\, \overline{S}_N(k) = \overline{p}\,] = p_{\pm} + C_0^{\pm}(\overline{p}), \qquad [6.85]$$

$$C(s) := E[\,S_N(k+1) \,|\, S_N(k) = s\,] = s + C_0(s), \; |s| \leq 1. \qquad [6.86]$$

REMARK 6.11.– *The following obvious relations take place:*

$$C(s) = C^+(\bar{p}) - C^-(\bar{p}), \; C^+(\bar{p}) + C^-(\bar{p}) = 1, \qquad [6.87]$$

Using [6.83], *condition* [6.84] *determines the frequency components* [6.85]*:*

$$C^{\pm}(\bar{p})) = \frac{1}{2}[1 \pm C(s)], \; s = p_+ - p_-, \; p_+ + p_- = 1. \qquad [6.88]$$

6.3.1. *Equilibrium state*

The equilibrium state for the probability frequencies $\rho_{\pm}$ is defined by the invariant states of the directing action

$$C_0^{\pm}(\bar{\rho}) = 0, \; \bar{\rho} := (\rho_+, \rho_-). \qquad [6.89]$$

Therefore, the equilibrium state of the directed force is determined by the equations

$$V_-\rho_- - V_+\rho_+ = 0, \; \rho_- + \rho_+ = 1. \qquad [6.90]$$

Therefore

$$C_0(\rho) = 0, \; \rho := \rho_+ - \rho_-. \qquad [6.91]$$

The equilibrium frequencies are determined by a solution of equations [6.90]. This solution is given by

$$\rho_{\pm} = V_{\mp}/V, \; V = V_- + V_+. \qquad [6.92]$$

The regression function of the increments of frequencies admits a representation in terms of frequency fluctuations.

COROLLARY 6.2.– *The directed forces are such that*

$$C_0^{\pm}(\bar{p}) = -V_0(\bar{p})(p_\pm - \rho_\pm),\ C_0(s) = -V(s)(s - \rho), \qquad [6.93]$$

$$V_0(\bar{p}) := Vp_+p_- = Vp_\pm(1 - p_\pm),\ V(s) = \frac{1}{4}V\,(1 - s^2). \qquad [6.94]$$

This follows from [6.78], [6.81] *and* [6.90]–[6.92].

A feature of the directed force [6.93] as a function of fluctuations is that the increments of stochastic experiments $S_N(k)$, $k \geq 0$, are acting (in the mean sense) in the opposite direction to the fluctuation $s - \rho$, i.e. the dynamics of the stochastic experiments are directed (in the mean sense) toward the point of equilibrium ρ.

Therefore, the directing action statistically describes the "force" that governs the dynamics of stochastic experiments and leads to the equilibrium state. This fact is described mathematically in the following result.

THEOREM 6.5.– *Assume that at least one of the following three initial states of stochastic experiments converges with probability 1:*

$$S_N^{\pm}(0) \xrightarrow{P1} \rho_\pm,\ S_N(0) \xrightarrow{P1} \rho = \rho_+ - \rho_-, \qquad [6.95]$$

as $N \to \infty$. *Then, their frequency components converge with probability 1 as well:*

$$S_N(k) \xrightarrow{P1} \rho,\ \textit{and}\ S_N^{\pm}(k) \xrightarrow{P1} \rho_\pm,\ N \to \infty, \qquad [6.96]$$

as $N \to \infty$ *for every finite* $k > 0$.

PROOF OF THEOREM 6.5.– In what follows, we denote the fluctuations of frequencies and binary stochastic experiments by

$$\widehat{S}_N^{\pm}(k) := S_N^{\pm}(k) - \rho_\pm,\ \widehat{S}_N(k) := S_N(k) - \rho, \qquad [6.97]$$

$$\widehat{s} := s - \rho,\ \rho = \rho_+ - \rho_-,\ \widehat{\rho}_\pm := p_\pm - \rho_\pm\ \ k \geq 0. \qquad [6.98]$$

As we have mentioned before, it is sufficient to prove the convergence of stochastic experiments for one of their frequency components. Consider the martingale difference for $k \geq 0$:

$$\widehat{\mu}_N(k+1) := \Delta\widehat{S}_N(k) - C_0(S_N(k)) = \Delta\widehat{S}_N(k) + V(S_N(k))\,\widehat{S}_N(k) \qquad [6.99]$$

The main assumption [6.78] implies that

$$E[\widehat{\mu}_N(k+1) \mid S_N(k)] = 0. \qquad [6.100]$$

Further, we need the following:

LEMMA 6.4.– *The quadratic characteristics of the martingale difference* [6.99] *are given by*

$$E[\widehat{\mu}_N^2(k+1) \,|\, S_N(k)] = B(S_N(k))/N, \qquad [6.101]$$
$$B(s) := 1 - C^2(s), \; C(s) = s + C_0(s).$$

Similarly, the quadratic characteristics of the frequency martingale difference

$$\widehat{\mu}_N^{\pm}(k+1) := \Delta \widehat{S}_N^{\pm}(k) - C_0^{\pm}(S_N(k)), \; k \geq 0 \qquad [6.102]$$

are given by

$$E[(\widehat{\mu}_N^{\pm}(k+1))^2 \,|\, \overline{S}_N(k)] = C^{\pm}(\overline{S}_N(k))\,[1 - C^{\pm}(\overline{S}_N(k))]\,/N. \qquad [6.103]$$

PROOF OF LEMMA 6.4.– As proved in [Lemma 2.1],

$$D[\,S_N^{\pm}(k+1) \,|\, \overline{S}_N(k) = \bar{p}\,] = \frac{1}{N}\, C^{\pm}(\bar{p})[1 - C^{\pm}(\bar{p})], \; k \geq 0. \qquad [6.104]$$

This implies that

$$E[\,(\widehat{\mu}_N^{\pm}(k+1))^2 \,|\, \overline{S}_N(k)\,] = E[\,(\mu_N^{\pm}(k+1))^2 \,|\, \overline{S}_N(k)\,]$$
$$= D[\,S_N^{\pm}(k+1) \,|\, \overline{S}_N(k)\,]. \qquad [6.105]$$

The relations for $\widehat{\mu}_N(k)$ are proved similarly. □

Consider the sum of martingale differences [6.99]

$$\widehat{M}_N(n) := \sum_{k=0}^{n} \widehat{\mu}_N(k+1), \; n \geq 0. \qquad [6.106]$$

Relations [6.101] imply that the quadratic characteristics of martingales [6.106] tend to zero with probability 1, as $N \to \infty$ if $n \geq 0$ is fixed. In particular, if $n = 0$, then

$$\widehat{M}_N(0) = \widehat{\mu}_N(1) = \Delta \widehat{S}_N(0) + V(S_N(0))\, \widehat{S}_N(0)$$
$$= \widehat{S}_N(1) - \widehat{S}_N(0) \cdot [1 - V(S_N(0))]. \qquad [6.107]$$

Thus

$$\widehat{S}_N(1) = \widehat{\mu}_N(1) + [1 - V(S_N(0))] \cdot \widehat{S}_N(0) \xrightarrow{P1} 0, N \to \infty \qquad [6.108]$$

according to the initial conditions [6.95] stated in Theorem 6.5.

Now, we prove the almost sure convergence [6.96], as well as [6.97], by induction if $k \geq 1$ is fixed. Theorem 6.5 is proved. □

6.3.2. *Approximation by a normal process of autoregression*

The existence of the equilibrium states [6.89], [6.91] of the stochastic experiments and the convergence proved in Theorem 6.5, allows us to use another approximation of stochastic experiments, with the help of the asymptotic behavior of the martingale component in the representation of increments of stochastic experiments [6.99], characterized by the first two moments (see [6.100]–[6.101]),

$$E[\widehat{\mu}_N(k+1) \mid S_N(k)] = 0, \qquad [6.109]$$

$$E[\widehat{\mu}_N^2(k+1) \mid S_N(k)] = B(S_N(k))/N, \ k \geq 0. \qquad [6.110]$$

The asymptotic behavior of the martingale difference in [6.99] can be obtained under the Lindeberg condition (see Shiryaev (2011))

$$\sum_{k=0}^{n-1} E\left[N\widehat{\mu}_N^2(k+1)\, I\Big\{ |\, \widehat{\mu}_N^2(k+1)| \geq \varepsilon/N \Big\} \mid S_N(k))\right] \xrightarrow{P1} 0. \qquad [6.111]$$

THEOREM 6.6.– *Assume that the conditions of Theorem 6.5 hold. Then, the conditions* [6.109]–[6.111] *imply that the sums of the normalized martingale difference* [6.99] *converge in distribution:*

$$\sqrt{N} \sum_{k=0}^{n-1} \widehat{\mu}_N(k+1) \xrightarrow{d} n\sigma W, \ N \to \infty, \ \forall n \geq 1, \qquad [6.112]$$

and

$$\sqrt{N} \sum_{k=0}^{n-1} \widehat{\mu}_N^{\pm}(k+1) \xrightarrow{d} n\breve{\sigma} W, \ N \to \infty, \ \forall n \geq 1. \qquad [6.113]$$

Here, W is a standard normal random variable and the variances σ^2 and $\breve{\sigma}^2$ are given by

$$\breve{\sigma} = \rho_+\rho_-, \ \sigma^2 = 4\breve{\sigma}^2 = 1 - \rho^2, \ \rho = \rho_+ - \rho_-.$$

REMARK 6.12.– *It is known (see Shiryaev (2011)) that the uniform convergence in the probability of the martingale difference*

$$E\left[\max_{1 \leq k \leq n} |\, \widehat{\mu}_N^{\pm}(k)|\right] \xrightarrow{P} 0, \ N \to \infty$$

implies the Lindeberg condition [6.111].

COROLLARY 6.3.– *The convergence of the sum of martingale differences* [6.99] *and their frequency components* [6.102] *implies the following convergence in distribution for every fixed* $k \geq 0$ *:*

$$\sqrt{N}[\Delta\widehat{S}_N(k) + V(S_N(k))\,\widehat{S}_N(k)] \xrightarrow{d} \sigma\widehat{W}(k+1),\ N \to \infty,\ k \geq 0, \qquad [6.114]$$

$$\sqrt{N}[\Delta\widehat{S}_N^{\pm}(k) + V(S_N(k))\,\widehat{S}_N^{\pm}(k)] \xrightarrow{d} \breve{\sigma}\widehat{W}(k+1),\ N \to \infty,\ k \geq 0. \qquad [6.115]$$

PROPOSITION 6.5.– *The limit relations* [6.114] *allow us to use the approximation of stochastic experiments by a normal process of autoregression defined by*

$$\Delta\widehat{\widetilde{S}}_N(k+1) = V(\widetilde{S}_N(k))\widehat{\widetilde{S}}_N(k) + (\sigma/\sqrt{N})\widehat{W}(k+1), \qquad [6.116]$$

$$\Delta\widehat{\widetilde{S}}{}_N^{\pm}(k+1) = V(\widetilde{S}_N^{\pm}(k))\widehat{\widetilde{S}}{}_N^{\pm}(k) + (\breve{\sigma}/\sqrt{N})\widehat{W}^{\pm}(k+1). \qquad [6.117]$$

In [6.114] and [6.116], as well as in [6.112]–[6.113], $\widehat{W}(k+1)$ and $\widehat{W}^{\pm}(k+1)$ are standard normal random variables.

REMARK 6.13.– *As a function of fluctuations, the regression function of the increments of the stochastic experiments admits the following representation:*

$$C_0(s) = -\frac{1}{4}V\left[\sigma^2 - (\widehat{s} + 2\rho)\,\widehat{s}\right]\widehat{s},\ \widehat{s} = s - \rho. \qquad [6.118]$$

Equality [6.118] *follows from the representation of the nonlinear factor* $V(s)$ [6.93]–[6.94] *as a function of fluctuations:*

$$1 - s^2 = \sigma^2 - (\widehat{s} + 2\rho)\widehat{s}.$$

Proposition 6.5 and representation [6.118] imply the following approximation of stochastic experiments by a linear normal process of autoregression.

PROPOSITION 6.6.– *The limit relations* [6.114] *imply the approximation of stochastic experiments by a linear normal process of autoregression defined by*

$$\Delta\widehat{\widetilde{S}}_N(k+1) = -\frac{V}{4}\sigma^2\widehat{\widetilde{S}}_N(k) + (\sigma/\sqrt{N})\widehat{W}(k+1), \qquad [6.119]$$

$$\Delta\widehat{\widetilde{S}}{}_N^{\pm}(k+1) = -V\breve{\sigma}^2\widehat{\widetilde{S}}{}_N^{\pm}(k) + (\breve{\sigma}/\sqrt{N})\widehat{W}^{\pm}(k+1). \qquad [6.120]$$

Stochastic experiments $\widehat{\widetilde{S}}_N(k)$ and $\widehat{\widetilde{S}}{}_N^{\pm}(k)$, $k \geq 0$ are well defined by difference stochastic equations [6.119] under the given initial values.

6.3.3. *Proof of Theorem 6.6*

Consider the fluctuations of the stochastic experiments centered at an equilibrium state ρ and normalized by the factor $\sqrt{N}$:

$$\zeta_N(k) := \sqrt{N}\widehat{S}_N(k) = \sqrt{N}[S_N(k) - \rho]. \qquad [6.121]$$

Furthermore, we use the following representation of the martingale difference [6.99] (see also [6.93]–[6.94]) with a separated linear component:

$$\sqrt{N}\,\widehat{\mu}_N(k+1) = \Delta\zeta_N(k) - \frac{1}{4}V\sigma^2\zeta_N(k) + \frac{1}{\sqrt{N}}R(\widehat{S}_N(k))\zeta_N^2(k), \quad [6.122]$$

where

$$R(\widehat{s}) = -\frac{1}{4}V(\widehat{s} + 2\rho)$$

The residual term on the right-hand side of [6.122] tends to zero in probability as $N \to \infty$, since the product of functions $R(\widehat{S}_N(k))\zeta_N^2(k)$ is bounded.

Consider the sum of martingale differences and its asymptotic representation

$$M_N(n) := \sum_{k=0}^{n-1} \sqrt{N}\,\widehat{\mu}_N(k+1) = \sum_{k=0}^{n-1}[\zeta_N(k+1) + V(S_N(k))\,\zeta_N(k)]. \qquad [6.123]$$

The quadratic characteristics of the martingale $M_N(n)$ are given by

$$\langle M_N\rangle_n = \sum_{k=0}^{n-1}[1 - C^2(S_N(k))], \; n \geq 0. \qquad [6.124]$$

The second term on the right-hand side of [6.124] tends to zero in probability as $N \to \infty$. By using Theorem 6.6, the quadratic characteristics converge with probability 1:

$$\langle M_N\rangle_n \xrightarrow{P1} n(1 - \rho^2) = n\sigma^2 \quad \text{as} \quad N \to \infty, \; n \geq 0. \qquad [6.125]$$

The central limit theorem for martingale differences (see, for example, Borovskikh and Korolyuk (1997)) implies the convergence in the distribution of martingale [6.123] to the sum of normal random variables $W(k)$, i.e. [6.112] holds.

Theorem 6.6 is proved. □

Taking into account the asymptotic representation of the martingale

$$M_N(n) = \sum_{k=0}^{n-1}[\Delta\zeta_N(k) - \frac{1}{4}V\sigma^2\zeta_N(k)] + \frac{1}{\sqrt{N}}\sum_{k=0}^{n-1} R(\widehat{S}_N(k))\zeta_N^2(k), \qquad [6.126]$$

we obtain the following result.

COROLLARY 6.4.– *The following convergence in distribution takes place:*

$$\sum_{k=0}^{n-1}[\Delta\zeta_N(k+1) - \frac{1}{4}V\sigma^2\zeta_N(k)] \xrightarrow{d} \sigma\sum_{k=0}^{n-1} W(k+1),\ N\to\infty,\ k\geq 1. \qquad [6.127]$$

Using the limit relation [6.114], we can introduce a new approximating model of normalized fluctuations of binary stochastic experiments in the following form, which is suitable for several applications.

PROPOSITION 6.7.– *The limit relations* [6.127] *imply that the normal process of autoregression* $\zeta_N(k),\ k\geq 0$ *is a solution of the following difference stochastic equation:*

$$\Delta\zeta_N(k) = -V(\rho)\zeta_N(k) + \sigma\Delta W(k), \qquad [6.128]$$

where

$$V(\rho) := \frac{1}{4}V(1-\rho^2) = \frac{1}{4}V\sigma^2,\ \Delta W(k) := W(k+1) - W(k), \sigma^2 = 1-\rho^2.$$

The existence of two models in the case of the persistent regression for binary stochastic experiments, described in Propositions 6.5 and 6.6, allows us to solve a problem of statistical analysis for these models, by using the initial values of the stochastic experiments, normalized by the sums of the sample values which satisfy the main assumption [6.78].

6.4. Difference diffusion models with two equilibriums

Now, we consider the dynamics of the statistical experiments and their evolutionary processes, which have two equilibriums. Unlike in the previous sections, we study the "zones of influence" of the two equilibriums generated by the principle of "stimulation and restraint". The universality of this principle is confirmed by real interpretations of the dynamics of economic processes (Maslov 2003, 2004, 2005a, 2005b), development of population genetic models (Korolyouk *et al.* 2014) and interpretation of learning processes (Korolyuk *et al.* 2016a).

As defined in the previous paragraphs, the dynamics of a discrete Markov diffusion is determined by the *evolutionary processes* $P_\pm(k),\ k\geq 0$, given by the conditional mathematical expectations:

$$P_\pm(k+1) := E[S_N^\pm(k+1)\,|\,S_N^\pm(k) = P_\pm(k)],\ k\geq 0, \qquad [6.129]$$

and also

$$C(k+1) := E[S_N(k+1)\,|\,S_N(k) = C(k)],\ k\geq 0. \qquad [6.130]$$

The relation between the evolutionary processes [6.129] and [6.130] is

$$C(k) = P_+(k) - P_-(k). \quad [6.131]$$

Given the obvious condition of balance

$$P_+(k) + P_-(k) \equiv 1, \ \ \forall k \geq 0, \quad [6.132]$$

the frequency evolutionary processes $P_\pm(k)$, $k \geq 0$ have a representation (see (0.4)):

$$P_\pm(k) = \frac{1}{2}[1 \pm C(k)], \ \ k \geq 0. \quad [6.133]$$

The difference evolution model is given by increments of frequency probabilities

$$\Delta P_\pm(k+1) = P_\pm(k+1) - P_\pm(k), \ k \geq 0. \quad [6.134]$$

as well as increments of the binary evolutionary process

$$\Delta C(k+1) = C(k+1) - C(k), \ k \geq 0, \quad [6.135]$$

The dynamics of the evolutionary processes [6.129] and [6.130] are given by the conditional mathematical expectations

$$V^\pm(p_\pm) := E[\Delta S_N^\pm(k+1) \mid S_N^\pm(k) = p_\pm], \ 0 \leq p_\pm \leq 1, \quad [6.136]$$

and

$$V_0(c) := -E[\Delta S_N(k+1) \mid S_N(k) = c], \ |c| \leq 1. \quad [6.137]$$

6.4.1. *The principle of "stimulation and restraint"*

The frequency probabilities $P_+(k)$ of the sample values $\delta_n^+(k)$, $k \geq 1$ are determined by *evolutionary processes*, whose dynamics by the stages $k \geq 0$) are determined by the *regression function of the increments*, which reflects the fundamental principle of interaction of the collective behavior of a set of objects – the principle "stimulation and restraint" (see Korolyouk 2015d).

From a mathematical point of view, the principle of "stimulation and restraint" is given by the linear function of the frequencies of values $p_\pm$ $(0 \leq p_\pm \leq 1, p_+ + p_- = 1)$ of the sampling values $\delta_n^\pm(k)$:

$$\pi_+ p_+ - \pi_- p_-$$

with two directing parameters $\pi_\pm$, which, without reducing the generality, also satisfy the properties of frequencies:

$$0 < \pi_\pm < 1, \ \pi_+ + \pi_- = 1.$$

It follows that the principle "stimulation and restraint" is expressed by frequency fluctuations

$$\pi_+ p_+ - \pi_- p_- = p_+ - \pi_- .$$

The linear component of the regression function of increments can also be represented in the following form:

$$\pi_- p_+ - \pi_+ p_- = p_+ - \pi_+ .$$

The product of two frequency fluctuations

$$(p_+ - \pi_+)(p_+ - \pi_-) = (p_- - \pi_-)(p_- - \pi_+)$$

generates a nonlinear regression function of increments with two equilibriums $\pi_\pm$. The completion of the construction of the frequency regression function of increments, taking into account the boundary absorption states 0, 1, determines the increments of the frequency probabilities:

$$V^\pm(p_\pm) = \mp V \cdot p_\pm(1 - p_\pm)(p_\pm - \pi_\pm)(p_\pm - \pi_\mp),\ V > 0.$$

A balance condition takes place:

$$V^+(p_+) + V^-(p_-) \equiv 0,\ p_+ + p_- = 1.$$

According to the principle of "stimulation and restraint", the linear regression function of increments can also be as follows:

$$\pi_\pm p_+ + \pi_\mp p_- = \pm\pi(p_+ - \overline{\pi}_\mp),\ \overline{\pi}_\mp := \mp\pi_\mp/\pi,\ \pi := \pi_+ - \pi_- .$$

Without a loss of generality, we assume $\pi := \pi_+ - \pi_- > 0$. Then, the equilibrium states $\overline{\pi}_\mp = \mp\pi_\mp/\pi$ take values outside of the interval $(0, 1)$:

$$\overline{\pi}_- := -\pi_-/\pi < 0,\ \overline{\pi}_+ := \pi_+/\pi > 1.$$

Hence, the equilibrium states determined by equilibriums $\overline{\pi}_\mp$ "stimulate" or "restrain" the dynamics of frequency probabilities, without dividing the range of values – the interval $(0, 1)$ on additional areas of influence. Such equilibriums are also considered in Maslov's economic studies (Maslov 2003, 2004, 2005a, 2005b).

The presence of two equilibriums in a regression function of increments, which determine *difference evolutionary equations* for increments of frequency probabilities and increments of binary evolutionary processes significantly, changes the dynamics of both the evolutionary processes and the statistical experiments themselves.

6.4.2. *Difference evolutionary model*

DEFINITION 6.1.– *The frequency probabilities $P_\pm(k)$, $k \geq 0$ are given by solutions of difference evolution equations*

$$\Delta P_\pm(k+1) = V^\pm(P_\pm(k)),\ k \geq 0, \qquad [6.138]$$

and also

$$\Delta C(k+1) = -V_0(C(k)),\ k \geq 0. \qquad [6.139]$$

The conditional mathematical expectations [6.129] and [6.130] are determined by the regression function of increments

$$V^\pm(p_\pm) = \mp V \cdot p_\pm(1-p_\pm)(p_\pm - \pi_\pm)(p_\pm - \pi_\mp),\ 0 \leq p_\pm \leq 1, \qquad [6.140]$$

and also

$$V_0(c) = -V_0 \cdot (1-c^2)(c+\pi)(c-\pi),\ \pi = \pi_+ - \pi_-, \qquad [6.141]$$
$$V_0 := V/4,\ |c| \leq 1.$$

The regression function of the increments of the binary evolutionary processes $C(k)$, $k \geq 0$ also has two equilibrium states $\mp\pi$.

Further analysis of the evolutionary process is given by the difference evolutionary equations [6.138] and [6.139], taking into account the regression function of increments [6.140] and [6.141], similarly to Korolyouk (2015a) and Korolyouk (2015b).

The regression functions of increments [6.140] and [6.141] allocate the zones specified by the equilibrium states $\pi_\pm$ and $\mp\pi$ (Figure 6.1).

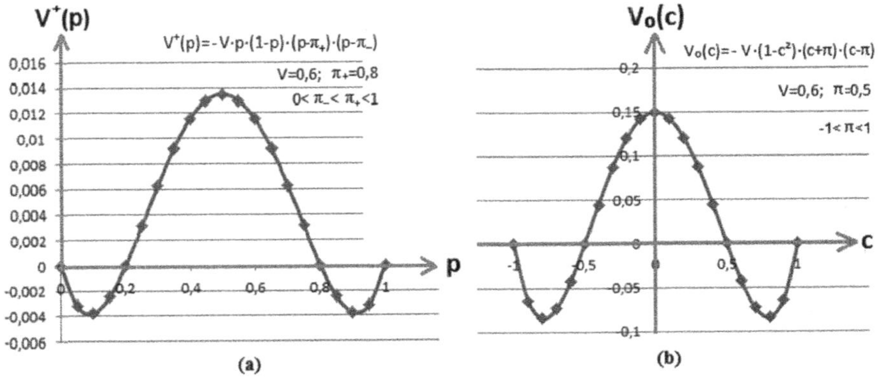

Figure 6.1. *Regression functions with two equilibriums: (a) frequency and (b) binary. For a color version of this figure, see www.iste.co.uk/koroliouk/random.zip*

The equilibrium states $\pi_\pm$ allocate three zones:

$$0 < p_+ < \pi_-, \ \pi_- < p_+ < \pi_+, \ \pi_+ < p_+ < 1. \qquad [6.142]$$

Similarly, there are three zones that define the binary regression function of increments $V_0(c)$:

$$-1 < c < -\pi, \ -\pi < c < \pi, \ \pi < c < 1. \qquad [6.143]$$

Now, using the classification of evolutionary process models, and taking into account the limit behavior of frequency probabilities $P_\pm(k)$, and binary evolutionary process $C(k)$ by $k \to \infty$, we formulate the classification theorem.

THEOREM 6.7.– *The frequency probabilities of alternatives $P_\pm(k)$, $k \geq 0$, given by the solution difference stochastic equation* [6.138], *identify areas of influence* [6.142] *by the following asymptotic behavior:*

– *In the attractive model:*

$$\lim_{k\to\infty} P_\pm(k) = \pi_\pm, \ \pi_- < P_\pm(0) \leq 1. \qquad [6.144]$$

– *In the repulsive model:*

$$\lim_{k\to\infty} P_+(k) = 0, \ 0 < P_+(0) < \pi_-; \qquad [6.145]$$

$$\lim_{k\to\infty} P_-(k) = 1, \ \pi_+ < P_-(0) < 1.$$

Similarly, binary evolutionary processes are determined by the zones of influence [6.143] of the equilibrium states $\mp\pi$, based on their asymptotic behavior.

THEOREM 6.8.– *Binary evolutionary process $C(k)$, $k \geq 0$, given by solutions of the difference evolutionary equation* [6.139], *determines the zones of the equilibrium states $\mp\pi$:*

– *In the attractive model:*

$$\lim_{k\to\infty} C(k) = \pi, \ -\pi < C(0) < 1;$$

– *In the repulsive model:*

$$\lim_{k\to\infty} C(k) = 0, \ 0 < C(0) < -\pi.$$

The proofs of Theorems [6.7] and [6.8] are similar. Therefore, we only consider the frequency positive probabilities of alternatives $P_\pm(k)$ $k \geq 0$, with the regression function of increments [6.140] (see Figure 6.1(a)):

$$V^+(p_+) = -V \cdot p_+(1 - p_+)(p_+ - \pi_-)(p_+ - \pi_+), \qquad [6.146]$$

takes the following values:

$$V^+(P_+(k)) < 0,\ 0 < P_+(0) < \pi_- \text{ or } \pi_+ < P_+(0) < 1; \qquad [6.147]$$
$$V^+(P_+(k)) > 0,\ \pi_- < P_+(0) < \pi_+.$$

The monotonicity of the increments [6.147] in the zones of influence of equilibriums and the boundedness of probabilities $P_+(k)\ :\ 0 \le P_+(k) \le 1$ allow us to conclude that the limits [6.144] and [6.145] exist for the frequencies

$$p_+^* = \lim_{k\to\infty} P_+(k).$$

The values of boundaries [6.144] and [6.145] follow from the obvious equation

$$0 = V^+(p_+^*) = p_+^*(1 - p_+^*)(p_+^* - \pi_-)(p_+^* - \pi_+).$$

For the initial value in the area of influence $\pi_- < P_+(0) < \pi_+$, the monotonic growth of increments [6.147] means that $\lim_{k\to\infty} P_+(k) = p_+^* = \pi_+$.

Similarly, the limit behavior of positive frequency probabilities $P_+(k),\ k \ge 0$ is determined in the areas of influence $\pi_+ < P_+(0) < 1$ and the condition $0 < P_+(0) < \pi_-$ is provided by a monotonous decline in increments. □

The classification of zones of influence of equilibriums of the regression function of increments allows us to make the preliminary conclusion concerning the real interpretation of equilibriums in modern economic space, and also in models of population genetics (Korolyouk *et al.* 2014), or in learning models (Korolyuk *et al.* 2016a).

6.4.3. *Interpretation of zones of influence of equilibriums $\pi_\pm$ in the economic space*

The dynamics of frequency probabilities $P_+(k),\ k \ge 0$ in the area of attractive equilibrium π_+

$$\lim_{k\to\infty} P_+(k) = \pi_+,\ \ P_+(0) > \pi_+$$

means that the presence of "relative" capital $CP_+(k) > C\pi_+$ leads to a general reduction of capital to the equilibrium state $C\pi_+$, which characterizes the objectivity of the principle of "stimulation and restraint".

However, attracting equilibrium π_+ in zone $\pi_- < P_+(0) < \pi_+$ also leads to an increase in capital to equilibrium $C\pi_+$, which also characterizes the objectivity of the principle of "stimulation and restraint".

Finally, in the zone of repulsive equilibrium, the dynamics of "relative" capital

$$\lim_{k\to\infty} P_+(k) = 0, \; 0 < P_+(0) < \pi_-$$

means that the subject of economic space with "relative" capital less than $C\pi_-$, loses capital.

There is a strategic problem: the reduction of the zone of "impoverishment" subjects the economic space and ideally – the elimination of the zone $(0, \; \pi_-)$.

6.4.4. *Difference stochastic model*

The difference evolutionary equations [6.138] and [6.139], together with the regression function of increments [6.140] and [6.141], generate deterministic evolutionary processes, expressed by the expected characteristics of statistical experiments.

The dynamics of the stochastic experiment are set by the difference stochastic equation using martingales (by $k \geq 0$)

$$\Delta\mu_N^{\pm}(k+1) := \Delta S_N^{\pm}(k+1) - V^{\pm}(S_N^{\pm}(k)), \; k \geq 0, \qquad [6.148]$$

and also

$$\Delta\mu_N(k+1) := \Delta S_N(k+1) + V_0(S_N(k)), \; k \geq 0. \qquad [6.149]$$

DEFINITION 6.2.– *The frequency statistical experiments are given by a solution of the difference stochastic equation*

$$\Delta S_N^{\pm}(k+1) = V^{\pm}(S_N^{\pm}(k)) + \Delta\mu_N^{\pm}(k+1), \; k \geq 0.$$

Binary statistical experiments are given by a solution of the difference stochastic equation

$$\Delta S_N(k+1) = -V_0(S_N(k)) + \Delta\mu_N(k+1), \; k \geq 0. \qquad [6.150]$$

The martingales [6.148] and [6.149] are characterized by the first two points:

$$E[\Delta\mu_N^{\pm}(k+1)] = 0, \; E[\Delta\mu_N(k+1)] = 0, \; k \geq 0, \qquad [6.151]$$

$$E[(\Delta\mu_N^{\pm}(k+1))^2 \mid S_N^{\pm}(k)] = V_+(S_N^{+}(k)) \cdot V_-(S_N^{-}(k))/N, \; k \geq 0,$$

$$E[(\Delta\mu_N(k+1))^2 \mid S_N(k)] = [1 - V^2(S_N(k))]/N, \; k \geq 0. \qquad [6.152]$$

Here, the regression functions

$$V_\pm(p_\pm) := p_\pm \mp V^{\pm}(p_\pm), \; 0 \leq p_\pm \leq 1,$$

$$V(c) := c - V_0(c), \; |c| \leq 1.$$

To calculate the second moments of the martingales [6.148] and [6.149], we will use the obvious relations:

$$E[(\delta_n^{\pm}(k+1))^2 \,|\, S_N^{\pm}(k) = p_{\pm}] = E[\delta_n^{\pm}(k+1) \,|\, S_N^{\pm}(k) = p_{\pm}]$$
$$= E[S_N^{\pm}(k+1) \,|\, S_N^{\pm}(k) = p_{\pm}] = V_{\pm}(p_{\pm}), \; k \geq 0, \quad [6.153]$$

as well as

$$E[(\delta_n(k+1))^2] \equiv 1, \; \forall k \geq 0,$$
$$E[\delta_n(k+1) \,|\, S_N(k) = c] = E[S_N(k+1) \,|\, S_N(k) = c] = V(c), \; |c| \leq 1.$$

Thus, we have

$$D[\delta_n(k+1) \,|\, S_N(k) = c] := E[(\delta_n(k+1))^2 \,|\, S_N(k) = c] -$$
$$- [E[\delta_n(k+1) \,|\, S_N(k) = c]]^2 = 1 - V^2(c).$$

Similarly,

$$D[\delta_n^{\pm}(k+1) \,|\, S_N^{\pm}(k) = p_{\pm}] = V_{\pm}(p_{\pm}) - V_{\pm}^2(p_{\pm}) = V_+(p_+)V_-(p_-).$$

Moreover, we will use the relation

$$D[S_N(k+1) \,|\, S_N(k) = c] = \frac{1}{N} D[\delta_n(k+1) \,|\, S_N(k) = c] = [1 - V^2(c)]/N,$$

which explains the presence of the multiplier N^{-1} in formulas [6.152].

It remains to note that by the definition of martingales [6.148] and [6.149], their conditional variances are determined by the equality (see [6.153])

$$E[(\Delta\mu_N(k+1))^2 \,|\, S_N(k)] = D[S_N(k+1) \,|\, S_N(k)].$$

Indeed, from definition [6.149], we have

$$E[(\Delta\mu_N(k+1))^2 \,|\, S_N(k)] = E[[\Delta S_N(k+1) + V_0(S_N(k))]^2 \,|\, S_N(k)]$$
$$= E[S_N^2(k+1) \,|\, S_N(k)] - [E[S_N(k+1) \,|\, S_N(k)]]^2 = D[S_N(k+1) \,|\, S_N(k)].$$

6.4.5. *Classification of equilibriums of the stochastic model statistical experiment*

The presence of the stochastic component $\Delta\mu_N(k+1)$ difference stochastic equation [6.150] for the binary statistical experiment significantly complicates the solution of the problem of its stability. It has been proposed to use the *stochastic approximation* method, developed by Robinson and Monroe (1951).

We enter the stochastic approximation parameters

$$a_k := a/k, \ k \geq 1,$$

satisfying the approximation conditions for $k \geq 0$:

$$\sum_{k=1}^{\infty} a_k = +\infty, \ \sum_{k=1}^{\infty} a_k^2 = a \sum_{k=1}^{\infty} \frac{1}{k^2} = a\frac{\pi^2}{6} < \infty \qquad [6.154]$$

We will describe the procedure of stochastic approximation for the difference stochastic equation [6.150] by the scheme

$$\Delta\alpha_N(k+1) = a_k[-V_0(\alpha_N(k)) + \sigma(\alpha_N(k))\Delta\mu_N(k+1)], \ k \geq 1. \quad [6.155]$$

The stochastic approximation normalization parameters satisfy the conditions [6.154].

REMARK 6.14.– *The discrete Markov diffusion* $\alpha_N(k)$, $k \geq 0$, *given by the solution of difference stochastic equation* [6.155], *is characterized by the first two moments:*

$$E[\Delta\alpha_N(k+1) \,|\, \alpha_N(k) = c] = -a_k V_0(c), \ k \geq 0,$$

$$E\Big[[\Delta\alpha_N(k+1)]^2 \,\Big|\, \alpha_N(k) = c\Big] = a_k^2\big[\sigma^2(c) + V_0^2(c)/N\big].$$

To classify the zones of influence of equilibriums $\pm\pi$ ($|\pi| < 1$) introduce classifiers of equilibriums:

$$\Lambda_\pm(c) := -V_0(1-c^2)(c \mp \pi), \ |c| \leq 1. \qquad [6.156]$$

The following property of classifiers [6.156] is used:

$$\Lambda_+(c) < 0, \ -\pi < c < 1 \ ; \ \ \Lambda_-(c) < 0, \ -1 < c < -\pi.$$

The stochastic approximation procedure uses the Nevelson–Hasminski theorem (Nevelson and Hasminskii 1976, Theorem 2.7.2) adapted for difference stochastic models:

THEOREM 6.9.– *Nevelson and Hasminskii (1976). Suppose there is a non-negative function* $\Lambda_0(c)$, $|c| \leq 1$, *with an equilibrium point* c_0*:* $\Lambda_0(c_0) = 0$, *satisfying the inequalities around the point* c_0, *by a fixed* $h > 0$*:*

$$\sup_{\frac{1}{h} \leq |c-c_0| \leq h} \Lambda_0(c)(c-c_0) < 0.$$

Then, the stochastic approximation procedure $\alpha_N(k),\ k \geq 0$ *converges, with probability 1, to the equilibrium point* c_0*:*

$$\alpha_N(k) \xrightarrow{P1} c_0,\ k \to \infty.$$

Now, the classification of zones of influence of equilibriums $\pm\pi$ is formulated as follows:

THEOREM 6.10.– *The dynamics of the statistical experiment* $S_N(k),\ k \geq 0$ *is characterized by the classifiers* [6.156]*:*

– *In the area of influence of the attractive equilibrium* π*:*

$$P \lim_{k\to\infty} \alpha_N(k) = \pi,\ -\pi < \alpha_N(0) < 1.$$

– *In the area of impact of the repulsive equilibrium* $-\pi$*:*

$$P \lim_{k\to\infty} \alpha_N(k) = 0,\ -1 < \alpha_N(0) < -\pi.$$

PROOF OF THE THEOREM 6.10.– The classifiers and the Nevelson–Hasminski theorem are used [6.9].

Calculating the value of generator DMD $\alpha_N(k),\ k \geq 0$ on test functions $\varphi_\pm(c) = (c \pm \pi)^2$ gives the following

$$\begin{aligned} L_\alpha\varphi_\pm(c) :&= E[\varphi_\pm(c + \Delta\alpha_N(k+1)) - \varphi_\pm(c) \,|\, \alpha_N(k) = c] \\ &= 2E[\Delta\alpha_N(k+1) \,|\, \alpha_N(k) = c](c \pm \pi) \\ &+ E[(\Delta\alpha_N(k+1))^2 \,|\, \alpha_N(k) = c], \end{aligned}$$

i.e. we have

$$L_\alpha\varphi_\pm(c) = -2a_k\Lambda_\pm(c)(c \mp \pi)^2 + a_k^2 B_N(c).$$

The use of [6.156] and of the Nevelson–Hasminski theorem complete the proof of Theorem 6.10. □

6.4.6. *Approximation of the stochastic component*

The martingale property (for $k \geq 0$) of stochastic components [6.148] and [6.149] allows us to characterize them in more detail. Namely, the normalized martingales [6.148] and [6.149] are determined by the Bernoulli distributions of possible frequency values $\nu_N^\pm$ at a given distribution $p_\pm$:

– in frequency representation

$$P\{\,\Delta\mu_N^{\pm}(p_\pm) = \nu_N^{\pm}/N \mid S_N^{\pm}(k) = p_\pm\,\}$$
$$= \frac{N!}{\nu_N^{+}!\nu_N^{-}!}(p_+)^{\nu_N^{+}}(p_-)^{\nu_N^{-}},\ \nu_N^{+} + \nu_N^{-} = N,\ 0 \le \nu_N^{\pm} \le N; \qquad [6.157]$$

– in binary representation

$$P\{\,\Delta\mu_N(c) = \nu_N/N \mid S_N(k) = c\,\}$$
$$= \frac{N!}{\nu_N^{+}!\nu_N^{-}!}(p_+)^{\nu_N^{+}}(p_-)^{\nu_N^{-}},\ \nu_N = \nu_N^{+} - \nu_N^{-},\ -N \le \nu_N \le N. \qquad [6.158]$$

The martingale property of the stochastic components [6.148] and [6.149], as well as their Bernoulli distributions [6.157] and [6.158], provide a "conditional form" of approximation using the evolutionary processes [6.132] and [6.133].

THEOREM 6.11.– *The normalized stochastic components* [6.157] *and* [6.158] *are approximated, in distribution, as* $N \to \infty$*, by a sequence of independent, normally distributed random variables* $W_{k+1}^{\pm}$, W_{k+1}, $k \ge 0$*:*

$$\sqrt{N}\Delta\mu_N^{\pm}(p_\pm) \xrightarrow{D} W_{k+1}^{\pm},\ E[W_{k+1}^{\pm}]^2 = \breve{\sigma}^2(p_+, p_-); \qquad [6.159]$$

$$\sqrt{N}\Delta\mu_N(c) \xrightarrow{D} W_{k+1},\ E[W_{k+1}]^2 = \sigma^2(c), \qquad [6.160]$$

which are characterized by the second moments

$$\breve{\sigma}^2(p_+, p_-) = V_+(p_+) \cdot V_-(p_-),\ \sigma^2(c) = 1 - V^2(c),$$

and are determined by regression functions

$$V_\pm(p_\pm) := p_\pm \mp V^{\pm}(p_\pm),\ V(c) := c - V_0(c). \qquad [6.161]$$

THE PROOF OF THEOREM 6.11.– The Bernoulli distributions [6.157] and [6.158] of the normalized martingales [6.148] and [6.149] in "conditional form" are generated by the normalized sums of sample values

$$\Delta\mu_N^{\pm}(p_\pm) = \frac{1}{\sqrt{N}}\sum_{n=1}^{N}\beta_n^{\pm}(p_\pm),\ \beta_n^{\pm}(p_\pm) = \delta_n^{\pm}(k+1) - V_\pm(p_\pm) \qquad [6.162]$$

$$\Delta\mu_N(c) = \frac{1}{\sqrt{N}}\sum_{n=1}^{N}\beta_n(c),\ \beta_n(c) = \delta_n(k+1) - V(c). \qquad [6.163]$$

Taking into account the definition of frequency martingales [6.148] and frequency regression functions [6.161], the frequency random variables in the sum [6.162] have the representation:

$$\beta_n^{\pm}(p_\pm) = \begin{cases} 1 - V_\pm(p_\pm), & \text{with probability } P_\pm = V_\pm(p_\pm), \\ -V_\pm(p_\pm), & \text{with probability } P_\mp = V_\mp(p_\mp). \end{cases} \qquad [6.164]$$

Similarly, taking into account the definitions of binary martingales [6.149] and binary regression functions [6.161], the binary random variables in the sum [6.163] have the representation:

$$\beta_n(c) = \begin{cases} 1 - V(c), & \text{with probability } P_+(c) = \frac{1}{2}[1 + V(c)], \\ -(1 + V(c)), & \text{with probability } P_-(c) = \frac{1}{2}[1 - V(c)]. \end{cases} \qquad [6.165]$$

The sample values [6.164] and [6.165] are characterized by the first two points:

$$E\beta_n^{\pm} = E\beta_n = 0,$$

$$E(\beta_n^{\pm})^2 = \breve{\sigma}^2(p_+, p_-) = V_+(p_+)\, V_-(p_-),\ E(\beta_n)^2 = \sigma^2(c) = 1 - V^2(c).$$

Therefore, the normalized martingales [6.159] and [6.160] are also characterized by the first two moments:

$$E\Delta\mu_N^{\pm}(p_\pm) = E\Delta\mu_N(c) = 0,$$

$$E(\Delta\mu_N^{\pm}(p_\pm))^2 = \breve{\sigma}^2(p_+, p_-) = V_+(p_+)\, V_-(p_-), \qquad [6.166]$$

$$E(\Delta\mu_N(c))^2 = \sigma^2(c) = 1 - V^2(c). \qquad [6.167]$$

The use of the central limit theorem for sums of independent, equally distributed random variables (Shiryaev 2011) completes the proof of Theorem 6.11. □

PROPOSITION 6.8.– *The statistical experiment with the normal approximation of the stochastic component is given by the following difference stochastic equation (Korolyouk* et al. *2016b; Korolyouk 2017):*

$$\Delta\zeta_N^{\pm}(k+1) = V^{\pm}(\zeta_N^{\pm}(k)) + \frac{1}{\sqrt{N}}\breve{\sigma}(\zeta_N^{\pm}(k))W^{\pm}(k+1),$$

$$\Delta\zeta_N(k+1) = -V_0(\zeta_N(k)) + \frac{1}{\sqrt{N}}\sigma(\zeta_N(k))W(k+1)$$

Here, $W^{\pm}(k+1)$, $W(k+1)$, $k \geq 0$ are the sequences of normally distributed standard random variables; the parameters $\breve{\sigma}^2$, σ^2 are defined in [6.166] *and* [6.167].

6.4.7. *Approximation of statistical experiment in discrete–continuous time*

The approximation of the normalized stochastic component, which is established in Theorem 6.1, gives rise to a new problem of statistical experiment approximation in discrete–continuous time.

DEFINITION 6.3.– *The difference stochastic equation in discrete–continuous time* $k = [Nt]$, $t \geq 0$ *is given by two normalized components:*

$$\Delta\zeta_N(t+1/N) = -\frac{1}{N}V_0(\zeta_N(t)) + \frac{1}{\sqrt{N}}\sigma(\zeta_N(t))\Delta\mu_N(t+1/N), \qquad [6.168]$$

$$t \geq 0.$$

The increments of the statistical experiment in discrete–continuous time are determined by the equality:

$$\Delta\zeta_N(t+1/N) := \zeta_N(t+1/N) - \zeta_N(t),\ t \geq 0.$$

The normalized stochastic martingale component is given by the conditional Bernoulli distribution

$$P\Big\{\frac{1}{\sqrt{N}}\Delta\mu_N(k+1) = \nu_N/N \,\Big|\, S_N(k) = C(k)\Big\}$$

$$= \frac{N!}{\nu_N^+!\nu_N^-!}(p_+)^{\nu_N^+}(p_-)^{\nu_N^-},\ \nu_N := \nu_N^+ - \nu_N^-.$$

Here, by definition,

$$P_\pm(k) := \frac{1}{2}[1 \pm C(k)],\ k = [Nt],\ t > 0.$$

Thus, the first two moments have the following values:

$$E\Delta\mu_N(t+1/N) \equiv 0,\ \forall t \geq 0,$$

$$E[(\Delta\mu_N(t+1/N))^2 \,|\, \zeta_N(t) = C(k)] = 1 - V^2(C(k)),\ k = [Nt].$$

So, we obtain the approximation theorem:

THEOREM 6.12.– *The normalized statistical experiments given by a solution of the difference stochastic equation* [6.168] *are approximated by the Ornstein–Uhlenbeck process given by the solution of the differential stochastic equation*

$$d\zeta_N(t) = -V_0(\zeta(t))dt + \sigma(\zeta(t))dW(t),\ t \geq 0, \qquad [6.169]$$

characterized by the generator

$$L_\zeta\varphi(c) = -V_0(c)\varphi'(c) + \frac{1}{2}\sigma^2(c)\varphi''(c),\ |c| \leq 1. \qquad [6.170]$$

PROOF OF THEOREM 6.12.– The martingale characterization of the Markov process is used $\zeta_N(t),\ t \geq 0$, given by the solution of the difference stochastic equation [6.168]:

$$\mu_N(t) := \varphi(\zeta_N(t)) - \varphi(\zeta_N(0)) - \int_0^{N[t/N]} L_\zeta^{(N)} \varphi(c)(\zeta_N(s))ds, \qquad [6.171]$$

$$k = [Nt].$$

The normalized generator is set as follows:

$$L_\zeta^N \varphi(c) = NE\Big[\varphi(c + \Delta\zeta_N(t)) - \varphi(c) \,|\, \zeta_N(t) = c\Big]. \qquad [6.172]$$

It is not difficult to make sure that the right-hand side of equation [6.171], taking into account [6.172], defines the martingale:

$$E\mu_N(t) = 0,\ E[\mu_N(t+s) \,|\, \mu_N(t)] = \mu_N(t),\ s > 0,\ t \geq 0.$$

Next, we use an approximation of the generator [6.172] on sufficiently smooth numerical test functions $\varphi(c) \in C_B^3(R)$, bounded and three times continuously differentiated with limited derivatives

$$L_\zeta^N \varphi(c) = L_\zeta \varphi(c) + R_N \varphi(c), \qquad [6.173]$$

with residual term:

$$R_N \varphi(c) \longrightarrow 0,\ N \to \infty,\ \varphi(c) \in C_B^3(R) \qquad [6.174]$$

The approximations [6.173] and [6.174] prove the convergence of the Markov process generators $\zeta_N(t),\ t \geq 0$, in a series scheme with the parameter $N \to \infty$, to the generator [6.170] of the Ornstein–Uhlenbeck process type [6.169], which ensures the convergence of finite-dimensional distributions of the normalized Markov discrete diffusion

$$\zeta_N(t) \Rightarrow \zeta(t),\ N \to \infty,$$

under the additional condition of the convergence of initial states

$$\zeta_N(0) \Rightarrow \zeta(0),\ N \to \infty.$$

Theorem 6.12 is proved. □

6.5. Multivariate statistical experiments with persistent nonlinear regression and equilibrium

Here, we consider multivariate statistical experiments with persistent nonlinear regression. Such stochastic sequences are determined by matrices of directing actions for frequencies of certain attributes. The conditions for the convergence of multivariate statistical experiments to the equilibrium state are studied. A stochastic approximation by the autoregressive process with normal perturbations is constructed for the discrete time model. The main fact is the approximation of statistical experiments with linear, as well as nonlinear, regression by a diffusion process with normal (Gaussian) stochastic components. The normal stochastic component can be viewed as a normal additive error of observations. It turns out that an appropriate linear regression may substitute a nonlinear regression in the approximation as the number of observations increases. We also describe statistical experiments with persistent regression by a generating matrix of numerical parameters, which influence the results of statistical experiments in the next step.

6.5.1. *Basic definitions and assumptions*

Multivariate statistical experiments are generated by samples of a fixed size N of jointly independent random variables $(\delta_r(k),\ 1 \leq r \leq N)$ being identically distributed for every fixed k and assuming a finite number M of values in a set $E = \{e_0, e_1, ..., e_M\}$, $M \geq 0$.

The multivariate statistical experiments are described by the family of frequencies

$$S_N^{(m)}(k+1) := \frac{1}{N}\sum_{r=1}^{N} I\{\delta_r(k) = e_m\},\ 0 \leq m \leq M,\ k \geq 0. \qquad [6.175]$$

Here, the symbol $I\{A\}$ stands for the indicator of a set A.

The balance condition is obvious:

$$\sum_{m=0}^{M} S_N^{(m)}(k) = 1,\ k \geq 0. \qquad [6.176]$$

The increments of frequencies

$$\Delta S_N^{(m)}(k+1) := S_N^{(m)}(k+1) - S_N^{(m)}(k),\ k \geq 0, \qquad [6.177]$$

admit the following balance condition:

$$\sum_{m=0}^{M} \Delta S_N^{(m)}(k) = 0,\ k \geq 1. \qquad [6.178]$$

The multivariate statistical experiments are characterized by the property of *persistent linear regression* that reflects a universal principle of *balanced reaction* at the next step of statistical experiments.

The influence of directing actions determining the choice of a characteristic is compensated by a delayed choice of a given characteristic.

The most general form of this principle corresponds to a construction of a regression function of increments expressed by means of a conditional mathematical expectation:

$$C_0^{(m)}(p) := E[\Delta S_N^{(m)}(k+1) \,|\, \overline{S}_N(k) = p], \; 0 \leq m \leq M, \; k \geq 0. \qquad [6.179]$$

Here, $\overline{S}_N(k) := (S_N^{(m)}(k), \; 0 \leq m \leq M)$ is a vector of states of statistical experiments at the k-th step,

$$\Delta\overline{S}_N(k+1) := \overline{S}_N(k+1) - \overline{S}_N(k), \; k \geq 0,$$

is the increment of the state of statistical experiments at the k-th step, and $p := (p_m, \; 0 \leq m \leq M)$ is the vector column of frequencies $S_N^m(k), \; 0 \leq m \leq M)$.

Consider the positive definite matrix of directing actions

$$\mathbb{V} = [V_{mn}, \; 0 \leq m, n \leq M] = (\overline{V}_m, \; 0 \leq m \leq M) \qquad [6.180]$$

where $\overline{V}_m$ is a vector row of the matrix $\mathbb{V}$.

PROPOSITION 6.9.– *The regression function of increments of frequencies* [6.179] *for multivariate statistical experiments can be written in the vector form as follows:*

$$\overline{C}_0(p) = G(p)\overline{1} - (M+1)\mathbb{V}p, \; G(p) := \overline{1}\mathbb{V}p, \qquad [6.181]$$

where $\overline{1}$ *is the vector of each, whose coordinate equals 1 (we treat* $\overline{1}$ *as a vector row if it is multiplied by a matrix on the left or a vector column otherwise).*

The multiplier $G(p)$ in the first term of [6.181] admits the representation

$$G(p) = \overline{1}\mathbb{V}p = \sum_{m,n=0}^{M} V_{mn}p_n = \sum_{m=0}^{M} \overline{V}_m p,$$

where $\overline{V}_m p$ denotes the scalar product of vectors:

$$\overline{V}_m p = \sum_{n=0}^{M} V_{mn}p_n, \; 0 \leq m \leq M.$$

Using the vector rows of matrix [6.180], we obtain an analogue of representation [6.181] for frequencies:

$$C_0^{(m)}(p) = \sum_{l=0}^{M} \overline{V}_l p - (M+1)\overline{V}_m p.$$

PROPOSITION 6.10.– *The regression functions of increments* [6.179] *are written as*

$$C_0^{(m)}(p) = -\sum_{l \neq m} [(\overline{V}_m - \overline{V}_l)p], \; 0 \leq m \leq M. \qquad [6.182]$$

Moreover, the balance condition

$$\overline{1}\,\overline{C}_0(p) = \sum_{m=0}^{M} C_0^{(m)}(p) = 0 \qquad [6.183]$$

holds.

Relation [6.182] provides an explicit expression of the (mean sense) *balance principle for reactions*: the regression function of increments of frequencies is such that the increment of the m-th frequency decreases by $\overline{V}_m p$; such a decrease is compensated by the term $\sum_{l \neq m} \overline{V}_l p$, by which the other frequencies increase according to the balance condition [6.183].

6.5.2. *Equilibrium and transformation of regression function*

Consider a state of equilibrium for a linear regression function (equilibrium) determined by the solutions of the following system of linear equations:

$$\overline{C}_0(\rho) = 0.$$

Using the decomposition [6.182] for the linear regression function, we obtain the equilibrium $\rho = (\rho_m, 1 \leq m \leq M)$ in the following form:

$$\overline{V}_l \rho = \overline{V}_m \rho, \; 0 \leq l, m \leq M.$$

The latter conditions can be rewritten in the form of a system of linear equations (see [6.182]) as follows:

$$\mathbb{V}\rho = C\overline{1} \qquad [6.184]$$

Moreover,

$$\overline{1}\mathbb{V}\rho = \sum_{m=0}^{M} \overline{V}_m \rho = (M+1) \cdot C, \; C = \text{const}. \qquad [6.185]$$

In what follows, we use the notation $\mathbb{V}^{-1} := [\overline{V}_{mn}]$ for the inverse matrix to the matrix of directing actions [6.180]. Here, $\overline{V}_{mn}$, $0 \leq m, n \leq M$ are the entries of the matrix $\mathbb{V}^{-1}$.

LEMMA 6.5.– *The state of equilibrium in the case of regression function* [6.182] *is given by*

$$\rho = (\rho_m = \overline{V}_m / \overline{V}, \ 0 \le m \le M), \qquad [6.186]$$

where $\overline{V}_m = \sum_{l=0}^{M} \overline{V}_{ml}$ *denotes the sum of entries of a row of the matrix* $\mathbb{V}^{-1} = [\overline{V}_{mn}]$, *and* $\overline{V} =: \sum_{m=0}^{M} \overline{V}_m = \sum_{m,n=0}^{M} \overline{V}_{mn}$ *denotes the sum of all entries of the matrix* $[\overline{V}_{mn}]$.

PROOF OF LEMMA 6.5.– Relation [6.184] implies

$$\rho = \mathbb{V}^{-1} C \overline{1} = C \mathbb{V}^{-1} \overline{1}. \qquad [6.187]$$

By definition,

$$\mathbb{V}^{-1} \overline{1} = (\overline{V}_m, \ 0 \le m \le M). \qquad [6.188]$$

Using [6.188], relation [6.187] can be rewritten as follows:

$$\rho = (\rho_m = C \overline{V}_m, \ 0 \le m \le M). \qquad [6.189]$$

Now, we use the property of the equilibrium

$$\sum_{m=1}^{M} \rho_m = 1$$

to evaluate the constant

$$C \sum_{m=1}^{M} \overline{V}_m = 1 \, ; \ C = 1/\overline{V}. \qquad [6.190]$$

The relation [6.185] follows from [6.189] and [6.190]. □

Let

$$\widehat{p} := p - \rho = (\widehat{p}_m = p_m - \rho_m, \ 0 \le m \le M) \qquad [6.191]$$

be the fluctuations of frequencies around the equilibrium.

REMARK 6.15.– *The regression function of increments* [6.181] *of multivariate statistical experiments only depends on the fluctuations of frequencies* [6.191], *namely*

$$\overline{C}_0(p) = \overline{C}_0(\widehat{p}) = (\overline{1} \mathbb{V} \widehat{p}) \overline{1} - (M+1) \overline{V}_m \widehat{p}, \ 0 \le m \le M. \qquad [6.192]$$

This property follows from the properties of linearity and [6.184].

Relation [6.192] *can be written in the component-wise form as follows:*

$$\overline{C}_0(\widehat{p}) = \left(C_0^{(m)}(\widehat{p}) = \sum_{l \ne m} [(\overline{V}_l - \overline{V}_m) \widehat{p}], \ 0 \le m \le M \right).$$

PROPOSITION 6.11.– *The regression functions for increments of fluctuations of frequencies* [6.192] *are given by*

$$\overline{C}_0(\widehat{p}) = -\widehat{\mathbb{V}}\widehat{p}, \qquad [6.193]$$

$$\widehat{\mathbb{V}} := \left(\widehat{\overline{V}}_m = \sum_{l \neq m} [\overline{V}_m - \overline{V}_l],\ 0 \leq m \leq M \right).$$

In other words,

$$\widehat{\overline{V}}_m := \left(\widehat{V}_{mn} = \sum_{l \neq m} [V_{ln} - V_{mn}],\ 0 \leq n \leq M \right),$$

where

$$\overline{V}_m := (V_{mn},\ 0 \leq n \leq M).$$

6.5.3. *Examples*

The diagonal matrices of directing actions [6.180], determining the regression functions of increments [6.193], are the simplest examples. Let a diagonal matrix of directing actions be given by

$$\mathbb{V} = [V_m \delta_{mn}\ ;\ \ 0 \leq m, n \leq M], \qquad [6.194]$$

where δ_{mn} is the Kronecker symbol:

$$\delta_{mn} = \begin{cases} 1 & \text{for } m = n, \\ 0 & \text{for } m \neq n. \end{cases}$$

EXAMPLE 6.1. ($M = 1$). Let the matrix of directing actions be determined by two parameters, V_m, $m = 1, 2$. According to relation [6.193],

$$C_0^{(m)}(\widehat{p}) = -(V_m - V_l)\widehat{p},\ l, m = 1, 2,\ l \neq m.$$

Thus,

$$C_0^{(1)}(\widehat{p}) = -V_1\widehat{P}_1 + V_2\widehat{P}_2,\ C_0^{(2)}(\widehat{p}) = V_1\widehat{P}_1 - V_2\widehat{P}_2.$$

In what follows, we denote by (+) the first attribute and by (-) the second one. Also, let

$$V_+ := V_1,\ V_- := V_2.$$

Then, the regression functions of the increments of binary attributes are given by the matrix

$$C_0(\widehat{p}) = -\widehat{\mathbb{V}}\widehat{p},\ \widehat{\mathbb{V}} = \begin{bmatrix} V_+ & -V_- \\ -V_+ & V_- \end{bmatrix}.$$

Hence,

$$C_0^+(\widehat{p}) = V_-\widehat{P}_- - V_+\widehat{P}_+,\ C_0^-(\widehat{p}) = V_+\widehat{P}_+ - V_-\widehat{P}.$$

The equilibrium of the regression function of increments is determined by relation [6.186]:

$$\rho_\pm = V_\pm^{-1}/\overline{V},\ \overline{V} := V_+^{-1} + V_-^{-1}.$$

Thus, the equilibrium of binary statistical experiments is such that

$$\rho_\pm = V_\mp/V,\ V := V_+ + V_-.$$

EXAMPLE 6.2. ($M = 2$). Let the statistical experiments be such that the corresponding matrix of directing actions [6.180] depends on three parameters, V_m, $m = 0, 1, 2$. Then, the matrix of directing actions determining the regression function for increments of fluctuations of frequencies [6.193] is of the following form:

$$\widehat{\mathbb{V}} = \begin{bmatrix} 2V_0 & -V_1 & -V_2 \\ -V_0 & 2V_1 & -V_2 \\ -V_0 & -V_1 & 2V_2 \end{bmatrix}.$$

Hence,

$$\begin{aligned} C_0(\widehat{p}) &= -\widehat{\mathbb{V}}\widehat{p} = \left(C_0^0(\widehat{p}),\ C_0^1(\widehat{p}),\ C_0^2(\widehat{p})\right) \\ &= (V_1\widehat{P}_1 + V_2\widehat{P}_2 - 2V_0\widehat{P}_0, V_0\widehat{P}_0 + V_2\widehat{P}_2 - 2V_1\widehat{P}_1, \\ &\quad V_0\widehat{P}_0 + V_1\widehat{P}_1 - 2V_2\widehat{P}_2), \end{aligned}$$

This implies the matrix representation of the regression function [6.193]:

$$C_0(\widehat{p}) = -\widehat{\mathbb{V}}\widehat{p},\ \widehat{\mathbb{V}} := [\widehat{V}_{mn};\ 0 \le m, n \le 2],$$

$$\widehat{V}_{mn} = -V_n,\ n \neq m,\ \widehat{V}_{mm} = 2V_m.$$

The equilibrium of statistical experiments is evaluated according to the relations [6.186]–[6.187]:

$$\overline{p} = \left(\rho^{(0)},\ \rho^{(1)},\ \rho^{(2)}\right);$$

$$\rho^{(0)} = V_1V_2/(V_0V_1 + V_0V_2 + V_1V_2),\ \ \rho^{(1)} = V_0V_2/(V_0V_1 + V_0V_2 + V_1V_2),$$

$$\rho^{(2)} = V_0V_1/(V_0V_1 + V_0V_2 + V_1V_2)\ ;\ \ \rho^{(0)} + \rho^{(1)} + \rho^{(2)} = 1.$$

EXAMPLE 6.3. ($M > 2$). Let the matrix of directing actions [6.194] be diagonal and depend on the parameters V_m, $0 \leq m \leq M$.

Then, the matrix of directing actions for fluctuations [6.193] is

$$\widehat{V}_{mn} = \begin{cases} MV_m & \text{for } m = n, \\ -V_n & \text{for } m \neq n. \end{cases}$$

The regression function of increments [6.193] is rewritten as follows:

$$C_0^{(m)}(\widehat{p}) = -\widehat{\mathbb{V}}\widehat{p} = -\sum_{l\neq m}[(V_m - V_l)]\widehat{p}$$
$$= \sum_{l\neq m} V_l\widehat{P}_l - MV_m\widehat{P}_m, \ 0 \leq m \leq M.$$

This implies the matrix representation of the regression function of increments [6.193], namely

$$C_0(\widehat{p}) = -\widehat{\mathbb{V}}\widehat{p}, \ \widehat{\mathbb{V}} := [\widehat{V}_{mn}; \ 0 \leq m, n \leq M],$$

$$\widehat{V}_{mn} = \begin{cases} MV_m & \text{for } m = n, \\ -V_n & \text{for } m \neq n. \end{cases}$$

The equilibrium of statistical experiments can be evaluated by relation [6.186], which was proved in Lemma 6.5:

$$\overline{\rho} = \left(\rho^{(0)}, \rho^{(1)}, \ldots, \rho^{(M)}\right) \ : \ \rho^{(m)} = \prod_{l\neq m} V_l \Big/ \sum_{l=0}^{M} \prod_{l\neq m} V_l, \ 0 \leq m \leq M.$$

6.5.4. *The state of equilibrium of multivariate statistical experiments*

An important property of multivariate statistical experiments with persistent regression is the existence of an equilibrium state $\rho = (\rho_m, \ 1 \leq m \leq M)$, for the regression function given by equality [6.186] (see Lemma 6.5).

The existence of a state of equilibrium ρ follows from the existence of a steady state of multivariate statistical experiments.

THEOREM 6.13.– *Assume that*

$$S_N^{(m)}(0) \xrightarrow{P1} \rho_m, \ N \to \infty, \ 1 \leq m \leq M.$$

Then, the statistical experiments converge:

$$S_N^{(m)}(k) \xrightarrow{P1} \rho_m, \ N \to \infty, \ 1 \leq m \leq M, \qquad [6.195]$$

for every finite $k > 0$.

PROOF.– Consider the martingale differences

$$\begin{aligned}\Delta\mu_N^{(m)}(k+1) :&= \Delta S_N^{(m)}(k+1) - E[\Delta S_N^{(m)}(k+1) \,|\, \overline{S}_N(k)] \\ &= S_N^{(m)}(k+1) - E[S_N^{(m)}(k+1) \,|\, \overline{S}_N(k)] \\ &= S_N^{(m)}(k+1) - C^{(m)}(\overline{S}_N(k)), \; k \geq 0.\end{aligned} \quad [6.196]$$

Here,

$$C^{(m)}(\overline{p}) := p + C_0^{(m)}(\overline{p}),$$

by definition.

The quadratic characteristic of the martingale difference [6.196] is easy to evaluate:

$$\begin{aligned}\langle\Delta\mu_N^{(m)}\rangle(k+1) :&= E\left[[S_N^{(m)}(k+1) - E[S_N^{(m)}(k+1) \,|\, \overline{S}_N(k)]]^2 \,|\, \overline{S}_N(k)\right] \\ &= C^{(m)}(\overline{S}_N(k))[1 - C^{(m)}(\overline{S}_N(k))]/N =: \sigma_m^2(\overline{S}_N(k)).\end{aligned} \quad [6.197]$$

Let

$$\Delta M_N^{(m)}(n) := \sum_{k=0}^{n} \Delta\mu_N^{(m)}(k+1), \; n \geq 0, \quad [6.198]$$

be the sum of the martingale differences [6.196].

Using definition [6.196], we can write the quadratic characteristic of the martingale [6.198] as follows:

$$\langle M_N^{(m)}\rangle(n) = \frac{1}{N}\sum_{k=0}^{n} \sigma_m^2(\overline{S}_N(k)), \; n \geq 0, \quad [6.199]$$

$$\sigma_m^2(\overline{p}) := C^{(m)}(\overline{p})[1 - C^{(m)}(\overline{p})].$$

Hence,

$$\langle M_N\rangle_n^{(m)} \xrightarrow{P1} 0,, \; N \to \infty, \; 0 \leq m \leq M$$

for all fixed $n \geq 0$. This implies the convergence of martingales [6.197] for every finite integer number $n \geq 0$:

$$\Delta\mu_N^{(m)}(n) \xrightarrow{P1} 0,, \; N \to \infty, \; 0 \leq m \leq M.$$

In particular,

$$\Delta\mu_N^{(m)}(0) = S_N^{(m)}(1) - C(S_N^{(m)}(0)) = S_N^{(m)}(1) - S_N^{(m)}(0) - C_0^{(m)}(S_N^{(m)}(0))$$

for $n = 0$. By the assumption of the theorem,

$$C_0^{(m)}(S_N^{(m)}(0)) \xrightarrow{P1} C_0^{(m)}(\rho_m) = 0, \; N \to \infty, \; 0 \leq m \leq M.$$

Therefore,

$$S_N^{(m)}(1) - \rho_m \xrightarrow{P1} 0, \; N \to \infty, \; 0 \leq m \leq M.$$

Using the induction, we prove convergence [6.195] for every $k \geq 1$. □

6.5.5. *Stochastic approximation of multivariate statistical experiments*

The stable state introduced in Theorem 6.13 does not have a characterization that can be used for the analysis of multivariate statistical experiments. An important result for the discrete time is the asymptotic approximation of multivariate statistical experiments by a normal autoregressive process. The corresponding result is stated below as $N \to \infty$.

THEOREM 6.14.– *Let the assumptions from Theorem 6.13 hold. Then, the normalized fluctuations converge in distribution, namely*

$$\sqrt{N}[\Delta\widehat{S}_N^{(m)}(k+1) + \mathbb{V}\widehat{S}_N^{(m)}(k)] \xrightarrow{D} \sigma_m W(k+1), \; N \to \infty, \; k \geq 0 \qquad [6.200]$$

for every finite $k \geq 0$. Here, $\{W(k), \; k \geq 1\}$ is a sequence of standard jointly independent Gaussian random variables.

Moreover,

$$\sigma_m^2 := \rho_m(1 - \rho_m), \; 0 \leq m \leq M.$$

PROPOSITION 6.12.– *The normal autoregressive process approximating multivariate statistical experiments by* [6.175]–[6.181] *is given by*

$$\Delta\widetilde{S}_N^{(m)}(k+1) + \mathbb{V}\widetilde{S}_N^{(m)}(k) = \frac{\sigma_m}{\sqrt{N}} W(k+1), \; k \geq 0. \qquad [6.201]$$

REMARK 6.16.– *The stochastic approximation* [6.201] *of the model* [6.176]–[6.179] *of multivariate statistical experiments preserves the property of the persistent linear regression:*

$$E[\Delta\widetilde{S}_N^{(m)}(k+1) \mid \widetilde{\widetilde{S}}_N(k)] = -\mathbb{V}\,\widetilde{S}_N^{(m)}(k), \; k \geq 0.$$

PROOF OF THEOREM 6.14.– Consider the normalized martingale defined by [6.198]:

$$\Delta\overline{M}_N^{(m)}(n) := \sqrt{N}\Delta M_N^{(m)}(n) = \sqrt{N}\sum_{k=0}^{n}\Delta\mu_N^{(m)}(k+1),\ n \geq 0. \qquad [6.202]$$

By relation [6.199], the quadratic characteristic of the normalized martingale [6.202] is of the form

$$\langle\Delta\overline{M}_N^{(m)}\rangle(n) = \sum_{k=0}^{n}\sigma_m^2(S_N^{(m)}(k)),\ n \geq 0. \qquad [6.203]$$

According to Theorem 6.13,

$$\sigma_m^2(S_N^{(m)}(k)) \xrightarrow{P1} \sigma_m^2 := \rho_m(1-\rho_m),\ N \to \infty,\ k \geq 0.$$

Thus, the quadratic characteristic of the normalized martingale [6.203] converges, i.e.

$$\langle\Delta\overline{M}_N^{(m)}\rangle(n) \xrightarrow{P1} (n+1)\sigma_m^2,\ N \to \infty,\ n \geq 0. \qquad [6.204]$$

Now, the central limit theorem (see, for example, Liptser and Shiryaev (2001)) implies that the predictable part of the normalized martingale [6.203] converges in distribution to the sum of normal random variables; i.e.

$$\sqrt{N}[\Delta S_N^{(m)}(k+1) + \mathbb{V}S_N^{(m)}(k)] \xrightarrow{D} W_{\sigma_m}(k+1),\ N \to \infty,\ k \geq 0. \qquad [6.205]$$

Moreover, the convergence of quadratic characteristics of the normalized martingales [6.204] means that

$$W_{\sigma_m}(k) = \sigma_m W(k),\ k \geq 0, \qquad [6.206]$$

where $W(k),\ k \geq 0$ is a sequence of jointly independent standard normal random variables. The limit variance of the normalized martingale equals the sum of variances of martingale differences.

The relations [6.205] and [6.206] mean that the convergence [6.200] holds for every finite $k \geq 0$. Theorem 6.14 is proved. □

6.6. Multivariant Wright–Fisher model

The population genetic models by Wright–Fisher are defined by regression functions, which are determined by a ratio of two quadratic forms (Ethier and Kurtz 1986, Chapter 10).

However, the equilibrium state, defined by the equilibrium point of the regression function, requires additional analysis (see, for example, Korolyuk and Korolyouk (1995) and Skorokhod *et al.* (2002)).

At the same time, the equilibrium is easily determined for the incremental regression function at each stage (Koroliouk *et al.* 2014; Koroliouk (2015a, 2015b)).

In the present consideration, the models of the population genetics of genotype interactions are determined by the difference evolution equations with regression functions of increments for the frequency probabilities of genotypes.

In this case, the equilibrium state of the probability frequency is given by the equilibrium of the regression function of increments, which is postulated by the form of such a function.

For multivariant Wright–Fisher generalized models, we define the equilibrium states, expressed by the fluctuations of probability relations, in distinction with the traditionally used fluctuations, expressed by the difference between the current value of a random process and its equilibrium value. Thus, the drift component of the gene frequency dynamic process, primarily expressed as a ratio of two quadratic forms, is transformed into a cubic parabola with a certain normalization factor.

6.6.1. *Regression function of increments*

The probabilities of genotype frequencies at each stage $k \geq 0$ are determined by the evolutionary process $P(k) = (P_m(k),\ 0 \leq m \leq M)$ with $M + 1$ $(M \geq 1)$ finite number of the state set $E = \{e_0, e_1, \dots e_M\}$.

The dynamics of the frequency probabilities at the next $k+1$-th stage $(k \geq 0)$ are given by the regression function (Ethier and Kurtz 1986, Chapter 10)

$$P_m(k+1) := W_m(p)/W(p),\ 0 \leq m \leq M,\ k \geq 0, \qquad [6.207]$$

$$W_m(p) := p_m \sum_{n=0}^{M} W_{mn} p_n,\ \ 0 \leq m \leq M, \qquad [6.208]$$

$$W(p) := \sum_{n=0}^{M} W_m(p). \qquad [6.209]$$

The probabilities of frequencies obey the usual restrictions $0 \leq p_m \leq 1$, $\sum_{n=0}^{M} p_n = 1$. The respective restrictions for the survival parameters are $0 \leq W_{mn} \leq 1, 0 \leq m, n \leq M$.

The increment of the probability at each stage

$$\Delta P_m(k+1) := P_m(k+1) - P_m(k),\ 0 \leq m \leq M,\ k \geq 0, \quad [6.210]$$

is given by the incremental regression function

$$\Delta P_m(k+1) = W_0^{(m)}(p),\ 0 \leq m \leq M, \quad [6.211]$$

$$W_0^{(m)}(p) = V_0^{(m)}(p)/W(p),\ V_0^{(m)}(p) := W_m(p) - p_m W(p),\ 0 \leq m \leq M. \quad [6.212]$$

Let us introduce new parameters of survival:

$$V_{mn} := 1 - W_{mn},\ 0 \leq m, n \leq M. \quad [6.213]$$

Then, the numerator of incremental regression function [6.212] is transformed into the form:

$$V_0^{(m)}(p) = p_m \left[\sum_{n=0}^{M} p_n (V_n, p) - (V_m, p) \right],\ 0 \leq m \leq M, \quad [6.214]$$

and the normalizing denominator [6.209] has the form:

$$W(p) = 1 - \sum_{n=0}^{M} p_n (V_n, p) \quad [6.215]$$

where the scalar product

$$(V_m, p) := \sum_{n=0}^{M} V_{mn} p_n,\ 0 \leq m \leq M. \quad [6.216]$$

Introduce the equilibriums of incremental regression functions [6.214] by the relations:

$$(V_m, \rho) = \pi,\ 0 \leq m \leq M,\ \pi := \prod_{n=0}^{M} \rho_n. \quad [6.217]$$

The normalized constant π is also generated by equilibriums $\rho = (\rho_m, 0 \leq m \leq M)$.

LEMMA 6.6.– *The equilibriums of the incremental regression functions* [6.212]–[6.216] *are given by the relation:*

$$\rho_m = \pi \overline{V}_m,\ 0 \leq m \leq M,\ \pi := \prod_{n=0}^{M} \rho_n, \quad [6.218]$$

where $\overline{V}_m := \sum_{n=0}^{M} \overline{V}_{mn}$, $0 \leq m \leq M$, *with the summands which are the elements of inverse matrix* $\mathbb{V}^{-1} := [\overline{V}_{mn}, 0 \leq m, n \leq M]$, *with respect to the directing parameter matrix* $\mathbb{V} = [V_{mn};\ 0 \leq m, n \leq M]$, *under the additional normalization condition* $\sum_{m=0}^{M} \overline{V}_m = \sum_{m,n=0}^{M} \overline{V}_{mn} = \pi^{-1}$.

PROOF.– The relation [6.217] means that

$$\mathbb{V}\rho = \pi\mathbf{1},\ \pi\mathbf{1} := (\pi,\ 0 \leq n \leq M).$$

Hence, the vector of equilibriums has the following representation:

$$\rho = \pi\mathbb{V}^{-1}\mathbf{1},\ \rho_m = \overline{V}_m\pi,\ 0 \leq m \leq M, \quad [6.219]$$

i.e. the assertion of Lemma [6.218]. □

COROLLARY 6.5.– *The equilibria* [6.218] *provide the equilibrium state of the probability frequency* [6.207]*:*

$$V_0^{(m)}(\rho) \equiv 0,\ 0 \leq m \leq M. \quad [6.220]$$

COROLLARY 6.6.– *The equilibria* [6.218] *generate a representation of the scalar products* [6.216] *by fluctuations of the probability relations:*

$$(V_m, p) = \pi p_m/\rho_m,\ 0 \leq m \leq M. \quad [6.221]$$

The normalizing constant π *is defined in* [6.217].

First of all, note that relation [6.218] coincides with the definition of the equilibrium [6.217], under the additional assumption that the directing parameter matrix $\mathbb{V} = V_m\delta_{mn};\ 0 \leq m, n \leq M$ is diagonal. Hence, we have the following:

LEMMA 6.7.– *The following relation takes place:*

$$\pi\rho^{-1} = \mathbb{V}\mathbf{\mathit{I}}, \quad [6.222]$$

which coincides with formula [6.221].

Now, the incremental regression functions [6.212]–[6.215] with the relations [6.220] generate the following.

PROPOSITION 6.13.– *The incremental regression functions with the Wright–Fisher normalization are given by the relations:*

$$W_0^{(m)}(p) = V_0^{(m)}(p)/W(p), \quad [6.223]$$

$$V_0^{(m)}(p) = \pi\rho_m[\sum_{m=0}^{M} p_n^2/\rho_n - p_m/\rho_m], \quad [6.224]$$

$$W(p) = 1 - \pi\sum_{m=0}^{M} p_n^2/\rho_n. \quad [6.225]$$

It is obvious that the global balance condition:

$$\sum_{m=0}^{M} V_0^{(m)}(p) = 0, \tag{6.226}$$

which in the scalar form is the following:

$$\sum_{m=0}^{M} p_m \sum_{n=0}^{M} p_n^2/\rho_n - \sum_{m=0}^{M} p_m^2/\rho_m \equiv 0. \tag{6.227}$$

6.6.2. *Equilibrium state*

The presence of equilibrium state is provided by the equilibrium point of the incremental regression function:

$$V_0^{(m)}(\rho) = 0,\ 0 \le m \le M,\ \rho = (\rho_m,\ 0 \le m \le M). \tag{6.228}$$

The normalizing Wright–Fisher factor has the form:

$$W(\rho) = 1 - \pi,\ \pi = \prod_{m=0}^{M} \rho_m. \tag{6.229}$$

The equilibrium generated by the state $\rho = (\rho_m,\ 0 \le m \le M)$ is interpreted by the convergence of evolutionary processes [6.207].

THEOREM 6.15.– *For any initial data:* $0 < P_m(0) < 1,\ 0 \le m \le M$, *the evolutionary processes* $P_m(k),\ 0 \le m \le M,\ k \ge 0$, *which are determined by the solutions of the difference evolutionary equation* [6.211], *with the incremental regression function* [6.223]–[6.225] *converging, by* $k \to \infty$, *to equilibrium*

$$\lim_{k\to\infty} P_m(k) = \rho_m,\ 0 \le m \le M. \tag{6.230}$$

PROOF.– The property of the main components is used, which is specified by the sum

$$\sum_{n=0}^{M} p_n^2/\rho_n = \sum_{n=0}^{M} p_n(p_n/\rho_n). \tag{6.231}$$

This means averaging the fluctuations of the probability relations p_n/ρ_n, $0 \le m \le M$ on the distribution of frequencies at the current stage. In this case, the fluctuations of the ratios are equal to one for $p_n = \rho_n$, $0 \le m \le M$, and at the same time, the main component of the incremental regression function is also equal to one.

Consequently, the possible values of the frequency probabilities can be split into three zones:

$$(+) \quad p_n < \rho_n \quad ; \qquad (-) \quad p_n > \rho_n \quad ; \qquad (0) \quad p_n = \rho_n \quad .$$

The signs of the incremental regression functions [6.222]–[6.224] in such a zone are the same:

in zone (+), the probabilities increase; in zone (-), they decrease.

Therefore, a limit [6.229] exists, whose value is ensured by the necessary condition for the existence of a limit:

$$\lim_{k\to\infty} \Delta P_m(k+1) = 0. \qquad [6.232]$$

□

6.7. Binary evolutionary process

The binary evolutionary process $P_\pm(k)$, $k \geq 0$ is determined by the following regression functions Ethier and Kurtz (1986):

$$P_\pm(k+1) = W_\pm(p)/W(p), \; k \geq 0, \qquad [6.233]$$

$$W_\pm(p) = P_\pm(W_\pm p_\pm + p_\mp), \qquad [6.234]$$

$$W(0) = W_+(p_+) + W_-(p_-) = W_+ p_+^2 + 2p_+p_- + W_- p_-^2. \qquad [6.235]$$

The frequency probabilities at the k-th stage satisfy the usual conditions $0 \leq p_\pm \leq 1$, $p_+ + p_- = 1$. The survival parameters are also limited by the relation $0 < W_\pm < 1$.

For probability increments

$$\Delta P_\pm(k+1) := P_\pm(k+1) - P_\pm(k), \; k \geq 0, \qquad [6.236]$$

the corresponding regression functions of increments can be represented as follows:

$$W_0^\pm(p_\pm) = W_\pm(p_\pm)/W(p) - p_\pm, \qquad [6.237]$$

or equivalently

$$W_0^\pm(p_\pm) = V_0^\pm(p_\pm)/W(p), \qquad [6.238]$$

$$V_0^\pm(p_\pm) = W_\pm(p_\pm) - p_\pm W(p). \qquad [6.239]$$

Now, introduce direction parameters, based on the survival ones: $V_\pm := 1 - W_\pm$. The relative equilibriums will be $\rho_\pm = V_\mp^{-1}$, with the normalization condition

$V_+ + V_- = 1$. Then, the numerators [6.233] of the regression function have the following form:

$$W_\pm(p_\pm) = p_\pm(1 - \pi p_\mp/\rho_\pm), \; \pi := \rho_+\rho_-. \qquad [6.240]$$

Therefore, the numerator [6.238] transforms into the following:

$$V_0^\pm(p_\pm) = p_\mp W_\pm(p_\pm) - p_\pm W_\mp(p_\mp) = p_+p_-(\rho_\pm p_\mp - \rho_\mp p_\pm). \qquad [6.241]$$

The linear component has the following representation:

$$\rho_+p_- - \rho_-p_+ = -(p_+ - \rho_+) = p_- - \rho_-. \qquad [6.242]$$

So, the regression functions of the increments of binary evolutionary processes are represented by the probability of fluctuations:

$$V_0^\pm(p_\pm) = -p_+p_-(p_\pm - \rho_\pm). \qquad [6.243]$$

The Wright–Fisher normalizing factor has the following representation:

$$W(p) = 1 - \pi[p_+^2/\rho_+ + p_-^2/\rho_-] = 1 - [\rho_-p_+^2 + \rho_+p_-^2]. \qquad [6.244]$$

The balance condition is also evident:

$$V_0^+(p_+) + V_0^-(p_-) \equiv 0. \qquad [6.245]$$

The evolutionary processes considered above serve as predictable components of stochastic models in population genetics and are represented by conditional mathematical expectations:

$$P_m(k+1) := E[S_N^{(m)}(k+1) \mid S_N(k) = P(k)], \; 0 \le m \le M, \; k \ge 0. \qquad [6.246]$$

The stochastic models in population genetics are determined by the averaged sums

$$S_N(k) := \frac{1}{N}\sum_{n=1}^{N} \delta_n(k), \; k \ge 0 \qquad [6.247]$$

of random sample variables $\delta_n(k)$, $1 \le n \le N$, which take values in a finite set, with $M+1$ $(M \ge 1)$ states $E = \{e_0, e_1, \ldots, e_M\}$ (see Korolyouk (2016b)). So, the stochastic models [6.247] are defined by the sum of two components:

$$S_N(k+1) = V(S_N(k)) + \Delta\mu_N(k+1), \; k \ge 0. \qquad [6.248]$$

The first, predictable component, is generated by the conditional mathematical expectations:

$$V_m(P_m(k)) = P_m(k) + V_0^{(m)}(P_m(k))/W(P_m(k)), \qquad [6.249]$$

$$0 \le m \le M, \; k \ge 0.$$

The second component forms a martingale difference

$$\Delta\mu_N(k+1) = S_N(k+1) - V(S_N(k)),\ k \geq 0, \tag{6.250}$$

characterized by the first moments:

$$E\Delta\mu_N^{(m)}(k+1) = 0,\ 0 \leq m \leq M,$$
$$E[(\Delta\mu_N^{(m)}(k+1))^2 \mid S_N(k)] = \sigma_m^2(S_N(k)),\ 0 \leq m \leq M,\ k \geq 0. \tag{6.251}$$

The conditional dispersion is determined by regression functions:

$$\sigma_m^2(p) = V_m(p)[1 - V_m(p)],\ 0 \leq m \leq M. \tag{6.252}$$

The asymptotic properties of the stochastic models [6.248]–[6.251] by $N \to \infty$, as well as by $k \to \infty$, are investigated in Korolyouk (2020). The algorithms of phase merging (Korolyuk and Limnios 2005) and statistical estimation of drift parameter (Korolyouk 2016a, 2020) can be directly applied.

7

Random Evolutionary Systems in Discrete–Continuous Time

7.1. Discrete Markov evolutions in an asymptotic diffusion environment

The asymptotic diffusion environment, defined by a solution of a difference evolutionary equation in an ergodic Markov environment, determines the limit diffusion environment for a discrete Markov evolution.

7.1.1. *Asymptotic diffusion perturbation*

The *Markov random environment* is given by a two-component Markov chain (x_n, τ_n), $n \geq 0$, homogeneous by the second component (Korolyuk and Limnios 2005):

$$\mathcal{P}\{x_{n+1} \in B, \; \tau_{n+1} \leq t + c \,|\, x_n = x, \; \tau_n = c\} = P(x,B)G_x(t), \; t \geq 0. \tag{7.1}$$

The semi-Markov kernel

$$Q(x,B,t) := P(x,B)G_x(t), \; x \in E, \; B \in \mathcal{E}, \; t \geq 0, \tag{7.2}$$

specifies the transition probabilities of a two-component Markov chain (x_n, τ_n), $n \geq 0$:

$$P(x,B) := \mathcal{P}\{x_{n+1} \in B \,|\, x_n = x\}, \; x \in E, \; B \in \mathcal{E}, \tag{7.3}$$

$$G_x(t) := \mathcal{P}\{\theta_x \leq t\} = \mathcal{P}\{\theta_{n+1} \leq t \,|\, x_n = x\}, \; x \in E, \; B \in \mathcal{E}, \; t \geq 0. \tag{7.4}$$

Here, the random variables

$$\theta_{n+1} := \tau_{n+1} - \tau_n, \; n \geq 0, \; \tau_0 = 0, \tag{7.5}$$

determine the sojourn time in the states $x \in E$ of the embedded Markov chain x_n, $n \geq 0$. In the particular case of exponentiality of distributions [7.4]

$$\overline{G}_x(t) = 1 - G_x(t) = e^{-q(x)t}, \ t \geq 0.$$

The Markov random environment is really Markov, i.e. x_{τ_n}, $n \geq 0$, is a Markov chain.

For arbitrary functions of the distribution of the sojourn time [7.4], the Markov random environment becomes a *semi-Markov* process.

The asymptotic diffusion is generated by the following *difference evolution equation*:

$$\Delta Y^\varepsilon(t + \varepsilon^2) = \varepsilon A_0(Y^\varepsilon(t);\, x) + \varepsilon^2 A(Y^\varepsilon(t);\, x), \ x \in E. \qquad [7.6]$$

The singular term $A_0(y;\, x)$ with scaling factor ε satisfies the balance condition:

$$\int_E \rho(dx) A_0(y;\, x) \equiv 0. \qquad [7.7]$$

The averaging [7.7] is determined by the stationary distribution of the embedded Markov chain x_n, $n \geq 0$:

$$\rho(B) = \int_E \rho(dx) P(x, B), \ B \in \mathcal{E}, \ \rho(E) = 1. \qquad [7.8]$$

It is known (Korolyuk and Limnios 2005) that the solution of difference evolutionary equation [7.6] and [7.7] generates a diffusion process, as $\varepsilon \to 0$.

According to the tradition of singular perturbation theory (Korolyuk and Limnios 2005), the discrete Markov process is considered in a discrete–continuous time with small parameter $\varepsilon \to 0$ ($\varepsilon > 0$): $t_k = [k\varepsilon^2]$, $k \geq 0$.

BASIC ASSUMPTION 7.1.– The discrete Markov process with asymptotic diffusion is given by the difference stochastic equation

$$\Delta\zeta^\varepsilon(t + \varepsilon^2) = -\varepsilon^2 V(Y^\varepsilon(t))\zeta^\varepsilon(t) + \varepsilon\sigma(Y^\varepsilon(t))\Delta\mu^\varepsilon(t), \ t \geq 0. \qquad [7.9]$$

The asymptotic diffusion $Y^\varepsilon(t)$, $t \geq 0$, is determined by a solution of the difference evolutionary equation [7.6] with the balance condition [7.7].

The discrete Markov process $\zeta^\varepsilon(t)$, $t \geq 0$, given by the difference stochastic equation solution [7.9], is supplemented by an asymptotic diffusion $Y^\varepsilon(t)$, $t \geq 0$, given by the difference evolutionary equation [7.6].

THEOREM 7.1.– *Under the conditions of uniform ergodicity of the embedded Markov chain x_n, $n \geq 0$, with the stationary distribution* [7.8], *the two-component process of the discrete Markov process* [7.9], *and the asymptotic diffusion $Y^\varepsilon(t)$, $t \geq 0$, converge in distribution, as $\varepsilon \to 0$, to an Ornstein–Uhlenbeck diffusion process with evolution:*

$$\{\zeta^\varepsilon(t),\, Y^\varepsilon(t)\} \xrightarrow{D} \{\zeta^0(t),\, Y^0(t)\},\ \varepsilon \to 0,\ 0 \leq t \leq T. \qquad [7.10]$$

The limit two-component diffusion process $\{\zeta^0(t),\, Y^0(t)\}$, $t \geq 0$, with evolution, is determined by the generator:

$$\mathfrak{L}^0(y)\varphi(c,\, y) = -\, V(y)\varphi'_c(c,\, y) + \frac{1}{2}\sigma^2(y)\varphi''_c(c,\, y)$$
$$+ \widehat{A}(y)\varphi'_y(c,\, y) + \frac{1}{2}\widehat{B}^2(y)\varphi''_y(c,\, y), \varphi \in C_0^2(R, \mathcal{B}). \qquad [7.11]$$

Here, by definition,

$$\widehat{A}(y) := \int_E \rho(dx)[A(y;\, x) + A_1(y;\, x)], \qquad [7.12]$$

$$A_1(y;\, x) := A_0(y;\, x)\, P\, R_0\, A'_{0y}(y;\, x), \qquad [7.13]$$

$$\widehat{B}^2(y) = \int_E \rho(dx) B(y;\, x),$$

$$B(y;\, x) = A_0(y;\, x)\, P\, [R_0 + \frac{1}{2}\mathbb{I}]\, A_0(y;\, x). \qquad [7.14]$$

The potential kernel (Korolyuk and Limnios 2005, section 5.2):

$$R_0 = (Q + \Pi)^{-1} - \Pi,\ Q := P - I,\ \Pi\varphi(x) := \int_E \rho(dx)\varphi(x). \qquad [7.15]$$

REMARK 7.1.– *With an additional condition for the continuity of parameters $V(y)$ and $G(y)$, $y \in R$, the Markov processes $\zeta^0(t)$, $Y^0(t)$, $t \geq 0$, given by the generator* [7.11], *are Feller processes (Liggett 2010, p. 93, Theorem 3.3).*

REMARK 7.2.– *The limit two-component diffusion process $(\zeta^0(t),\ Y^0(t),\ t \geq 0)$, determined by the generator* [7.11], *has a stochastic presentation by the difference stochastic equation:*

$$d\zeta^0(t) = -V(Y^0(t))\zeta^0(t)dt + \sigma(Y^0(t))dW(t),$$
$$dY^0(t) = \widehat{A}(Y^0(t))dt + \widehat{B}(Y^0(t))dW_0(t). \qquad [7.16]$$

The parameters of the limit diffusion $\zeta^0(t)$, $t \geq 0$, depend on the diffusion process $Y^0(t)$, $t \geq 0$, which is unequally determined by the averaged values $\widehat{A}$, $\widehat{B}$ [7.12]–[7.15] *that satisfy the uniform Lipshitz conditions (Gikhman and Skorohod 1979, Chapter 2).*

PROOF OF THEOREM 7.1.– The basic idea of the proof is formulated at the beginning of section 2.1.2. The operator characterization of the Markov process, determined by the generator on the class of real-valued functions, acting on the set of values of the Markov process, is used (Korolyuk and Limnios 2005).

First of all, we used the operator characterization of the extended three-component Markov chain

$$\{\zeta^\varepsilon(t_n),\, Y^\varepsilon(t_n),\, x^\varepsilon(t_n) := x_{t_n/\varepsilon^2}\},\ t_n = n\varepsilon^2,\ \varepsilon > 0, \qquad [7.17]$$

in the following form.

LEMMA 7.1.– *The extended Markov chain is determined by the generator*

$$\mathbb{L}^\varepsilon(x)\varphi(c,\, y,\, x) = \varepsilon^{-2}[\Gamma^\varepsilon(y)A^\varepsilon(x)P - \mathbb{I}]\varphi(c,\, y,\, x). \qquad [7.18]$$

Here, the following operators are used:

$$\begin{aligned} \Gamma^\varepsilon(y)\varphi(c) &:= E[\varphi(c + \Delta\zeta^\varepsilon(t;\, y)) \,|\, \zeta^\varepsilon(t;\, y) = c], \\ A^\varepsilon(x)\varphi(y) &:= E[\varphi(y + \Delta Y^\varepsilon(t;\, x)) \,|\, Y^\varepsilon(t;\, x) = y], \\ P\varphi(x) &:= \int_E P(x,\, dz)\varphi(z). \end{aligned} \qquad [7.19]$$

The statement of Lemma 7.1 follows from the definition of the generator of the three-component Markov chain [7.17].

A significant step of the proof of Theorem 7.1.1 is presented in the next lemma.

LEMMA 7.2.– *The generator* [7.18] *and* [7.19] *of the three-component Markov chain* [7.17], *acting on real-valued functions* $\varphi(c,\, y,\, x)$, *which have finite derivatives up to the third order, has the following asymptotic representation:*

$$\mathbb{L}^\varepsilon(x)\varphi(c,\, y,\, x) = \left[\varepsilon^{-2}\mathbb{Q} + \varepsilon^{-1}\mathbb{A}_0(x)P + \mathbb{A}(x)P + \mathbb{L}^0(y)P + \mathbb{R}_\varepsilon(y;\, x)\right]\varphi(c,\, y,\, x); \qquad [7.20]$$

$$\mathbb{A}_0(x)\varphi(y) = A_0(y;\, x)\varphi'(y),\ \mathbb{A}(x)\varphi(y) = A(y;\, x)\varphi'(y); \qquad [7.21]$$

$$\mathbb{L}^0(y)\varphi(c) = -V(y)c\varphi'(c) + \frac{1}{2}\sigma^2(y)\varphi''(c). \qquad [7.22]$$

The residual term $\mathbb{R}_\varepsilon(y;\, x)$ *is negligible:*

$$\mathbb{R}_\varepsilon(y;\, x)\varphi(c,\, y,\, x) \to 0,\ \varepsilon \to 0,\ \varphi \in C^3(R^2).$$

PROOF OF LEMMA 7.2.– We use the following transformation of the generator [7.18]:

$$\mathbb{L}^\varepsilon(x)\varphi(c,\, y,\, x)$$

$$= \varepsilon^{-2}\Big[\mathbb{Q} + (A^\varepsilon - \mathbb{I})P + (\Gamma^\varepsilon - \mathbb{I})P + \mathbb{R}_\varepsilon(y;\, x)\Big]\varphi(c,\, y,\, x), \qquad [7.23]$$

The residual term is:

$$\mathbb{R}_\varepsilon(y;\, x)\varphi(c,\, y,\, x) = (\Gamma^\varepsilon - \mathbb{I})(A^\varepsilon - \mathbb{I})P\varphi(c,\, y,\, x). \qquad [7.24]$$

Then we calculate

$$\varepsilon^{-2}[\Gamma^\varepsilon(y) - \mathbb{I}]\varphi(c) = \varepsilon^{-2}E[\varphi(c + \Delta\zeta^\varepsilon(t;\, y) \mid \zeta^\varepsilon(t;\, y) = c]$$

$$= [\mathbb{L}^0(y) + \mathbb{R}_\varepsilon(y;\, c)]\varphi(c); \qquad [7.25]$$

The limit generator $\mathbb{L}^0(y)$ is defined in [7.22].

The next transformation

$$\varepsilon^{-2}[A^\varepsilon(x) - \mathbb{I}]\varphi(y) = \varepsilon^{-2}E[\varphi(y + \Delta y^\varepsilon(t;\, x)) - \varphi(y) \mid y^\varepsilon(t;\, x) = y]$$

$$= \varepsilon^{-2}\left[E[\Delta y^\varepsilon(t;\, x)]\varphi'(y) + \frac{1}{2}E[\Delta y^\varepsilon(t;\, x)]^2\varphi''(y) + \mathbb{R}_\varepsilon(y;\, c)\right]$$

$$= \left[\varepsilon^{-1}\mathbb{A}_0(y;\, x) + \mathbb{A}(y;\, x)\right]\varphi'(y) + \frac{1}{2}\mathbb{A}_0^2(y;\, x)\varphi''(y) + \mathbb{R}_\varepsilon\varphi(y). \qquad [7.26]$$

allows us to obtain the asymptotic expansion for Lemma 7.2:

$$\mathbb{L}^\varepsilon(x)\varphi(c,\, y,\, x) = [\varepsilon^{-2}\mathbb{Q} + \varepsilon^{-1}\mathbb{A}_0(x)\, P + \mathbb{A}(x)\, P$$

$$+ \frac{1}{2}(\mathbb{A}_0(x)\, P)^2 + \mathbb{L}^0(y)]\varphi(c,\, y,\, x) + \mathbb{R}_\varepsilon(y;\, x)\varphi(c,\, y,\, x). \qquad [7.27]$$

□

Next, we use the solution of the singular perturbation problem for the truncated operator.

LEMMA 7.3.– *The solution of the singular perturbation problem for the truncated operator is realized on the perturbed test functions as follows:*

$$\mathbb{L}_0^\varepsilon(x)\varphi^\varepsilon(c,\, y,\, x) = \Big[\varepsilon^{-2}\mathbb{Q} + \varepsilon^{-1}\mathbb{A}_0(x)\, P + \mathbb{A}(x)\, P + \frac{1}{2}(\mathbb{A}_0(x)\, P)^2$$

$$+ \mathbb{L}^0(y)\Big][\varphi(c,\, y) + \varepsilon\varphi_1(c,\, y,\, x) + \varepsilon^2\varphi_2(c,\, y,\, x)]$$

$$= \mathfrak{L}^0(y)\varphi(c,\, y) + \mathbb{R}_\varepsilon(x)\varphi(c,\, y). \qquad [7.28]$$

The corresponding averaged parameters are determined by formulas [7.12]–[7.15].

The limit operator is calculated as follows:

$$\mathfrak{L}^0(y)\varphi(c,\,y) = \mathbb{L}^0(y) + \widehat{A}(y)\varphi'_y(c,\,y) + \frac{1}{2}\widehat{B}^2(y)\varphi''_y(c,\,y). \qquad [7.29]$$

PROOF OF LEMMA 7.3.– For solving the singular perturbation problem for the truncated operator, we consider the following asymptotic expansion in powers of ε:

$$\begin{aligned}\mathbb{L}_0^\varepsilon(x)\varphi^\varepsilon(c,\,y,\,x) &= \varepsilon^{-2}\mathbb{Q}\varphi(c,\,y) + \varepsilon^{-1}[\mathbb{Q}\varphi_1(c,\,y,\,x)\\ &+ \mathbb{A}_0(x)\varphi(c,\,y)] + \Big[\mathbb{Q}\varphi_2(c,\,y,\,x) + \mathbb{A}_0(x)\,P\,\varphi_1(c,\,y,\,x) \qquad [7.30]\\ &+ [\mathbb{A}(x) + \frac{1}{2}(\mathbb{A}_0(x)P)^2 + \mathbb{L}^0(y)]\varphi(c,\,y)\Big] + \mathbb{R}_\varepsilon(x)\varphi(c,\,y).\end{aligned}$$

Obviously, $\mathbb{Q}\varphi(c,\,y) = 0$.

The balance condition [7.7] is then used. The solution of the equation:

$$\mathbb{Q}\varphi_1(c,\,y,\,x) + \mathbb{A}_0(x)\varphi(c,\,y) = 0$$

is given by the formula (Korolyuk and Limnios 2005, section 5.4):

$$\varphi_1(c,\,y,\,x) = \mathbb{R}_0\mathbb{A}_0(x)\varphi(c,\,y).$$

Substituting this result into the following equation, we have:

$$\mathbb{Q}\varphi_2(c,\,y,\,x) + \big[\mathbb{B}(x) + \mathbb{A}(x) + \mathbb{L}^0(y)\big]\varphi(c,\,y) = \mathfrak{L}^0(y)\varphi(c,\,y). \qquad [7.31]$$

Here, by definition,

$$\mathbb{B}(x) := \mathbb{A}_0(x)PR_0\mathbb{A}_0(x) + \frac{1}{2}(\mathbb{A}_0(x)P)^2. \qquad [7.32]$$

The limit operator is calculated by using the balance condition:

$$\mathfrak{L}^0(y)\varphi(c,\,y) = \Big\{\Pi\Big[\mathbb{B}(x) + \mathbb{A}(x)\Big]\Pi + \mathbb{L}^0(y)\Big\}\varphi(c,\,y). \qquad [7.33]$$

We recall the projector's operation:

$$\Pi\mathbb{B}(x) = \int_E \rho(dx)B(y,\,x),$$

$$B(y,\,x) = A_0(y,\,x)PR_0A_0(y,\,x) + \frac{1}{2}(A_0(y,\,x))^2.$$

Taking into account the definition of evolution operators [7.21], the limit generator is determined by formula [7.30]. □

The proof of Theorem 7.1 is complete. □

REMARK 7.3.– *Using the definition of the potential operator R_0 (see Korolyuk and Limnios 2005, section 5.5), we present to the reader a further analysis of the formula for the diffusion parameter* $\mathbb{B}(x)$.

7.2. Discrete Markov process with asymptotically small diffusion

The discrete Markov process in discrete–continuous time is given by a solution of the difference stochastic equation in a series scheme, approximated by the Ornstein–Uhlenbeck process with continuous time and asymptotically small diffusion (see section 7.2.1).

The exponential generator for the discrete Markov process is derived from the method of large deviations analysis for Markov processes, similar to that in Feng and Kurtz (2006) (see section 7.2.2).

The rate functional for the discrete Markov process is introduced using the method developed in Freidlin and Wentzell (2012) (see section 7.2.3).

7.2.1. *Asymptotically small diffusion*

The discrete Markov process in discrete–continuous time is given in a series scheme with a small parameter of the series $\varepsilon \to 0$ $(\varepsilon > 0)$ by a solution of the following difference stochastic equation:

$$\Delta_\varepsilon \zeta_\varepsilon(t) = -\varepsilon^3 V(\zeta_\varepsilon(t)) + \varepsilon^2 \sigma \Delta_\varepsilon \mu(t), \; t = k\varepsilon^3, \; k \geq 0. \qquad [7.34]$$

The discrete–continuous time means the following scaling of increments:

$$\Delta_\varepsilon \alpha(t) := \alpha(t + \varepsilon^3) - \alpha(t), \; \alpha(t) = \zeta_\varepsilon(t) \text{ or } \mu(t). \qquad [7.35]$$

The stochastic component $\mu(t)$, $t \geq 0$, is a martingale difference with the characteristics

$$E\mu(t) = 0, \; E[\Delta_\varepsilon \mu(t)]^2 = 1. \qquad [7.36]$$

The discrete Markov processes [7.34]–[7.36] are given by the following transition probabilities (Korolyuk and Limnios 2005):

$$P_\varepsilon \varphi(c) = E\big[\varphi(c + \Delta_\varepsilon \zeta_\varepsilon(t)) \,|\, \zeta_\varepsilon(t) = c\big]. \qquad [7.37]$$

The generating operator (generator) of the discrete Markov process [7.34]–[7.37] has the following expression:

$$Q_\varepsilon \varphi(c) = E\big[\varphi(c + \Delta_\varepsilon \zeta_\varepsilon(t)) - \varphi(c) \,|\, \zeta_\varepsilon(t) = c\big]. \qquad [7.38]$$

The approximation of the discrete Markov process with asymptotically small diffusion, given by the difference stochastic equation [7.34]–[7.38], uses the martingale characterization (Korolyuk and Limnios 2005):

$$\mu^\varepsilon(t) = \varphi(\zeta_\varepsilon(t)) - \varphi(\zeta_\varepsilon(0)) - \int_0^{\varepsilon^3[t/\varepsilon^3]} L_\varepsilon \varphi(\zeta_\varepsilon(u))du, \; t = k\varepsilon^3. \qquad [7.39]$$

The upper integration limit in [7.39] has the asymptotics $\lim_{\varepsilon \to 0} \varepsilon^3[t/\varepsilon^3] = t$.

The normalized generator of the discrete Markov process L_ε in the martingale characterization [7.39], due to the discrete–continuous time normalization $k = [t/\varepsilon^3]$, is determined by the formula:

$$L_\varepsilon \varphi(c) = \varepsilon^{-3} Q_\varepsilon \varphi(c). \qquad [7.40]$$

Next, we establish that the discrete Markov process [7.34]–[7.38] is approximated by *asymptotically small diffusion*.

THEOREM 7.2.– *The normalized generators* [7.40] *have the following asymptotic representation, as* $\varepsilon \to 0$*:*

$$L_\varepsilon \varphi(c) = L_\varepsilon^0 \varphi(c) + \varepsilon R_\varepsilon \varphi(c), \qquad [7.41]$$

by the limit generator of the Ornstein–Uhlenbeck process:

$$L_\varepsilon^0 \varphi(c) = -V(c)\varphi^{'}(c) + \varepsilon \frac{\sigma^2}{2} \varphi^{''}(c), \qquad [7.42]$$

and the residual term is negligible:

$$R_\varepsilon \varphi(c) \to 0, \; \varepsilon \to 0, \; \varphi(c) \in C^3(\mathbb{R}). \qquad [7.43]$$

PROOF OF THEOREM 7.2.– The generating operator

$$Q_\varepsilon \varphi(c) = (P_\varepsilon - \mathbb{I})\varphi(c), \qquad [7.44]$$

by virtue of definition [7.37] of discrete Markov process transition probabilities, has the following Taylor decomposition:

$$\begin{aligned} Q_\varepsilon \varphi(c) &= E\big[\Delta_\varepsilon \zeta_\varepsilon(t)\varphi'(c) + \frac{1}{2}\left(\Delta_\varepsilon \zeta_\varepsilon(t)\right)^2 \varphi''(c) + \varepsilon^4 R_\varepsilon(c)\varphi\big] \\ &= -\varepsilon^3 V(c)\varphi'(c) + \frac{1}{2}\varepsilon^4\sigma^2\varphi''(c) + \varepsilon^4 R_\varepsilon \varphi(c) \end{aligned} \qquad [7.45]$$

on sufficiently smooth test functions $\varphi(c) \in C^3(\mathbb{R})$.

Substituting the decomposition [7.45] into [7.40], we obtain:

$$L_\varepsilon \varphi(c) = -V(c)\varphi'(c) + \frac{1}{2}\varepsilon\sigma^2\varphi''(c) + \varepsilon R_\varepsilon \varphi(c). \tag{7.46}$$

The generator of asymptotically small diffusion is determined by the first two components of decomposition [7.46]:

$$L_\varepsilon^0 \varphi(c) := -V(c)\varphi'(c) + \frac{1}{2}\sqrt{\varepsilon}\sigma^2\varphi''(c). \tag{7.47}$$

The generator L_ε^0 defines an Ornstein–Uhlenbeck-type continuous process $\zeta_\varepsilon^0(t)$, $t \geq 0$, with asymptotically small diffusion:

$$d\zeta_\varepsilon^0(t) = -V(\zeta_\varepsilon^0(t)) + \sqrt{\varepsilon}\sigma dW(t), \tag{7.48}$$

where the stochastic component is characterized as follows:

$$E[dW(t)] = 0, \; E[dW(t)]^2 = dt. \tag{7.49}$$

The approximation of the discrete Markov process by asymptotically small diffusion means the following:

$$\left(L_\varepsilon - L_\varepsilon^0\right)\varphi(c) = \varepsilon R_\varepsilon \varphi(c), \; |\, \varepsilon R_\varepsilon \varphi(c)\,| = o(\varepsilon), \; \varepsilon \to 0. \tag{7.50}$$

The proof of Theorem 7.2 is complete. □

Consider the martingale characterization of asymptotically small diffusion in the following form:

$$\mu_0^\varepsilon(t) = \varphi(\zeta_\varepsilon^0(t)) - \varphi(\zeta_\varepsilon^0(0)) - \int_0^{\varepsilon^3[t/\varepsilon^3]} L_\varepsilon^0 \varphi(\zeta_\varepsilon^0(u))du. \tag{7.51}$$

According to the model limit theorem (Korolyuk and Limnios 2005, section 6.3, Theorem 6.3), the following asymptotic relation, in probability, holds:

$$\mu^\varepsilon(t) - \mu_0^\varepsilon(t) \xrightarrow{P} 0, \; \varepsilon \to 0, \tag{7.52}$$

provided by

$$R_\varepsilon \varphi(c) \to 0, \; \varepsilon \to 0, \; \varphi \in C^2(\mathbb{R}). \tag{7.53}$$

The approximation [7.48] means an essential extension of the results for Ornstein–Uhlenbeck processes to the class of discrete Markov processes given by the solutions of the difference stochastic equation [7.34]. This is also the novelty of the method.

7.2.2. *Exponential generator of the discrete Markov process*

The discrete Markov process approximation by an Ornstein–Uhlenbeck process with asymptotically small diffusion generates an exponential martingale characterization (Feng and Kurtz 2006, Chapter 1, section 2.3) with the following scaling:

$$\mu_e^\varepsilon(t) = \exp\Big\{\varphi(\zeta^\varepsilon(t)) - \varphi(\zeta^\varepsilon(0)) - \int_0^{\varepsilon^3[t/\varepsilon^3]} H^\varepsilon\varphi(\zeta^\varepsilon(c))ds\Big\},\ t = k\varepsilon^3. \tag{7.54}$$

The corresponding *exponential generator* (Borovskikh and Korolyuk 1997, section 2.5) is:

$$H^\varepsilon\varphi(c) := \varepsilon^{-2}\ln\Big[1 + e^{-\varphi(c)/\varepsilon}Q_\varepsilon\, e^{\varphi(c)/\varepsilon}\Big]. \tag{7.55}$$

The exponential generator is generated by the nonlinear exponential semigroup

$$H_t^\varepsilon\varphi(c) = \ln E\Big[e^{\varphi(\zeta^\varepsilon(t))/\varepsilon}\,|\,\zeta^\varepsilon(0) = c\Big]. \tag{7.56}$$

The generating operator of the discrete Markov process [7.38] has the form:

$$Q_\varepsilon\varphi(c) = E[\varphi(c + \Delta_\varepsilon\zeta^\varepsilon(t)) - \varphi(c)\,|\,\zeta^\varepsilon(t) = c]. \tag{7.57}$$

In accordance with the method of large deviations analysis for Markov processes developed in Feng and Kurtz (2006), the first stage of the large deviations approach for the Markov process $\zeta^\varepsilon(t)$, $t \geq 0$, is solved by the limit theorem (Feng and Kurtz 2006, Chapter 3).

THEOREM 7.3.– *The exponential discrete Markov process generator* [7.55] *is calculated by the passage to limit*

$$\lim_{\varepsilon\to 0} H^\varepsilon\varphi(c) = H^0\varphi(c),\ \varphi(c) \in C^3(\mathbb{R}). \tag{7.58}$$

The limit exponential discrete Markov process generator [7.58] *is determined by the relation*

$$H^0\varphi(c) = -V(c)\varphi'(c) + \frac{1}{2}\sigma^2[\varphi'(c)]^2 \tag{7.59}$$

PROOF OF THEOREM 7.3.– The main idea is based on the asymptotic relation

$$e^{-\varphi(c)/\varepsilon}Q_\varepsilon\, e^{\varphi(c)/\varepsilon} = \varepsilon^2[H^0\varphi(c) + R_\varepsilon\varphi(c)], \tag{7.60}$$

Hereinafter, $R_\varepsilon\varphi$ means a negligible term, i.e.

$$R_\varepsilon\varphi \to 0,\ \varepsilon \to 0,\ \varphi \in C^3(R).$$

The justification of formula [7.59] is based on the asymptotic representation

$$e^{-\varphi(c)/\varepsilon}\, Q_\varepsilon\, e^{\varphi(c)/\varepsilon} = E[\exp \Delta_\varepsilon\varphi(c) - 1 \mid \zeta^\varepsilon(t) = c],$$
$$\Delta_\varepsilon\varphi(c) := [\varphi(c + \Delta_\varepsilon\zeta^\varepsilon(t)) - \varphi(c)]/\varepsilon. \quad [7.61]$$

With Taylor's formula, we have:

$$\Delta_\varepsilon\varphi(c) = \Delta_\varepsilon\zeta^\varepsilon(t)\varphi'(c)/\varepsilon + \frac{1}{2}(\Delta_\varepsilon\zeta^\varepsilon(t))^2\varphi''(c)/\varepsilon + R_\varepsilon\varphi(c). \quad [7.62]$$

The negligible term $R_\varepsilon\varphi(c)$, together with the second term [7.62], are of the order $O(\varepsilon^3)$. Hence, only the first term in [7.62] gives the asymptotic contribution.

Using the Taylor expansion up to the second order:

$$e^\Delta - 1 = \Delta + \frac{1}{2}\Delta^2 + O(\Delta^3), \quad [7.63]$$

we obtain the asympotic representation of formula [7.61], which can be computed as follows:

$$E[\exp \Delta_\varepsilon\varphi(c) - 1 \mid \zeta^\varepsilon(t) = c] =$$
$$= E_c[\Delta_\varepsilon\zeta^\varepsilon(t)/\varepsilon]\varphi'(c) + \frac{1}{2}E_c[(\Delta_\varepsilon\zeta^\varepsilon(t)/\varepsilon)^2](\varphi')^2(c) + R_\varepsilon\varphi. \quad [7.64]$$

The first term on the right-hand side of [7.64] becomes:

$$E_c[\Delta_\varepsilon\zeta^\varepsilon(t)/\varepsilon]\varphi'(c) = -\varepsilon^2 V(c)\varphi'(c) + \varepsilon^3 R_\varepsilon\varphi(c). \quad [7.65]$$

The second term on the right-hand side of [7.64] becomes:

$$E_c[(\Delta_\varepsilon\zeta^\varepsilon(t)/\varepsilon)\varphi'(c)]^2 = \varepsilon^2\sigma^2(\varphi'(c))^2 + \varepsilon^3 R_\varepsilon\varphi(c). \quad [7.66]$$

So,

$$E_c[e^{\Delta_\varepsilon\varphi(c)} - 1] = \varepsilon^2[-V(c)\varphi'(c) + \frac{1}{2}\sigma^2(\varphi'(c))^2] + \varepsilon^3 R_\varepsilon\varphi(c). \quad [7.67]$$

Finally, we have:

$$e^{-\varphi(c)/\varepsilon}\, Q_\varepsilon\, e^{\varphi(c)/\varepsilon} = \varepsilon^2[H^0\varphi(c) + \varepsilon R_\varepsilon\varphi(c)], \quad [7.68]$$

where the limit exponential generator is

$$H^0\varphi(c) := -V(c)\varphi'(c) + \frac{1}{2}\sigma^2(\varphi'(c))^2. \quad [7.69]$$

The proof of Theorem 7.3 is complete. □

REMARK 7.4.– *The exponential generator* [7.59] *sets the martingale (exponential) characterization of the Ornstein–Uhlenbeck process*

$$d\zeta_\varepsilon^0(t) = -V(\zeta_\varepsilon^0(t))dt + \sqrt{\varepsilon}\sigma dW(t), \; t \geq 0, \qquad [7.70]$$

set by the generator [7.40].

LEMMA 7.4.– *The exponential generator of the Ornstein–Uhlenbeck process* [7.70], *set by the generator* [7.40]–[7.43], *has a limit representation, as* $\varepsilon \to 0$*:*

$$\lim_{\varepsilon\to 0} H_\zeta^\varepsilon \varphi(c) = \lim_{\varepsilon\to 0} e^{-\varphi(c)/\varepsilon}\, \varepsilon L_\varepsilon^0 \; e^{\varphi(c)/\varepsilon} = H^0\varphi(c), \qquad [7.71]$$

where H^0 *is defined in formula* [7.59].

REMARK 7.5.– *The assertions of Theorem 7.3 and Lemma 7.5 mean that the exponential generator* [7.59] *raises the problem of large deviations for the discrete Markov process* [7.34]–[7.36], *as well as for Ornstein–Uhlenbeck processes* [7.70].

7.2.3. *Rate functional of the discrete Markov process*

In Freidlin and Ventzell (2012, Chapter 4), the problem of leaving a region of a dynamical system with Gaussian (asymptotically small) perturbances [7.70] is investigated with the application of the *rate functional*, which is given *a priori* by the equality:

$$S_T(\varphi) = \frac{1}{2\sigma^2}\int_0^T [\dot\varphi_t + V(\varphi_t)]^2 dt. \qquad [7.72]$$

Following Feng and Kurtz (2006, Part 1), the rate functional of the Ornstein–Uhlenbeck process [7.70], which is characterized by the exponential generator [7.59], is realized by a solution of the variational problem (Frechet–Legendre transform):

$$\mathcal{L}(c,q) = \sup_p \Big\{ pq - H^0(c,p) \Big\}. \qquad [7.73]$$

The rate functional of the Markov process $\zeta^\varepsilon(t)$, $t \geq 0$, is determined by the integral on test functions $\varphi_t \in C_T^1(\mathbb{R})$, $\dot\varphi_t := d\varphi_t/dt$:

$$S_T(\varphi) = \int_0^T \mathcal{L}(\varphi_t, \dot\varphi_t)dt, \qquad [7.74]$$

The function of two variables $H^0(c,p)$ is determined by the limit exponential generator [7.59]:

$$H^0(c, \dot\varphi) \equiv H^0\varphi(c). \qquad [7.75]$$

Thereby

$$H^0(c,\dot{\varphi}) = -V(c)\dot{\varphi} + \frac{1}{2}\sigma^2\dot{\varphi}^2. \qquad [7.76]$$

The fact of coincidence of the exponential generator of the Ornstein–Uhlenbeck process with asymptotically small diffusion [7.48] and [7.49] with the exponential generator of the discrete Markov process [7.54]–[7.58] provides the following result.

THEOREM 7.4.– *The rate functionals of the Ornstein–Uhlenbeck process* [7.70] *and the discrete Markov process* [7.34]–[7.36] *coincide and are determined by the integral* [7.72].

PROOF 7.18.– The variational problem [7.73], taking into account [7.74], involves the Frechet–Legendre transformation in [7.73] for the integral function [7.74]:

$$\mathcal{L}(c,q) = [q + V(c)]^2. \qquad [7.77]$$

Substituting the test function $\varphi_t = c$, $\dot{\varphi}_t = q$, we obtain the assertion of Theorem 7.4. □

7.3. The problem of discrete Markov random evolution leaving an interval

As an example of the application of discrete Markov random evolution approximation by the Ornstein–Uhlenbeck process with asymptotically small diffusion, we consider the problem of random evolution leaving an interval. In this case, we use the exponential generator defined by a variational problem (the Frechet–Legendre transformation).

We will construct the rate functional for discrete Markov evolution, which coincides with the rate functional of the Ornstein–Uhlenbeck process.

The rate functional is defined by the solution of the variational problem (Feng and Kurtz 2006, Chapter 2).

The problem of discrete Markov evolution leaving an interval can be solved by the method outlined in Freidlin and Ventzell (2012) for Ornstein–Uhlenbeck processes. The discrete Markov evolution is considered in discrete–continuous time and is specified by the solutions of the difference stochastic equation in a series scheme, which ensures the approximation by the Ornstein–Uhlenbeck process with continuous time with asymptotically small diffusion. Difference stochastic equations for discrete Markov evolution were also considered in the scheme of diffusion approximation of statistical experiments in Korolyouk (2020).

The discrete Markov diffusion, generated by the solutions of the difference stochastic equation [7.34], approximates the real statistical experiments with normalization (Korolyouk 2015a):

$$\zeta^{\varepsilon}(t) = S_N(k) - \rho,\ t = k\varepsilon^3,\ k \geq 0. \qquad [7.78]$$

Given the value range of real statistical experiments $0 \leq S_N(k) \leq 1$, we determine the sojourn interval for approximating stochastic equations as follows:

$$(-\rho,\ 1-\rho),\ 0 \leq \rho \leq 1. \qquad [7.79]$$

Now we have the opportunity to use the results of Freidlin–Wentzell's monograph adapted to our problem (a variant of Theorem 1.2 (Freidlin and Ventzell 2012, Chapter 4)).

THEOREM 7.5.– *The probability of the statistical experiment* [7.78] *and* [7.79] *leaving the interval* $(-\rho, 1-\rho)$ *is given by the following asymptotic relation (Korolyouk 2016c):*

$$\lim_{\varepsilon\to 0} \varepsilon^2 \ln P\{\zeta^{\varepsilon}(t) \in (-\rho, 1-\rho) \,|\, \zeta^{\varepsilon}(0) = c\} = -\min_{\varphi \in \overline{H}(c)} S_T(\varphi), \qquad [7.80]$$

in which

$$\overline{H}(c) := \{\varphi_t : \varphi_0 = c,\ \varphi_t \in [-\rho, 1-\rho],\ 0 \leq t \leq T\}$$

In addition, the following asymptotic relation holds:

$$\lim_{\varepsilon\to 0} \varepsilon^2 \ln P\{\tau^{\varepsilon} \leq t\} = -\min_{\varphi \in H(c)} S_T(\varphi), \qquad [7.81]$$

for the exit time of the interval:

$$\tau^{\varepsilon} := \min_{\varphi \in H(c)} \{t : \zeta^{\varepsilon}(t) \notin (-\rho, 1-\rho)\}.$$

Here, by definition, the set of test functions is:

$$H(c) := \{\varphi_t : \varphi_0 = c,\ \varphi_t \notin (-\rho, 1-\rho),\ 0 \leq t \leq T\}.$$

For a dynamic system with the linear velocity

$$dS(t)/dt = -VS(t),\ -\infty \leq t \leq T, \qquad [7.82]$$

the potential $U(c)$ is determined by the relation (Freidlin and Ventzell 2012, section 4.2, p. 149)

$$U(c) = \frac{1}{2} V s^2,\ s \in R. \qquad [7.83]$$

We denote the solution of the dynamic system

$$d\widehat{S}(t)/dt = V\widehat{S}(t),\ \widehat{S}(-\infty) = 0,\ \widehat{S}(T) = c,\ -\infty \le t \le T. \qquad [7.84]$$

The following equality is obvious:

$$\widehat{S}_0(t) = ce^{V(t-T)}. \qquad [7.85]$$

There is a variant of Theorem 7.3.1 (Freidlin and Ventzell 2012, section 4.3, p. 162):

THEOREM 7.6.– *The discrete Markov process* [7.78] *and* [7.79] *is characterized by the potential* [7.83] *of the dynamic system* [7.82], *namely there is an inequality*

$$S_T(\varphi_t) \ge \frac{1}{\sigma^2} Vs^2,\ \forall\{\varphi_t\ :\ \varphi_0 = 0, \varphi_T = c\}. \qquad [7.86]$$

PROOF 7.19.– The inequality [7.86] follows from the obvious identity

$$S_T(\varphi_t) = \frac{1}{\sigma^2}\Big[\frac{1}{2}\int_0^T [\dot{\varphi}_t - V\varphi_t]^2 dt + Vs^2\Big]. \qquad [7.87]$$

Thus, for the test functions $\varphi_t\ :\ \varphi_0 = 0, \varphi_T = c$, there is an inequality [7.86]. □

REMARK 7.6.– *Given the solution* [7.84] *of the dynamic system* [7.82], *we can also argue the validity of inequality*

$$S_T(\varphi_t) \ge \widehat{S}_T(\widehat{\varphi_t}), \qquad [7.88]$$

where the test function $\widehat{\varphi}_t$ *is determined by a solution of the dynamic system* $\widehat{\varphi}_t$*:*

$$\dot{\widehat{\varphi}}_t - V\widehat{\varphi}_t = 0,\ \widehat{\varphi}_{-\infty} = 0,\ \widehat{\varphi}_T = c. \qquad [7.89]$$

Therefore, the unique extremal of the rate function [7.74] *is given by a solution of the dynamic system* [7.89], *namely:*

$$\widehat{\varphi_t} = ce^{V(t-T)},\ -\infty \le t \le T. \qquad [7.90]$$

Thus, the inequality [7.88] *is established.*

The asymptotic analysis of exit from the interval $(-\rho,\ 1 - \rho)$ of a statistical experiment [7.54]–[7.58] can be realized using the approximation of the process $\zeta^{\varepsilon}(t)$, $t \ge 0$, by the Ornstein–Uhlenbeck process $\zeta^0_{\varepsilon}(t)$, $t \ge 0$ with the generator [7.47].

Consider the distribution functions of the Ornstein–Uhlenbeck process (see Freidlin and Ventzell 2012, section 4.1):

$$U^{\varepsilon}_t(c) = P\{\zeta^0_{\varepsilon}(t) \in (-\rho, 1-\rho) \,|\, \zeta^0_{\varepsilon}(0) = c\}, \qquad [7.91]$$

$$\overline{U}^{\varepsilon}_t(c) = P\{\tau^{\varepsilon} \le t) \,|\, \zeta^0_{\varepsilon}(0) = c\}. \qquad [7.92]$$

Using Theorem 3.1 (Freidlin and Ventzell 2012, section 4.3, p. 162), we can refine the results of Theorem 1.2 (Freidlin and Ventzell 2012, section 4.1, p. 145).

We can make sure that the distribution function [7.91] is determined by the solution of the boundary problem for a differential equation of the second order with a small parameter ε in the term of the highest derivative (see, for example, Freidlin and Ventzell (2012, Chapter 4)):

$$\partial U_t^\varepsilon(c)/\partial t = L_\varepsilon^0 U_t^\varepsilon(c), U_0^\varepsilon(c) = 1, s \in (-\rho, 1-\rho), \quad U_0^\varepsilon(c_0) = 0, s_0 \in \{-\rho, 1-\rho\}, \tag{7.93}$$

In particular, the probability of leaving the interval (ρ_-, ρ_+), $\rho_- := -\rho$, $\rho_+ := 1-\rho$

$$U_\pm^\varepsilon(c) := P\{\zeta_\varepsilon^0(\tau^\varepsilon) = \rho_\pm \,|\, \zeta_\varepsilon^0(0) = c\}, \ \rho_- < s < \rho_+ \tag{7.94}$$

is determined by a solution of the boundary elliptic problem:

$$L_\varepsilon^0 U_\pm^\varepsilon(c) = 0, \ U_\pm^\varepsilon(\rho_\pm) = 1, \ U_\pm^\varepsilon(\rho_\mp) = 0. \tag{7.95}$$

The solution of the boundary problem [7.95] is given by the relationships:

$$U_+^\varepsilon(c) = \int_{\rho_-}^{s} \exp[\frac{V}{\varepsilon\sigma^2}r^2]dr \Big/ \bar{U}_\varepsilon, \tag{7.96}$$

$$U_-^\varepsilon(c) = \int_{s}^{\rho_+} \exp[\frac{V}{\varepsilon\sigma^2}r^2]dr \Big/ \bar{U}_\varepsilon. \tag{7.97}$$

Here, the normalizing constant

$$\bar{U}_\varepsilon = \int_{\rho_-}^{\rho_+} \exp[\frac{V}{\varepsilon\sigma^2}r^2]dr. \tag{7.98}$$

COROLLARY 7.1.– *The probabilities of leaving the interval* (ρ_-, ρ_+) *are characterized by the limit behavior:*

$$\lim_{\varepsilon\to 0} P\{\zeta_\varepsilon^0(\tau^\varepsilon) = \rho_+ \,|\, \zeta_\varepsilon^0(0) = c\} = \begin{cases} 1, & \rho > 1/2, \\ 0, & \rho < 1/2. \end{cases} \tag{7.99}$$

In particular, we can see that there is an asymptotic probability behavior [7.93] *as* $\rho = 1/2$*:*

$$\lim_{\varepsilon\to 0} P\{\zeta_\varepsilon^0(\tau^\varepsilon) = \pm 1/2 \,|\, \zeta_\varepsilon^0(0) = c\} = 1/2. \tag{7.100}$$

8

Diffusion Approximation of Random Evolutions in Random Media

Discrete–continuous time is determined by the connection of discrete instants of time $k \geq 0$, $k \in \mathbb{N} := \{0, 1, ...\}$ with continuous time $t \geq 0$, $t \in \mathbb{R}_+ := [0, +\infty)$ by the formula

$$k = [Nt],\ t \geq 0. \qquad [8.1]$$

The integer part of the product means that

$$[Nt] = k \ \mathit{for}\ t_k^N := k/N,\ k \geq 0. \qquad [8.2]$$

The transition from the discrete instants of time $k \geq 0$ to the continuous time $t \geq 0$ means that when increasing the sample size $N \to \infty$, the intervals between adjacent moments of continuous time tend to 0:

$$\Delta_N := t_{k+1}^N - t_k^N = 1/N \to 0,\ N \to \infty. \qquad [8.3]$$

However, the condition

$$t_k^N = k/N \to t,\ N \to \infty, \qquad [8.4]$$

also means that $k \to \infty$.

8.1. Binary discrete Markov evolutions

As previously discussed, *statistical experiments* are defined by the averaged sums of sample variables $(\delta_n(k), 1 \leq n \leq N)$, $k \geq 0$, jointly independent for fixed $k \geq 0$ and identically distributed for different $n \in [1, N]$, taking two values ± 1

$$S_N(k) := \frac{1}{N}\sum_{n=1}^{N} \delta_n(k), \ -1 \leq S_N(k) \leq 1, \ k \geq 0. \qquad [8.5]$$

The binary statistical experiment [8.5] can also specify the difference between the frequency statistical experiments (see section 1.1):

$$S_N(k) = S_N^+(k) - S_N^-(k), \ k \geq 0. \qquad [8.6]$$

In the general case, both values are arbitrary $\{e_0, e_1\}$; the binary statistical experiment [8.5] can be considered in the vector form:

$$S_N(k) = (S_N^+(k), \ S_N^-(k)), \ k \geq 0. \qquad [8.7]$$

Evolutionary processes play a fundamental role, given by the conditional mathematical expectations:

$$P_\pm(k+1) := E[S_N^\pm(k+1) \mid S_N^\pm(k) = P_\pm(k)], \ k \geq 0. \qquad [8.8]$$

Obviously, there is a global balance condition:

$$P_+(k) + P_-(k) \equiv 1, \ \forall k \geq 0. \qquad [8.9]$$

The dynamics of the evolutionary process [8.8] are determined for increments

$$\Delta P_\pm(k+1) = P_\pm(k+1) - P_\pm(k), \ k \geq 0, \qquad [8.10]$$

which, evidently, also satisfy the global balance condition:

$$\Delta P_+(k) + \Delta P_-(k) \equiv 0, \ \forall k \geq 0. \qquad [8.11]$$

The evolutionary process increments [8.10] are given by the conditional mathematical expectations:

$$V_0^\pm(p_\pm) = E[\Delta S_N^\pm(k+1) \mid S_N^\pm(k) = p_\pm], \ k \geq 0. \qquad [8.12]$$

The *regression functions of increments* [8.12] also satisfy the global balance condition:

$$V_0^+(p_+) + V_0^-(p_-) \equiv 0, \ \forall p_\pm, p_+ + p_- = 1. \qquad [8.13]$$

The choice of regression functions of increments [8.12] and the corresponding evolutionary process [8.8] is a fundamental problem in studying the dynamics of evolutionary process, and is realized in a form of basic assumptions.

The basic principle of the regression functions of increments [8.12] selection is based on the *linear dependence* of regression functions of increments from the fluctuations:

$$\widehat{P}_{\pm}(k) := P_{\pm}(k) - \rho_{\pm},$$

where $\rho_{\pm}$ are equilibrium states: $V_0^{\pm}(\rho_{\pm}) = 0$. However, the nonlinear component of the regression functions of increments should provide a *balanced impact* on the dynamics of evolutionary process [8.8].

BASIC ASSUMPTION 8.1.– The regression function of increments of binary evolutionary process [8.8] is given by the relation

$$V_0^{\pm}(p_{\pm}) := -V_0 p_+ p_- (p_{\pm} - \rho_{\pm}). \quad [8.14]$$

The frequency evolutionary process [8.8] is given by the solutions of the *difference evolutionary equation*

$$\Delta P_{\pm}(k+1) = -V_0^{\pm}(P_{\pm}(k)),\ k \geq 0. \quad [8.15]$$

8.1.1. *Discrete Markov evolutions*

The study of statistical experiments in section 2.1 justifies the introduction of *discrete Markov evolutions* in discrete–continuous time:

$$\zeta_N^{\pm}(t) := \sqrt{N}[S_N^{\pm}(k) - \rho_{\pm}],\ k = [Nt],\ t \geq 0. \quad [8.16]$$

The normalized fluctuations of discrete Markov diffusion [8.16] mean that

$$S_N^{\pm}(k) = \rho_{\pm} + \zeta_N^{\pm}(t)/\sqrt{N},\ t \geq 0. \quad [8.17]$$

The global balance condition for discrete Markov evolution [8.16] is also evident:

$$\zeta_N^{+}(t) + \zeta_N^{-}(t) \equiv 0,\ \forall t \geq 0. \quad [8.18]$$

Thus, discrete Markov evolution [8.16] is uniquely determined by one of the components $\zeta_N^{\pm}(t)$, $t \geq 0$.

In the rest of this chapter, we will use the notation $\zeta(t) = \zeta^{+}(t)$.

The dynamics of discrete Markov evolution [8.16] is determined by a solution of the difference stochastic equation given by two components: the predictable component defined by regression functions of increments [8.14] and the stochastic component, with the normalized quadratic characteristic.

BASIC ASSUMPTION 8.2.– Discrete Markov evolution [8.16] is determined by a solution of the difference stochastic equation

$$\Delta\zeta_N(t) = -\frac{1}{N}V_0 S_N^+(k)S_N^-(k)\zeta_N(t) + \frac{1}{\sqrt{N}}\sigma(S_N^\pm(k))\Delta\mu_N(t), \quad t \geq 0. \qquad [8.19]$$

The quadratic characteristic (see Lemma 1.3) has the form:

$$\sigma^2(p) = V^+(p_+)V^-(p_-) \qquad [8.20]$$

The regression functions are expressed as:

$$\begin{aligned} V^\pm(p_\pm) &= p_\pm - V_0 p_+ p_-(p_\pm - \rho_\pm) \\ &= \rho_\pm + (1 - V_0 p_+ p_-)(p_\pm - \rho_\pm). \end{aligned} \qquad [8.21]$$

The difference stochastic equation [8.19] is basic in the asymptotic analysis of discrete Markov evolution [8.16] in discrete–continuous time.

The nonlinear factors of the predictable component and the quadratic characteristic in discrete–continuous time converge, in probability, as follows:

$$V_0(S_N(k)) := V_0 S_N^+(k)S_N^-(k) \xrightarrow{P} V_0\rho_+\rho_-, \ k \to \infty, \qquad [8.22]$$

$$\sigma^2(S_N(k)) \xrightarrow{P} \rho_+\rho_-, \ k \to \infty, \qquad [8.23]$$

The convergences [8.22] and [8.23] of difference stochastic equation components with nonlinear regression functions of increments generate a linear approximation of the difference stochastic equation [8.19]:

BASIC ASSUMPTION 8.3.– Discrete Markov evolution [8.16] is approximated by a solution of the next linear difference stochastic equation

$$\Delta\zeta_N^0(t) = -\frac{1}{N}V_0\pi\zeta_N^0(t) + \frac{1}{\sqrt{N}}\breve{\sigma}\Delta\mu_N^0(t), \ t \geq 0, \qquad [8.24]$$

where $\pi := \rho_+\rho_- = \breve{\sigma}^2$.

The discrete Markov evolution $\Delta\zeta_N^0(t), t \geq 0$, given by a solution of the difference stochastic equation [8.24], can be approximated, as $N \to \infty$, in discrete–continuous time, by a *diffusion process with evolution.*

THEOREM 8.1.– *The discrete Markov evolution* $\Delta\zeta_N^0(t)$, $t \geq 0$, *determined by a solution of the difference stochastic equation* [8.24], *converges in distribution:*

$$\zeta_N^0(t) \xrightarrow{D} \zeta^0(t), \ N \to \infty, \ 0 \leq t \leq T, \qquad [8.25]$$

to a diffusion process with evolution $\zeta^0(t), t \geq 0$, *given by the following generator, which acts on the class of finite real-valued functions* $\varphi(c) \in C^3(R)$, $(R := (-\infty, +\infty))$, *thrice continuously differentiated with bounded derivatives:*

$$L^0\varphi(c) = -V_0\pi c\varphi'(c) + \frac{1}{2}\breve{\sigma}^2\varphi''(c), \qquad [8.26]$$

by the additional initial conditions:

$$\zeta_N^0(0) \xrightarrow{P} \zeta^0(0),\ N \to \infty,\ E|\zeta_N^0(t)| < +\infty. \qquad [8.27]$$

The diffusion process with evolution $\zeta^0(t),\ t \geq 0$ *is an Ornstein–Uhlenbeck process with continuous time, determined by a solution of the difference stochastic equation*

$$d\zeta^0(t) = -V_0\pi\zeta^0(t)dt + \breve{\sigma}dW(t),\ t \in [0, T]. \qquad [8.28]$$

The Brownian motion $W(t),\ t \geq 0$ *has the standard parameters:*

$$EdW(t) = 0,\ E(dW(t))^2 = dt. \qquad [8.29]$$

8.1.2. *Justification of diffusion approximation*

The limit theorems for Markov random processes are based on the operator characterization of the Markov process (by means of the generator) on a class of real-valued functions with argument in the set of Markov process values (Korolyuk and Limnios 2005). The convergence of generators on a sufficiently rich class of real-valued functions ensures the convergence of finite-dimensional distributions (Ethier and Kurtz 1986; Skorokhod 2008).

Following Skorokhod (2008) (see also Korolyuk and Limnios (2005)), a generator of discrete Markov evolution with linear regression functions of increments [8.24] is defined as follows:

$$L_N^0\varphi(c) = NE[\varphi(c + \Delta\zeta_N^0(t)) - \varphi(c) \mid \zeta_N^0(t) = c]. \qquad [8.30]$$

PROOF OF THEOREM 8.1.– An essential stage of the proof of Theorem 8.1 is provided in the following lemma:

LEMMA 8.1.– *There is a convergence of generators* [8.30]*:*

$$\lim_{N\to\infty} L_N^0\varphi(c) = L^0\varphi(c),\ \varphi(c) \in C^3(R) \qquad [8.31]$$

in the class $C^3(R)$ *of real-valued finite functions, three times continuously differentiable with bounded derivatives. The limit generator*

$$L^0\varphi(c) = -V_0\pi c\varphi'(c) + \frac{1}{2}\breve{\sigma}^2\varphi''(c),\ \pi = \breve{\sigma}^2 = \rho_+\rho_-, \qquad [8.32]$$

defines the limit process of Ornstein–Uhlenbeck type [8.28].

PROOF OF LEMMA 8.1.– Using the representation [8.24] of the discrete Markov process of fluctuations $\zeta_N^0(t),\ t \geq 0$, let us calculate the first two moments of increments:

$$E[\Delta\zeta_N^0(t)|\zeta_N^0(t) = c] = -V_0\pi c/N, \tag{8.33}$$

$$E[(\Delta\zeta_N^0(t))^2|\zeta_N^0(t) = c] = \breve{\sigma}^2/N + o(1/N)\ \forall c,\ \breve{\sigma}^2 = \rho_+\rho_-. \tag{8.34}$$

We now apply the Taylor formula in representation [8.30] of the generator L_N^0, using a test function $\varphi(c) \in C^3(R)$:

$$L_N^0\varphi(c) = N\Big[E[\Delta\zeta_N^0(t) \,|\, \zeta_N^0(t) = c]\varphi'(c)$$

$$+ E[(\Delta\zeta_N^0(t))^2 \,|\, \zeta_N^0(t) = c]\frac{1}{2}\varphi''(c)\Big] + R_N\varphi(c). \tag{8.35}$$

Here, the residual term, under the condition of Lemma 8.1, is negligible:

$$R_N\varphi(c) \to 0,\ N \to \infty,\ \varphi(c) \in C^3(R). \tag{8.36}$$

Using representations [8.33]–[8.35] of the first two moments of increments, we obtain the statement [8.31] of Lemma 8.1 in asymptotic form:

$$L_N^0\varphi(c) = L^0\varphi(c) + R_N\varphi(c).$$

The representations [8.33]–[8.35] imply the assertions of Lemma 8.1. □

The proof of Theorem 8.1 is based on Theorem 1 of A.V. Skorokhod (2008, section 2.1), from which follows the convergence of finite-dimensional distributions of normalized discrete Markov diffusion fluctuations. □

REMARK 8.1.– *The discrete Markov evolution* $\zeta_N^+(t),\ t \geq 0$, *with the linear regression functions of increments:*

$$V_0^+(\zeta_N(t)) = -V_0\zeta_N^+(t),\ t \geq 0, \tag{8.37}$$

obviously also has the approximation by a diffusion process with evolution, with the generator

$$L^+\varphi(c) = -V_0c\varphi'(c) + \frac{1}{2}\breve{\sigma}^2\varphi''(c),\ \breve{\sigma}^2 = \rho_+\rho_-. \tag{8.38}$$

Similarly, the following convergence takes place.

LEMMA 8.2.– *There is a convergence of generators*

$$\lim_{N\to\infty} L_N^+\varphi(c) = L^+\varphi(c),\ \varphi(c) \in C^3(R)$$

on a class of real-valued finite functions $C^3(R)$, three times continuously differentiated with bounded derivatives. The limit generator L^+ is given by the representation [8.38].

The diffusion approximation justification for a discrete Markov evolution $\zeta_N^0(t)$ with the linear difference stochastic equation [8.24] obviously provides the statement of Theorem 8.1 for discrete Markov evolution frequency $\zeta_N^+(t)$ with linear regression functions of increments [8.37].

Let us introduce the generator of discrete Markov evolution, given by a solution of the difference stochastic equation [8.19] with nonlinear components [8.20] and [8.21]:

$$L_N\varphi(c) = E[\varphi(c + \Delta\zeta_N(t)) - \varphi(c) \mid \zeta_N(t) = c]. \qquad [8.39]$$

COROLLARY 8.1.– *The convergence in probability* [8.22] *and* [8.23] *of the coefficients of equation* [8.19] *provides the asymptotic representation of the discrete Markov evolution generators:*

$$L_N\varphi(c) = L^0\varphi(c) + R_N\varphi(c),$$

$$R_N\varphi(c) \to 0,\ N \to \infty,\ \varphi(c) \in C^3(R).$$

The convergence in probability [8.22] and [8.23] of the coefficients of equation [8.19] is provided by the following lemma:

LEMMA 8.3.– *The discrete Markov evolution $\zeta_N^+(t)$, $t \geq 0$, given by a solution of the difference stochastic equation* [8.19] *with uniformly bounded initial values:*

$$E|\zeta_N^+(0)| \leq C < +\infty,\ N \to \infty, \qquad [8.40]$$

has a uniformly bounded second moment:

$$E|\zeta_N^+(t)|^2 \leq C < +\infty,\ 0 \leq t \leq T,\ N \to \infty. \qquad [8.41]$$

PROOF OF LEMMA 8.3.– The following semimartingale representation is used:

$$\zeta_N^+(t) = \zeta_N^+(0) + \frac{1}{N}V_N(t) + \frac{1}{\sqrt{N}}M_N(t),\ t \geq 0. \qquad [8.42]$$

The evolutionary component (see [8.14]) has the following form:

$$V_N^+(t) = -\sum_{k=0}^{[Nt]} V_0 S_N^+(k) S_N^-(k) \zeta_N(t_k^N),\ t_k^N := k/N. \qquad [8.43]$$

The stochastic component is:

$$M_N^+(t) = \sum_{k=0}^{[Nt]} \sigma(S_N(k)) \Delta\mu_N(k+1),\ t \geq 0. \qquad [8.44]$$

The quadratic characteristic (see [8.21]) is:

$$\sigma^2(p) := V^+(p)V^-(p),\ V^{\pm}(p) := p_{\pm} - V_0 p_+ p_-(p_{\pm} - \rho_{\pm}), \quad [8.45]$$

which has the following representation:

$$\sigma^2(S_N(k)) = V^+(S_N(k))V^-(S_N(k)), \quad [8.46]$$

$$V^{\pm}(S_N(k)) = \rho_{\pm} + [1 - V_0 S_N^+(k) S_N^-(k)]\zeta_N(t_k^N)/\sqrt{N}. \quad [8.47]$$

Following Liptser (1994), let us introduce the notations:

$$\zeta_N^*(t) := \sup_{0\le s\le t} |\zeta_N(s)|,\ t \ge 0. \quad [8.48]$$

Similarly to Liptser (1994), we have the following estimations:

$$\sup_{0\le s\le t} |V_N(s)|^2 \le N V_0^2 [\zeta_N^*(t)]^2,\ t \ge 0, \quad [8.49]$$

$$\sup_{0\le s\le t} |M_N(s)|^2 \le N C_1 \zeta_N^*(t),\ t \ge 0. \quad [8.50]$$

Therefore, we have the estimate:

$$E \sup_{0\le s\le t} |\zeta_N(t)|^2 \le C_0 + C_1 \int_0^T \zeta_N^*(t)dt + C_2 \int_0^T |\zeta_N^*(t)|^2 dt. \quad [8.51]$$

Given the boundedness of the integral

$$\int_0^T \zeta_N^*(t)dt \le C < +\infty,\ N \to \infty, \quad [8.52]$$

we apply Gronwall–Bellman's lemma

$$E \sup_{0\le s\le t} |\zeta_N^*(t)|^2 \le C_{01} + C_2 \int_0^t E|\zeta_N^*(s)|^2 ds,\ 0 \le t \le T, \quad [8.53]$$

which justifies the assertion of Lemma 8.3. □

8.2. Multivariate random evolutionary systems in discrete–continuous time

The diffusion approximation of multivariant discrete Markov evolutions in discrete–continuous time $k = [Nt],\ t \in R_+ := [0, +\infty)$ is realized for the normalized discrete Markov diffusion fluctuations:

$$\zeta_N^{(m)}(t) := \sqrt{N}[S_N^{(m)}(k) - \rho_m],\ 0 \le m \le M,\ k = [Nt],\ t \ge 0. \quad [8.54]$$

The discrete Markov evolution connection with discrete Markov diffusion [8.54] means that

$$S_N^{(m)}(k) = \rho_m + \varsigma_N^{(m)}(t)/\sqrt{N},\ 0 \leq m \leq M,\ k = [Nt],\ t \geq 0. \qquad [8.55]$$

The global balance condition for frequency discrete Markov diffusion:

$$\sum_{m=0}^{M} S_N^{(m)}(k) \equiv 1,\ \forall k \geq 0, \qquad [8.56]$$

generates the global balance condition for discrete Markov evolution:

$$\sum_{m=0}^{M} \varsigma_N^{(m)}(t) \equiv 0,\ \forall t \geq 0. \qquad [8.57]$$

In the asymptotic analysis of discrete Markov evolution [8.54] by $N \to \infty$, the approach given in section 1.3 for multivariant statistical experiments is used.

8.2.1. *Evolutionary process in discrete–continuous time*

The models of multivariant discrete Markov diffusion are defined by two components: predictable and stochastic components.

The predictable component is given by conditional mathematical expectations (section 1.3):

$$V_0^{(m)}(P(k)) := E[\Delta S_N^{(m)}(k+1) \mid S_N(k) = P(k)],\ k \geq 0, \qquad [8.58]$$

The regression functions of increment dependence on the vector parameter $p = (p_m\,,\ 0 \leq m \leq M)$ are postulated in Proposition 1.6:

$$V_0^{(m)}(p) := V_0 \pi p_m \left[\sum_{n=0}^{M} p_n^2/\rho_n - p_m/\rho_m \right],\ 0 \leq m \leq M. \qquad [8.59]$$

Using the parameters $\pi_m = \pi/\rho_m$, $0 \leq m \leq M$, $\pi := \prod_{m=0}^{M} \rho_m$, the canonical regression functions of increments [8.59] transform into a fluctuation representation

$$V_0^{(m)}(p, \widehat{p}) := V_0 p_m \left[\sum_{n=0}^{M} p_n \pi_n \widehat{p}_n - \pi_m \widehat{p}_m \right],\ 0 \leq m \leq M. \qquad [8.60]$$

PROPOSITION 8.1.– *(Basic assumption 8.4). The frequency evolutionary process*

$$P_m(k+1) := E[\Delta S_N^{(m)}(k+1) \mid S_N(k) = P(k)],\ 0 \leq m \leq M,\ k \geq 0, \quad [8.61]$$

is determined by the solutions of the difference evolutionary equation with nonlinear regression functions of increments [8.60]*:*

$$\Delta \widehat{P}_m(k+1) = -V_0^{(m)}(P(k), \widehat{P}(k)),\ 0 \leq m \leq M,\ k \geq 0. \quad [8.62]$$

8.2.2. *Difference stochastic equation*

Consider the convergence to the equilibrium state, defined by the equilibrium of regression functions of increments:

$$\rho := (\rho_m,\ 0 \leq m \leq M),\ \sum_{m=0}^{M} \rho_m = 1. \quad [8.63]$$

DEFINITION 8.1.– *The stochastic components of discrete Markov evolution are determined by the following martingale differences:*

$$\Delta \mu_N^{(m)}(k+1) := \Delta S_N^{(m)}(k+1) - E[\Delta S_N^{(m)}(k+1) \mid S_N(k)], \quad [8.64]$$
$$k \geq 0.$$

The conditional mathematical expectations in [8.58] are determined by regression functions of increments [8.60] with the normalized fluctuation evolution process:

$$V_0^{(m)}(S_N(k), \zeta_N(t)) := V_0 \cdot [S_N^{(m)}(k) \sum_{n=0}^{M} \pi_n S_N^{(n)}(k) \zeta_N^{(n)}(t)$$
$$-\pi_m S_N^{(m)}(k) \zeta_N^{(m)}(t)],\ 0 \leq m \leq M,\ k \geq 0. \quad [8.65]$$

PROPOSITION 8.2.– *(Basic assumption 8.5). The frequency discrete Markov evolution* [8.54] *is determined by a solution of the following difference stochastic equation:*

$$\Delta \zeta_N^{(n)}(t+\Delta) = \frac{1}{N} V_0^{(m)}(S_N(k), \zeta_N(t))$$
$$+ \frac{1}{\sqrt{N}} \sigma_m(S_N(k)) \Delta \mu_N^{(n)}(t+\Delta),\ 0 \leq m \leq M,\ k \geq 0. \quad [8.66]$$

The stochastic component is distinguished by the conditional quadratic characteristic:

$$\sigma_m^2(S_N(k)) := E[(\Delta \mu_N^{(m)}(t+\Delta))^2 \mid S_N(k)]$$
$$= V^{(m)}(S_N^{(m)}(k))[1 - V^{(m)}(S_N^{(m)}(k))],\ 0 \leq m \leq M,\ k \geq 0. \quad [8.67]$$

LEMMA 8.4.– *The predictable component, along with the quadratic characteristic of the stochastic component of the difference stochastic equation* [8.66], *converge in probability, by* $N \to \infty$:

$$V_0^{(m)}(S_N(k), \zeta_N(t)) \xrightarrow{P} -V_0 \pi \zeta_N^{(m)}(t), \ k \geq 0,$$

$$\sigma_m^2(S_N(k)) \xrightarrow{P} \rho_m(1-\rho_m), \ 0 \leq m \leq M, \ k \geq 0, \qquad [8.68]$$

PROOF OF LEMMA 8.4.– There is a convergence in the probability of the predictable component

$$V_0^{(m)}(S_N(k), \zeta_N(t)) \xrightarrow{P} -V_0 \pi \zeta_N^{(m)}(t), \ N \to \infty, \ 0 \leq m \leq M, \ k \geq 0, \quad [8.69]$$

and also convergence in the probability of the quadratic characteristic (see [8.67])

$$\sigma_m^2(S_N(k)) = \sigma_m^2 \left(\rho + \frac{1}{\sqrt{N}} \zeta_N(t) \right) \xrightarrow{P} \rho_m(1-\rho_m), \ N \to \infty. \qquad [8.70]$$

By virtue of [8.65], the predictable component has the expression:

$$V_0 \Big[\rho_m \sum_{n=0}^{M} \rho_n \pi_n \zeta_N^{(n)}(t) - \pi_m \zeta_N^{(m)}(t) \Big]. \qquad [8.71]$$

Considering the identity $\rho_n \pi_n = \pi$, along with the global balance condition [8.57], the nonlinear term in [8.71] disappears:

$$V_0 \sum_{n=0}^{M} \rho_n \pi_n \zeta_N^{(n)}(t) = V_0 \pi \sum_{n=0}^{M} \zeta_N^{(n)}(t) \equiv 0. \qquad [8.72]$$

Therefore, the following term remains:

$$-V_0 \rho_m \pi_m \zeta_N^{(m)}(t) = -V_0 \pi \zeta_N^{(m)}(t).$$

Thus, the nonlinear predictable component of the difference stochastic equation [8.71] converges, in probability, to a linear component [8.68], which is identical for all m: $0 \leq m \leq M$.

The convergence, in probability, of the stochastic component of the difference stochastic equation [8.68] is obvious because of [8.70]. The proof of Lemma 8.4 is complete. □

An important consequence of Lemma 8.4 is a linear approximation of nonlinear discrete Markov evolution, formulated as follows:

PROPOSITION 8.3.– *(Basic assumption 8.6). The frequency nonlinear discrete Markov evolutions are approximated by the solutions of the linear difference stochastic equation*

$$\Delta\zeta_N^{(m)}(t+\Delta) = -\frac{1}{N}V_0\pi\zeta_N^{(m)}(t) + \frac{1}{\sqrt{N}}\breve{\sigma}_m\Delta\mu_N^{(m)}(t+\Delta),$$
$$0 \le m \le M,\ \breve{\sigma}_m^2 = \rho_m(1-\rho_m). \qquad [8.73]$$

8.2.3. *Diffusion approximation of random evolutionary systems in discrete–continuous time*

Using the Ornstein–Uhlenbeck continuous process, we study the approximation of multivariate nonlinear discrete Markov evolutions, determined by the solutions of the difference stochastic equation [8.66] in a discrete–continuous time $k = [Nt]$, $t \in R_+ := [0, +\infty)$, by $N \to \infty$, hence, also by $k \to \infty$.

In this case, we substantially use linear discrete Markov evolution [8.73], which approximates nonlinear discrete Markov evolution [8.66].

THEOREM 8.2.– *By the condition of convergence, in probability, of initial conditions*

$$\zeta_N^{(m)}(0) \xrightarrow{P} \zeta^{(m)}(0)\ , 0 \le m \le M\ , N \to \infty, \qquad [8.74]$$

the convergence, in distribution, of linear solutions of the difference stochastic equation [8.73] *takes place*

$$\zeta_N^{(m)}(t) \xrightarrow{D} \zeta^{(m)}(t)\ , 0 \le m \le M\ , 0 \le t \le T\ , N \to \infty. \qquad [8.75]$$

The limit diffusion process with evolution $\zeta^{(m)}(t)$, $0 \le t \le T$ is set by the generator

$$L^{(m)}\varphi(c) = V_0\pi c\varphi'(c) + \frac{1}{2}\breve{\sigma}_m^2\varphi''(c),\ \varphi(c) \in C_0^\infty(R) \qquad [8.76]$$

on the class $C_0^\infty(R)$ of real-valued functions, bounded together with their derivatives and having bounded supports.

The parameters of the generator satisfy the relations:

$$V_0 > 0\ , \pi = \prod_{m=0}^{M} \rho_m\ , \breve{\sigma}_m^2 = \rho_m(1-\rho_m)\ , 0 \le m \le M.$$

CONCLUSION 8.1.– The limit diffusion process with evolution is given by the differential stochastic equation

$$d\zeta^{(m)}(t) = V_0\pi\zeta^{(m)}(t)dt + \breve{\sigma}_m dW_m(t)\;, 0 \le m \le M,\; t \ge 0. \qquad [8.77]$$

The evolutional component is identical for all m: $0 \le m \le M$.

The stochastic components are defined as the standard process of Brownian motion with dispersion $\breve{\sigma}_m^2 = \rho_m(1-\rho_m)$, $0 \le m \le M$:

$$EdW_m(t) = 0\;, E(dW_m(t))^2 = dt\;, 0 \le m \le M.$$

PROOF OF THEOREM 8.2.– We use the operator and martingale characterization of discrete Markov evolution, which is determined by the solution of the difference stochastic equation [8.64]–[8.67], taking into account the asymptotic properties of the parameters.

The generator of discrete Markov evolution $\zeta(t)$, $t \ge 0$ is the following (see section 2.1):

$$\mathbb{L}_N\varphi(c) = NE[\varphi(c + \Delta\zeta_N(t+\Delta)) - \varphi(c) \mid \zeta_N(t) = c], \qquad [8.78]$$

acting on the class of functions $\varphi(c) \in C_0^\infty(R)$.

The martingale characterization of discrete Markov evolution is due to the ratio:

$$\int_0^{[Nt]/N} \mathbb{L}_N\varphi(\zeta_N(s))ds$$

$$= \sum_{k=0}^{t_N-1} E[\varphi(c + \Delta\zeta_N(t_k+\Delta)) - \varphi(c) \mid \zeta_N(t_k) = c] \qquad [8.79]$$

$$= E\sum_{k=0}^{t_N-1}[\varphi(\zeta_N(t_{k+1})) - \varphi(\zeta_N(t_k))] = E[\varphi(\zeta_N(t_N)) - \varphi(\zeta_N(0))].$$

Here, by definition,

$$t_N := [Nt]/N = k_N/N \to T,\; N \to \infty.$$

The key point in the proof of Theorem 8.2 is formulated in the following.

LEMMA 8.5.– *The following asymptotic representation of generator* [8.78] *takes place:*

$$\mathbb{L}_N\varphi(c) = \mathbb{L}\varphi(c) + \mathbb{R}_N\varphi(c),\; \varphi(c) \in C_0^\infty(R), \qquad [8.80]$$

on the class of functions $\varphi(c) \in C_0^\infty(R)$, *with a negligible term:*

$$\mathbb{R}_N\varphi(c) \to 0, \ N \to \infty, \ \varphi(c) \in C_0^\infty(R) \tag{8.81}$$

The limit diffusion process is given by generator [8.76].

PROOF OF LEMMA 8.5.– Applying the Taylor expansion up to the third order:

$$\begin{aligned}\mathbb{L}_N\varphi(c) &= NE[\Delta\zeta_N(t+\Delta) \,|\, \zeta_N(t) = c]\varphi'(c) \\ &\quad + \frac{1}{2}NE[(\Delta\zeta_N(t+\Delta))^2 \,|\, \zeta_N(t) = c]\varphi''(c) + \mathbb{R}_N\varphi(c),\end{aligned} \tag{8.82}$$

the first two moments of discrete Markov evolution increments are calculated, taking into account [8.64]–[8.67]:

$$\begin{aligned}NE[\Delta\zeta_N^{(m)}(t+\Delta) \,|\, \zeta_N(t) = c] &= V_0^{(m)}(\rho + c/\sqrt{N}), \\ NE[(\Delta\zeta_N^{(m)}(t+\Delta))^2 \,|\, \zeta_N^{(m)}(t) = c] &= \sigma_m^2(\rho + c/\sqrt{N}).\end{aligned} \tag{8.83}$$

Next, the asymptotic properties of predictable and stochastic components are used:

$$V_0^{(m)}(\rho + c/\sqrt{N}) \to -V_0\pi c, \ N \to \infty, \tag{8.84}$$

$$\begin{aligned}&\sigma_m^2(\rho + c/\sqrt{N}) \to \sigma_m^2, \ N \to \infty, \\ &\sigma_m^2 := \rho_m(1-\rho_m), \ 0 \le m \le M.\end{aligned} \tag{8.85}$$

Combining the asymptotic representations [8.82]–[8.85], we obtain the statement [8.80]–[8.81] of Lemma 8.5. □

The martingale characterization of discrete Markov evolution [8.79] implies that

$$\mu_N^{(m)}(t) = \int_0^{[Nt]/N} \mathbb{L}_N\varphi(\zeta_N^{(m)}(s))ds - \varphi(\zeta_N(t_N)) + \varphi(\zeta_N(0)), \ t \ge 0. \tag{8.86}$$

This, together with the asymptotic representation of the discrete Markov evolution generator [8.80] and [8.81], ensures the fulfillment of the statement of Theorem 8.2.□

8.3. Discrete random evolutionary systems in a Markov random environment

Consider a discrete Markov evolution in a *Markov random environment* in which drift parameters depend on the states of ergodic Markov chains, which determines the change of states of a Markov random environment (Korolyouk 2015c).

The Markov random environment is considered under two assumptions:

1. *A discrete Markov random environment*, given by a *Markov chain* (MC) x_k, $k \geq 0$ in a measurable state space $(E, \mathcal{E})$ with transition probabilities

$$P(x, B) = \mathcal{P}\{x_{k+1} \in B \,|\, x_k = x\}, \; x \in E, \; B \in \mathcal{E}. \qquad [8.87]$$

2. *A continuous Markov random environment*, given by a *Markov process* $x(t)$, $t \geq 0$ in a measurable state space $(E, \mathcal{E})$ with transition probabilities [8.87] of the embedded Markov chain $x_k = x(\tau_k)$, $k \geq 0$, and by distributions of the sojourn times θ_{k+1}, $k \geq 0$:

$$\mathcal{P}\{\theta_{k+1} \geq t \,|\, x_k = x\} = \exp[-q(x)t], \; t \geq 0. \qquad [8.88]$$

Therefore, the renewal moments are determined as follows:

$$\tau_{k+1} = \tau_k + \theta_{k+1}, \; k \geq 0. \qquad [8.89]$$

8.3.1. *Discrete and continuous Markov random environments*

The diffusion approximation for discrete Markov evolution in a Markov random environment is investigated using the singular perturbation method (Korolyuk and Limnios 2005, Chapter 5). By tradition, the generator of a two-component Markov process is given by a singular perturbation with a small series parameter $\varepsilon > 0$ ($\varepsilon \to 0$). We next use the small parameter $\varepsilon^2 = 1/N$, or, elsewhere, $N = \varepsilon^{-2}$.

The transition to the small parameter ε clearly uses the singular perturbation technique.

BASIC ASSUMPTION 8.7.– The statistical experiments in a discrete Markov random environment with discrete–continuous time $t = k\varepsilon^2$, $k \geq 0$ are given by the solution of the difference stochastic equation:

$$\Delta\zeta_\varepsilon(t) = -\varepsilon^2 V(x_k)\zeta_\varepsilon(t) + \varepsilon\,\sigma(x_k)\Delta\mu_\varepsilon(t), \; t = k\varepsilon^2, \; k \geq 0. \qquad [8.90]$$

The parameters $V(x) > 0$ and $\sigma(x)$, $x \in E$ are bounded real-valued functions of the states $x \in E$ of the Markov chain x_k, $k \geq 0$, which determines the Markov random environment.

Here, by definition,

$$\Delta\zeta_\varepsilon(t) := \zeta_\varepsilon(t + \varepsilon^2) - \zeta_\varepsilon(t), \; t \geq 0.$$

The stochastic component $\Delta\mu_\varepsilon(t)$, $t \geq 0$ is given by martingale differences, characterized by the first two moments:

$$E[\Delta\mu_\varepsilon(t)] = 0, \; E[[\Delta\mu_\varepsilon(t)]^2 \,|\, \zeta_\varepsilon(t)] = 1. \qquad [8.91]$$

The discrete Markov evolution in a Markov random environment, given by the solutions of difference stochastic equations [8.90] and [8.91], has an approximation in a series scheme, with a series parameter $\varepsilon \to 0$, by a normal Ornstein–Uhlenbeck process type with continuous time, under additional conditions.

THEOREM 8.3.– *Let the following conditions be fulfilled:*

C1: *The Markov chain with transition probabilities* [8.87] *is uniformly ergodic with the stationary distribution* $\rho(B)$, $B \in \mathcal{E}$*:*

$$\rho(B) = \int_E \rho(dx)P(x, B), \ \rho(\mathcal{E}) = 1; \tag{8.92}$$

C2: *The martingale differences satisfy the conditions of the central limit theorem (Borovskikh and Korolyuk 1997);*

C3: *There is the convergence and uniform boundedness of the initial conditions:*

$$\zeta_\varepsilon(0) \xrightarrow{P} \zeta(0), \ \varepsilon \to 0, \ E|\zeta_\varepsilon(0)| \le C < +\infty. \tag{8.93}$$

There is then a convergence, in the distribution of the discrete Markov evolution:

$$\zeta_\varepsilon(t) \xrightarrow{D} \zeta(t), \ \varepsilon \to 0, \ t \in [0, T]. \tag{8.94}$$

The limit process $\zeta(t)$, $t \ge 0$, *is set by the generator*

$$\mathbb{L}\varphi(s) = -\widehat{V} s\, \varphi'(s) + \frac{1}{2}\widehat{\sigma}^2 \varphi''(s). \tag{8.95}$$

The drift parameters $\widehat{V}$ *and dispersions* $\widehat{\sigma}^2$ *are calculated by the averaging formulas:*

$$\widehat{V} = \int_E \rho(dx)V(x), \ \widehat{\sigma}^2 = \int_E \rho(dx)\sigma^2(x). \tag{8.96}$$

REMARK 8.2.– *The limit process* $\zeta(t)$, $t \ge 0$ *with the generator* [8.95] *and* [8.96] *is a normal Ornstein–Uhlenbeck diffusion process, determined by the stochastic differential equation:*

$$d\zeta(t) = -\widehat{V}\zeta(t)dt + \widehat{\sigma}dW(t), \ t \ge 0. \tag{8.97}$$

Here, $dW(t)$, $t \ge 0$ *is a standard process of Brownian motion with parameters:*

$$E[dW(t)] = 0, \ E[dW(t)]^2 = dt.$$

In contrast to discrete Markov evolution in a discrete Markov random environment, in which the discrete Markov evolution dynamics is determined at each (discrete–continuous) time instant by a solution of the difference stochastic equation [8.90], the discrete Markov evolution dynamics in a continuous Markov random environment is determined at the increment instants of the Markov process.

An extension of discrete Markov evolution in a continuous Markov random environment is assumed, according to the scheme:

$$\zeta_\varepsilon(t) = \zeta_\varepsilon(t_k),\ t_k = k\varepsilon^2 \leq t < t_{k+1} = t_k + \varepsilon^2,\ k \geq 0.$$

BASIC ASSUMPTION 8.8. Discrete Markov evolution in a continuous Markov random environment is given by the solutions of the difference stochastic equation

$$\Delta\zeta_\varepsilon(\tau_{k+1}) = -\varepsilon^2 V(x_k)\zeta_\varepsilon(\tau_k) + \varepsilon\,\sigma(x_k)\Delta\mu_\varepsilon(\tau_{k+1}). \tag{8.98}$$

The stochastic component $\Delta\mu_\varepsilon(\tau_k)$, $k \geq 0$ is characterized by the first two moments:

$$E[\Delta\mu_\varepsilon(\tau_{k+1}) \mid \zeta_\varepsilon(\tau_k)] = 0,\ E[[\Delta\mu_\varepsilon(\tau_{k+1})]^2 \mid \zeta_\varepsilon(\tau_k)] = 1. \tag{8.99}$$

Consider the accompanying functions, relative to the Markov processes $x(t)$, $t \geq 0$:

$$\nu(t) := \max\{k : \tau_k \leq t\},\ t \geq 0;$$

$$\tau(t) := \tau_{\nu(t)},\ \theta(t) := \tau_{\nu(t)+1} - \tau(t).$$

The discrete Markov evolution in discrete–continuous time $t = k\varepsilon^2$, $k \geq 0$ is given by the sum of increments:

$$\zeta_\varepsilon(t) = \zeta_\varepsilon(0) + \sum_{k=1}^{\nu(t/\varepsilon^2)} \Delta\zeta_\varepsilon(\tau_k),\ t = k\varepsilon^2,\ k \geq 0. \tag{8.100}$$

THEOREM 8.4.– *Let the following conditions be fulfilled:*

C1: *The Markov process $x(t)$, $t \geq 0$ be uniformly ergodic with the stationary distribution $\pi(B)$, $B \in \mathcal{E}$:*

$$\pi(dx)q(x) = q\rho(dx),\ q = \int_E \pi(dx)q(x); \tag{8.101}$$

C2: *The martingale differences $\Delta\mu_\varepsilon(\tau_k)$, $k \geq 1$ satisfy the central limit theorem conditions (Borovskikh and Korolyuk 1997);*

C3: There is a convergence of initial conditions:

$$\zeta_\varepsilon(0) \xrightarrow{P} \zeta(0),\ \varepsilon \to 0,\ E|\zeta_\varepsilon(0)| \le C < +\infty. \qquad [8.102]$$

There is then the convergence, in distribution, of discrete Markov evolutions:

$$\zeta_\varepsilon(t) \xrightarrow{D} \zeta(t),\ \varepsilon \to 0,\ t \in [0, T]. \qquad [8.103]$$

The limit process $\zeta(t)$, $t \ge 0$ is set by the generator

$$\mathbb{L}\varphi(s) = -\widehat{V} s\, \varphi'(s) + \frac{1}{2}\widehat{\sigma}^2 \varphi''(s). \qquad [8.104]$$

The drift parameter $\widehat{V}$ and dispersion $\widehat{\sigma}^2$ are calculated by the corresponding averages:

$$\widehat{V} = q \int_E \rho(dx) V(x),\ \widehat{\sigma}^2 = q \int_E \rho(dx) \sigma^2(x). \qquad [8.105]$$

REMARK 8.3.– *The limit process $\zeta(t)$, $t \ge 0$ with the generator* [8.104] *and* [8.105] *is a normal diffusion process of Ornstein–Uhlenbeck type, determined by the stochastic differential equation* [8.97].

The basic idea of proving limit Theorems 8.3 and 8.4 for Markov processes is the application of limit theorems for random processes, in particular, for random evolutionary systems, based on the convergence of generators for the corresponding Markov processes (Ethier and Kurtz 1986; Korolyuk 1999; Korolyuk and Limnios 2005; Skorokhod 2008). The convergence of generators, on a sufficiently rich class of real-valued functions, ensures the convergence of finite-dimensional distributions (Skorokhod 2008).

The presence of a random Markovian environment necessitates the technique of the *singular perturbation problem* for the reducible-invertible operator that determines a uniformly ergodic Markov chain (or Markov process) (Korolyuk and Limnios 2005).

First, the characterization of the extended Markov process with an additional component, which sets the Markov random environment, is used.

LEMMA 8.6.–

A: For a discrete Markov random environment, the following two-component Markov chain:

$$\left(\zeta_\varepsilon^d(t; x),\ x_\varepsilon(t) := x(t/\varepsilon^2)\right),\ t = k\varepsilon^2,\ k \ge 0, \qquad [8.106]$$

is determined by the generator

$$\mathcal{L}_\varepsilon^d(x)\varphi(c, x) = \varepsilon^{-2}\left[\Gamma_\varepsilon^d(x)\mathbb{P} - \mathbb{I}\right]\varphi(c, x), \qquad [8.107]$$

where the operators $\Gamma_\varepsilon^d(x)$ *and* $\mathbb{P}$ *are transition probabilities of the Markov chain component* [8.106], *defined by the following formulas:*

$$\Gamma_\varepsilon^d(x)\varphi(c) := E\left[\varphi\left(c+\Delta\zeta_\varepsilon^d(t;x)\right)\Big|\,\zeta_\varepsilon^d(t)=c\,,\,x_\varepsilon(t)=x\right],\ t\geq 0. \quad [8.108]$$

$$\mathbb{P}\varphi(x) := \int_E P(x,dy)\varphi(y); \quad [8.109]$$

B: *For a continuous Markov random environment, the two-component Markov process*

$$\left(\zeta_\varepsilon^c(t;x),\,x_\varepsilon(t)=x(k)\right),\ k=t/\varepsilon^2,\ t\geq 0, \quad [8.110]$$

is determined by the generator

$$\mathcal{L}_\varepsilon^c(x)\varphi(c,x) = q(x)\mathcal{L}_\varepsilon^d(x)\varphi(c,x). \quad [8.111]$$

PROOF OF LEMMA 8.6.–

A: The assertion A follows from the definition of the Markov chain generator in discrete–continuous time with the elementary interval $\Delta = \varepsilon^2$ (see Nevelson and Hasminskii (1976, section 2.1)).

B: The assertion B follows from the well-known presentation for the generator of the Markov process with continuous time, taking into account the transition probabilities of the embedded Markov chain. □

A significant step in the proof of Theorems 8.3 and 8.4 is realized in the next lemma.

LEMMA 8.7.– *The generator* [8.107] *of a two-component Markov chain* [8.106] *has an asymptotic expansion on the real-valued test functions* $\varphi(c,x)$ *with three bounded derivatives by c:* $\varphi(c,\cdot)\in C^3(R)$*:*

$$\mathcal{L}_\varepsilon^d(x)\varphi(c,x) = \left[\varepsilon^{-2}\mathbb{Q}+\mathbb{L}^0(x)\mathbb{P}\right]\varphi(c,x)+\mathbb{R}_\varepsilon(x)\varphi(c,x). \quad [8.112]$$

Here, by definition,

$$\mathbb{Q}\varphi(\cdot,x) := [\mathbb{P}-\mathbb{I}]\varphi(\cdot,x), \quad [8.113]$$

$$\mathbb{L}^0(x)\varphi(c) := -V(x)c\varphi'(c)+\frac{1}{2}\sigma^2(x)\varphi''(c). \quad [8.114]$$

The residual term

$$|\mathbb{R}_\varepsilon(x)\varphi(c,x)| \to 0, \varepsilon\to 0,\ \varphi(c,x)\in C^3(R). \quad [8.115]$$

PROOF OF LEMMA 8.7.– This is based on the following transformation of the generator [8.107]:

$$\mathcal{L}^d_\varepsilon(x)\varphi(c,x) = \varepsilon^{-2}[\mathbb{Q} + \mathbb{L}^d_\varepsilon(x)\mathbb{P}]\varphi(c,x). \quad [8.116]$$

Here, by definition,

$$\mathbb{L}^d_\varepsilon(x)\varphi(c,\cdot) := \varepsilon^{-2}[\Gamma^d_\varepsilon(x) - \mathbb{I}]\varphi(c,\cdot). \quad [8.117]$$

Then, we use the Taylor formula for test functions $\varphi(c)$ up to the third order inclusively:

$$\mathbb{L}^d_\varepsilon(x)\varphi(c) = \mathbb{L}^0(x)\varphi(c) + \mathbb{R}_\varepsilon(x)\varphi(c) \quad [8.118]$$

with the residual term that satisfies the condition [8.115].

The generator representation [8.112] conducts to a singular perturbation problem for the truncated operator (see Korolyuk and Limnios (2005, Chapter 5)):

$$\mathcal{L}^0_\varepsilon\varphi(c,x) := [\varepsilon^{-2}\mathbb{Q} + \mathbb{L}^0(x)\mathbb{P}]\varphi(c,x). \quad [8.119]$$

LEMMA 8.8.– *On the perturbed test functions*

$$\varphi_\varepsilon(c,x) = \varphi(c) + \varepsilon^2\varphi_1(c,x), \quad [8.120]$$

the solution of the singular perturbation problem (Korolyuk and Limnios 2005, Proposition 5.1) for the truncated operator [8.119] *is determined by the relation:*

$$\mathcal{L}^0_\varepsilon(x)\varphi_\varepsilon(c,x) = \mathbb{L}^0\varphi(c) \; + \; \mathbb{R}_\varepsilon(x)\,\varphi(c). \quad [8.121]$$

Here, the limit operator

$$\mathbb{L}^0\varphi(c) = -\widehat{V}c\varphi'(c) + \frac{1}{2}\widehat{\sigma}^2\varphi''(c). \quad [8.122]$$

The residual term $\mathbb{R}_N(x)$ *satisfies the condition* [8.115].

PROOF OF LEMMA 8.8.– On the perturbed test functions

$$\varphi(c) + \varepsilon^2\varphi_1(c,x),$$

we can calculate the generator

$$\begin{aligned}\mathcal{L}^0_\varepsilon(x)\varphi_\varepsilon(c,x) &= [\varepsilon^{-2}\mathbb{Q} + \mathbb{L}^0(x)\mathbb{P}][\varphi(c) + \varepsilon^2\varphi_1(c,x)]\\ &= \varepsilon^{-2}\mathbb{Q}\varphi(c) + [\mathbb{Q}\varphi_1(c,x) + \mathbb{L}^0(x)\mathbb{P}\varphi(c)] + \varepsilon^2\mathbb{L}^0(x)\mathbb{P}\varphi_1(c,x).\end{aligned} \quad [8.123]$$

Taking into account that $\mathbb{Q}\varphi(c) = 0$, as well as the representation

$$\mathbb{Q}\varphi(x) := [\mathbb{P} - \mathbb{I}]\varphi(x),$$

the first singular term is a generating operator of an ergodic Markov chain (see **C1** in Theorem 8.3), using the condition of solvability of equation

$$\mathbb{Q}\varphi_1(c, x) + \mathbb{L}^0(x)\mathbb{P}\varphi(c) = \mathbb{L}^0\varphi(c), \qquad [8.124]$$

which ensures the existence of a test function $\varphi_1(c, x)$ (see Korolyuk and Limnios (2005, Proposition 5.1)).

The averaged operator $\mathbb{L}^0$ is determined by the relation

$$\mathbb{L}^0 = \Pi\mathbb{L}^0(x)\mathbb{P}\Pi = \Pi\mathbb{L}^0(x)\Pi. \qquad [8.125]$$

The calculation, by formula [8.125], defines the limit operator [8.95] and [8.96]. □

At the end of the proof of Theorem 8.3, we apply the model limit theorem (Korolyuk and Limnios 2005, section 6.3), which ensures the convergence of finite-dimensional distributions.

PROOF OF THEOREM 8.4.– This is completed with calculations of solvability conditions for the generator $\mathbb{Q}^c = q(x)\mathbb{Q}$ of equation

$$\mathbb{Q}^c\varphi_1(c, x) + q(x)\mathbb{L}^0(x)\mathbb{P}\varphi(c) = \mathbb{L}^c\varphi(c), \qquad [8.126]$$

where

$$\mathbb{L}^c = \Pi q(x)\mathbb{L}^0(x)\mathbb{P}\Pi. \qquad [8.127]$$

The relation [8.127] with the projector

$$\Pi\varphi(x) := \int_E \pi(dx)\varphi(x), \qquad [8.128]$$

gives the limit operator [8.104] with parameters [8.105].

REMARK 8.4.– *The existence of perturbation* $\varphi_1(c, x)$, *which satisfies equation* [8.124], *provides the condition for solvability*

$$\Pi[\mathbb{L} - \mathbb{L}^0(x)\mathbb{P}]\varphi(c) = 0.$$

There is then a solution of equation [8.124]

$$\varphi_1(s, x) = \mathbb{R}_0^d\widetilde{\mathbb{L}}(x)\varphi(s), \ \widetilde{\mathbb{L}}(x) := \mathbb{L} - \mathbb{R}_0^d\mathbb{L}^0(x),$$

where $\mathbb{R}_0^d$ *is a generalized-inverse operator*

$$\mathbb{Q}^d \mathbb{R}_0^d = \mathbb{R}_0^d \mathbb{Q}^d = \Pi - \mathbb{I}.$$

In this case, the residual term in the expansion [8.123] *has the form*

$$\varepsilon^2 \mathbb{C}^0(x) \mathbb{P} \mathbb{R}_0^d \widetilde{\mathbb{L}}(x) \varphi(c) \to 0, \ \varepsilon \to 0.$$

8.4. Random evolutionary systems in a balanced Markov random environment

The *discrete Markov evolution with discrete–continuous time* $t = [k\varepsilon^2]$, $k \geq 0$ is given by the solutions of difference stochastic equations with two components: predictable components and stochastic components (martingale differences). The regression function of increments, as well as the diffusion coefficient of the stochastic component, depend on the states of the embedded Markov chain in a homogeneous (in time) uniformly ergodic Markov process, which describes the states of the random environment. The balanced Markov random environment is characterized by zero means of the regression increments over the ergodic distribution of the Markov process.

Here, the approximation, by the Ornstein–Uhlenbeck-type diffusion process in a series scheme, is obtained with a small parameter ε ($\varepsilon \to 0$). The drift and diffusion parameters are determined by averaging over the stationary distribution of the embedded Markov chain, taking into account the balance condition.

8.4.1. *Basic assumptions*

The Markov statistical experiments with discrete–continuous time are given by the solutions of difference stochastic equations with the parameters of predictable and stochastic components, which depend on the states of the Markov random environment.

The Markov random environment is considered under two assumptions:

1. *The discrete Markov random environment*, given by an ergodic *Markov chain* x_k, $k \geq 0$ in a measurable state space $(E, \mathcal{E})$ with transition probabilities:

$$P(x, B) = \mathcal{P}\{x_{k+1} \in \mathcal{B} \,|\, x_k = x\}, \ x \in E, \ \mathcal{B} \in \mathcal{E}. \qquad [8.129]$$

2. *The continuous Markov random environment*, given by an ergodic *Markov process* $x(t)$, $t \geq 0$ in a measurable state space $(E, \mathcal{E})$ with transition probabilities [8.129] of an embedded Markov chain $x_k = x(\tau_k^\varepsilon)$, $k \geq 0$, and distributions of sojourn times in the states θ_{k+1}, $k \geq 0$:

$$\mathcal{P}\{\theta_{k+1} \geq t \,|\, x_k = x\} = \exp[-q(x)t], \ t \geq 0. \qquad [8.130]$$

Therefore, the renewal instants have the following form:

$$\tau_{k+1}^{\varepsilon} = \tau_k^{\varepsilon} + \varepsilon^2 \theta_{k+1}, \ k \geq 0. \qquad [8.131]$$

The drift parameter $V(x)$, $x \in E$ can take both positive and negative values with the balance condition for a random environment:

$$\int_E \rho(dx)V(x) = 0. \qquad [8.132]$$

Consequently, in a balanced environment, the stationary averaged drift increment is zero. Hence, the component normalization of the difference stochastic equation for statistical experiments is different.

BASIC ASSUMPTION A IN A DISCRETE MARKOV RANDOM ENVIRONMENT.– Discrete Markov evolution in a balanced Markov random environment [8.129] with discrete–continuous time $t = k\varepsilon^2$, $k \geq 0$ is given by the solution of the difference stochastic equation (compare with [8.90]):

$$\Delta\zeta^{\varepsilon}(t) = \varepsilon\,[V(x_k)\zeta^{\varepsilon}(t) + \sigma(x_k)\Delta\mu^{\varepsilon}(t)], \ k \geq 0, \qquad [8.133]$$

under additional conditions

$$E[\Delta\mu^{\varepsilon}(t)] = 0, \ E[[\Delta\mu^{\varepsilon}(t)]^2 \,|\, \zeta^{\varepsilon}(t)] = 1, \ t \geq 0. \qquad [8.134]$$

The drift parameters $V(x)$ and $\sigma(x)$, $x \in E$ are bounded real-valued functions on the states $x \in E$ of the Markov chain x_k, $k \geq 0$, which determines the Markov random environment.

Here, by definition,

$$\Delta\zeta^{\varepsilon}(t+\varepsilon^2) := \zeta^{\varepsilon}(t+\varepsilon^2) - \zeta^{\varepsilon}(t), \ t = [k\varepsilon^2].$$

BASIC ASSUMPTION B IN A CONTINUOUS MARKOV RANDOM ENVIRONMENT.– We define the extension of discrete Markov evolution in a continuous Markov random environment as a stepwise process:

$$\zeta^{\varepsilon}(t) = \zeta^{\varepsilon}(\tau_k^{\varepsilon}), \ \tau_k^{\varepsilon} \leq t < \tau_{k+1}^{\varepsilon}, \ \ k \geq 0.$$

Discrete Markov evolution in a continuous Markov random environment is given by the solutions of the difference stochastic equations

$$\Delta\zeta^{\varepsilon}(t) = \varepsilon\,[V(x_k)\zeta^{\varepsilon}(\tau_k^{\varepsilon}) + \sigma(x_k)\Delta\mu^{\varepsilon}(\tau_{k+1}^{\varepsilon})], \ k \geq 0, \qquad [8.135]$$

with additional conditions:

$$E[\Delta\mu^{\varepsilon}(\tau_{k+1}^{\varepsilon}) \,|\, \zeta^{\varepsilon}(\tau_k^{\varepsilon})] = 0, \ E[[\Delta\mu^{\varepsilon}(\tau_{k+1}^{\varepsilon})]^2 \,|\, \zeta^{\varepsilon}(\tau_k^{\varepsilon})] = 1, \ k \geq 0. \qquad [8.136]$$

The discrete Markov evolutions, defined by the solutions of the difference stochastic equations [8.133] and [8.134], as well as [8.135] and [8.136], under additional conditions, have an approximation in a series scheme with a parameter $\varepsilon \to 0$, by a normal Ornstein–Uhlenbeck process with continuous time. At the same time, the algorithm for calculating the drift parameter and the dispersion of the stochastic component in the limit process is significantly altered.

THEOREM 8.5.– *Let the following conditions be fulfilled:*

C0: *Balance condition*

$$\int_E \rho(dx)V(x) = 0. \tag{8.137}$$

C1A: *Markov chain x_k, $k \geq 0$ with transitional probabilities* [8.129] *is uniformly ergodic with a stationary distribution $\rho(B)$, $B \in \mathcal{E}$:*

$$\rho(B) = \int_E \rho(dx)P(x, B), \ \rho(E) = 1; \tag{8.138}$$

C1B: *Markov chain $x(t)$, $t \geq 0$ with transitional probabilities* [8.129] *is uniformly ergodic with a stationary distribution $\pi(B)$, $B \in \mathcal{E}$:*

$$\pi(dx)q(x) = q\rho(dx), \ q = \int_E \pi(dx)q(x); \tag{8.139}$$

C2: *Martingale differences $\Delta\mu^\varepsilon(t)$, $t \geq 0$ satisfy the condition*

$$E[\Delta\mu^\varepsilon(t)]^3 < C < +\infty, \ \forall t \in [0, T];$$

C3: *There is a convergence of initial conditions:*

$$\zeta^\varepsilon(0) \xrightarrow{P} \zeta(0), \ \varepsilon \to 0, \ E\,|\zeta(0)| \leq C < +\infty, \tag{8.140}$$

There is then a convergence, in distribution, of discrete Markov evolution:

$$\zeta^\varepsilon(t) \xrightarrow{D} \zeta(t), \ \varepsilon \to 0, \ t \in [0, T]. \tag{8.141}$$

In this case:

A: *In a discrete Markov random environment, the limit process $\zeta(t)$, $t \geq 0$ is set by the generator*

$$\mathbb{L}_d^0\varphi(c) = \widehat{V_0}s\varphi'(c) + \frac{1}{2}\widehat{\sigma_0^2}(c)\varphi''(c). \tag{8.142}$$

The drift parameter $\widehat{V_0}$ and dispersions $\widehat{\sigma_0^2}$ are calculated by the averaging formulas:

$$\widehat{V_0} = \widehat{V_0^2} - \widehat{V^2}, \ \widehat{\sigma_0^2}(c) = \widehat{\sigma^2} + (2\widehat{V_0^2} - \widehat{V^2})c^2, \qquad [8.143]$$

where

$$\widehat{V_0^2} := \int_E \rho(dx)V(x)\mathbb{R}_0 V(x), \ \widehat{V^2} := \int_E \rho(dx)V^2(x),$$

$$\widehat{\sigma^2} := \int_E \rho(dx)\sigma^2(x). \qquad [8.144]$$

The potential operator $\mathbb{R}_0$ is reducibly invertible with respect to the Markov chain generator $\mathbb{Q} := \mathbb{P} - \mathbb{I}$ (Korolyuk 1999):

$$\mathbb{Q}\,\mathbb{R}_0 = \mathbb{R}_0\,\mathbb{Q} = \Pi - \mathbb{I} \qquad [8.145]$$

The projector Π on the subspace of zeros of operator $\mathbb{Q}$ is determined as follows:

$$\Pi\varphi(x) = \int_E \rho(dx)\varphi(x). \qquad [8.146]$$

***B:** In a continuous Markov random environment, the limit process $\zeta(t)$, $t \geq 0$ is set by the generator*

$$\mathbb{L}_c^0\varphi(c) = q\,\mathbb{L}_d^0\varphi(c). \qquad [8.147]$$

REMARK 8.5.– *The limit process $\zeta(t)$, $t \geq 0$, with the generator given by equations* [8.142]–[8.144] *or by relation* [8.147], *is a normal Ornstein–Uhlenbeck diffusion process, given by the following stochastic differential equation:*

$$d\zeta(t) = -\widehat{V_0}\zeta(t)dt + \widehat{\sigma_0}dW(t), \ t \geq 0. \qquad [8.148]$$

Here, $W(t)$, $t \geq 0$ is a standard process of Brownian motion with parameters:

$$E[W(t)] = 0, \ E[[W(t)]^2] = t.$$

The main idea of proving the limit theorem, Theorem 8.5, for Markov processes is to apply limit theorems for random processes, in particular, for random evolutionary processes, based on the convergence of generators, defining the corresponding Markov processes (Korolyuk 1999; Korolyuk and Limnios 2005; Skorokhod 2008). The convergence of finite-dimensional distributions is ensured by the convergence of generating operators (generators) on a sufficiently large class of real-valued test functions.

In the rest of this chapter, we use the class $C_b^3(R)$ of real-valued functions $\varphi(c)$ with finite support (different from 0 on finite intervals) and three times continuously differentiated with bounded derivatives. In addition, using the test functions $\varphi(c, x)$, $x \in \mathcal{E}$, which depend on the additional argument x, the condition $\varphi(c, \cdot) \in C_b^3(R)$ means the existence of a function $\varphi(c) \in C_b^3(R)$, which limits, at the top, the function $\varphi(c, x) : |\varphi(c, x)| \leq \varphi(c)$.

The presence of the random Markov environment necessitates the use of singular perturbation theory for the reducibly invertible operator, which determines a uniformly ergodic Markov chain (or Markov process) (Korolyuk and Limnios 2005).

First, the characterization of an extended Markov process with an additional component, which sets the Markov random environment, is used.

LEMMA 8.9.–

A: *In the case of a discrete Markov random environment, the two-component Markov chain*

$$\{\zeta^\varepsilon(t),\, x^\varepsilon(t) := x(t/\varepsilon^2)\},\; t = k\varepsilon^2,\; \varepsilon > 0 \qquad [8.149]$$

is set by the generator

$$\mathbb{L}_d^\varepsilon(x)\varphi(c, x) = \varepsilon^{-2}[\mathbb{C}^\varepsilon(x)\mathbb{P} - \mathbb{I}]\varphi(c, x). \qquad [8.150]$$

where the operators $\mathbb{P}$ *and* $\mathbb{C}^\varepsilon(x)$ *are defined as follows:*

$$\mathbb{P}\varphi(x) := \int_E P(x, dy)\varphi(y), \qquad [8.151]$$

$$\mathbb{C}^\varepsilon(x)\varphi(c) := E[\varphi(c + \Delta\zeta^\varepsilon(t)) \,|\, \zeta^\varepsilon(t) = c,\, x^\varepsilon(t) = x],\; t \geq 0. \qquad [8.152]$$

B: *In the case of a continuous Markov random environment, the two-component Markov chain*

$$\{\zeta^\varepsilon(t),\, x^\varepsilon(t) := x(t/\varepsilon^2)\},\; t = k\varepsilon^2,\; k \geq 0, \qquad [8.153]$$

is set by the generator

$$\mathbb{L}_c^\varepsilon\varphi(c, x) = q(x)\mathbb{L}_d^\varepsilon(x)\varphi(c, x). \qquad [8.154]$$

PROOF OF LEMMA 8.9.–

A: The statement *A* follows from the definition of the Markov chain generator in discrete–continuous time with scaling ε^2 (see Nevelson and Hasminskii (1976, section 2.1)).

B: The statement *B* follows from the well-known definition of the generator of the Markov process with continuous time, taking into account the transition probabilities of the embedded Markov chain. □

The essential step of the proof of Theorem 8.3 is realized in the next lemma.

LEMMA 8.10.– *The generator* [8.150] *of the two-component Markov chain* [8.149] *has the following asymptotic decomposition on real-valued test functions* $\varphi(c,\cdot) \in C_b^3(R)$*:*

$$\mathbb{L}_d^\varepsilon(x)\varphi(c,x) = \Big[\varepsilon^{-2}\mathbb{Q} + \varepsilon^{-1}\mathbb{C}_1(x)\mathbb{P} + \mathbb{C}_2(x)\mathbb{P}\Big]\varphi(c,x) + \varepsilon\mathbb{R}^\varepsilon(x)\mathbb{P}\varphi(c,x). \qquad [8.155]$$

Here

$$\mathbb{Q}\varphi(\cdot,x) = [\mathbb{P} - \mathbb{I}]\varphi(\cdot,x), \qquad [8.156]$$

$$\mathbb{C}_1(x)\varphi(c,\cdot) = V(x)\,c\,\varphi'(c,\cdot),$$
$$\mathbb{C}_2(x)\varphi(c,\cdot) = \frac{1}{2}[V^2(x)c^2 + \sigma^2(x)]\varphi''(c,\cdot), \qquad [8.157]$$

and the residual term with bounded $\mathbb{R}^\varepsilon(x)$ *(see* [8.162]*):*

$$\varepsilon\mathbb{R}^\varepsilon\varphi(c,x) \to 0,\ \varepsilon \to 0,\ \varphi(c,\cdot) \in C_b^3(R). \qquad [8.158]$$

PROOF OF LEMMA 8.10.– Consider the following transformation of the generator [8.150]:

$$\mathbb{L}_d^\varepsilon(x)\varphi(c,x) = \varepsilon^{-2}[\mathbb{Q} + \mathbb{C}_0^\varepsilon(x)\mathbb{P}]\varphi(c,x). \qquad [8.159]$$

Here, by definition,

$$\mathbb{C}_0^\varepsilon(x) := \mathbb{C}^\varepsilon(x) - \mathbb{I}. \qquad [8.160]$$

Furthermore, taking into account the basic assumption [8.133], using the Taylor expansion of functions $\mathbb{C}_0^\varepsilon(x)\varphi(c)$ by the degrees of $\Delta\zeta_x^\varepsilon(t)$ in the neighborhood of the point $\zeta_x^\varepsilon(t) = c$, up to the third order, we obtain:

$$\begin{aligned}\mathbb{C}_0^\varepsilon(x)\varphi(c) &= E\varphi\Big(c + \Delta\zeta_x^\varepsilon(t+\varepsilon^2)\Big) - \varphi(c)\\ &= E\varphi\Big(c + \varepsilon\Big[V(x)\cdot c + \sigma(x)\Delta\mu^\varepsilon(t+\varepsilon^2)\Big]\Big) - \varphi(c) \qquad [8.161]\\ &= \varepsilon V(x)c\varphi'(c) + \frac{1}{2}\varepsilon^2\Big[V^2(x)c^2 + \sigma^2(x)\Big]\varphi''(c) + \varepsilon^3\mathbb{R}^\varepsilon(x)\varphi(c),\end{aligned}$$

where

$$\mathbb{R}^\varepsilon(x)\varphi(c) = \frac{1}{6}E\Big[(\Delta\zeta_x^\varepsilon)^3\,\varphi'''(c)\Big] = \frac{1}{6}\Big(V^3(x)c^3 + 3V(x)c\sigma^2(x) + \sigma^3(x)E[\Delta\mu^\varepsilon(x)]^3\Big)E\varphi'''(c+\theta\Delta\zeta_x^\varepsilon(t)). \qquad [8.162]$$

The residual term in [8.162], considering the basic assumption **A** and the condition **C2** of Theorem 8.4, is bounded on the class of functions $C_b^3(R)$. Consequently, the condition [8.158] is fulfilled.

The next stage of the proof of Theorem 8.4 is the use of the singular perturbation technique for the truncated operator (see Korolyuk and Limnios (2005, Chapter 5)):

$$\overline{\mathbb{L}_d^{\varepsilon}}(x)\varphi(c,x) = [\varepsilon^{-2}\mathbb{Q} + \varepsilon^{-1}\mathbb{C}_1(x)\mathbb{P} + \mathbb{C}_2(x)\mathbb{P}]\varphi(c,x) \qquad [8.163]$$

LEMMA 8.11.– *The solution of the singular perturbation problem for the truncated operator* [8.163] *on perturbed test functions*

$$\varphi^{\varepsilon}(c,x) = \varphi(c) + \varepsilon\varphi_1(c,x) + \varepsilon^2\varphi_2(c,x), \qquad [8.164]$$

is given by the following relation:

$$\overline{\mathbb{L}_d^{\varepsilon}}(x)\varphi(c,x) = \mathbb{L}_d^0\varphi(c) + \mathbb{R}_2^{\varepsilon}(x)\varphi(c). \qquad [8.165]$$

Here, the limit operator $\mathbb{L}_d^0$ *is determined by relations* [8.142]–[8.144], *and the residual term* $\mathbb{R}_2^{\varepsilon}(x)$ *is negligible:*

$$\mathbb{R}_2^{\varepsilon}(x)\varphi(c) \to 0,\ \varepsilon \to 0,\ \varphi(c) \in C_b^3(R). \qquad [8.166]$$

PROOF OF LEMMA 8.11.– Consider the relationship

$$\begin{aligned}
&\overline{\mathbb{L}_d^{\varepsilon}}(x)\varphi(c,x)\\
&= \left[\varepsilon^{-2}\mathbb{Q} + \varepsilon^{-1}\mathbb{C}_1(x)\mathbb{P} + \mathbb{C}_2(x)\mathbb{P}\right][\varphi(c) + \varepsilon\varphi_1(c,x) + \varepsilon^2\varphi_2(c,x)]\\
&= \varepsilon^{-2}\mathbb{Q}\varphi(c) + \varepsilon^{-1}[\mathbb{Q}\varphi_1(c,x) + \mathbb{C}_1(x)\varphi(c)] \qquad [8.167]\\
&+ [\mathbb{Q}\varphi_2(c,x) + \mathbb{C}_1(x)\mathbb{P}\varphi_1(c,x) + \mathbb{C}_2(x)\varphi(c)] + \mathbb{R}_2^{\varepsilon}(x)\varphi(c).
\end{aligned}$$

The first term on the right-hand side of [8.167] is zero because $\mathbb{Q}\varphi(c) \equiv 0$. The second term on the right-hand side of [8.167] is zero due to the balance condition [8.137]:

$$\mathbb{Q}\varphi_1(c,x) + \mathbb{C}_1(x)\varphi(c) = 0. \qquad [8.168]$$

Therefore, the solution of equation [8.168] is given by the following relation (see Korolyuk and Limnios (2005, section 5.2)):

$$\varphi_1(c,x) = \mathbb{R}_0\mathbb{C}_1(x)\varphi(c). \qquad [8.169]$$

Next, we substitute [8.169] into the third term of the right-hand side of [8.167]. The solution of the singular perturbation problem:

$$\mathbb{Q}\varphi_2(c,x) + \mathbb{C}_1(x)\mathbb{P}\varphi_1(c,x) + \mathbb{C}_2(x)\varphi(c) = \mathbb{L}^0\varphi(c), \qquad [8.170]$$

is then given by the next relation (see Korolyuk and Limnios (2005, section 5.2, Proposition 5.2)):

$$\mathbb{L}^0\varphi(c) = \Pi\,[\,\mathbb{C}_1(x)\mathbb{P}\mathbb{R}_0\mathbb{C}_1(x) + \mathbb{C}_2(x)\,]\,\varphi(c). \quad [8.171]$$

The following transformation of operator [8.171] specifies the form of the limit operator in Theorem 8.5. Given the definition of operators [8.157], we calculate the first term in [8.171]:

$$\begin{aligned}\Pi\mathbb{C}_1(x)\mathbb{P}\mathbb{R}_0\mathbb{C}_1(x)\varphi(c) &= \Pi\mathbb{C}_1(x)\mathbb{P}\mathbb{R}_0V(x)c\varphi'(c)\\ &= \Pi V(x)c\mathbb{P}\mathbb{R}_0V(x)[c\varphi'(c)]' \\ &= \Pi V(x)\mathbb{P}\mathbb{R}_0V(x)[c\varphi'(c) + c^2\varphi''(c)].\end{aligned} \quad [8.172]$$

Let us calculate the parameter represented by the composition of the operators on the right-hand side of [8.172]:

$$\widehat{V_0} := \Pi V(x)\mathbb{P}\mathbb{R}_0V(x) = \Pi V(x)\mathbb{Q}\mathbb{R}_0V(x) + \Pi V(x)\mathbb{R}_0V(x). \quad [8.173]$$

Using equality [8.145] as well as the balance condition [8.137], we obtain the relations [8.143] and [8.144].

Therefore, the right part of [8.173] has a representation $\widehat{V_0} = \widehat{V_0^2} - \widehat{V^2}$; that is, the first term in [8.171] has the form:

$$[\Pi\,\mathbb{C}_1(x)\mathbb{P}\mathbb{R}_0\mathbb{C}_1(x)]\,\varphi(c) = \widehat{V_0}\,[c\varphi'(c) + c^2\varphi''(c)]. \quad [8.174]$$

The calculation of the second term in [8.171] gives the following expression:

$$\Pi\,\mathbb{C}_2(x)\,\varphi(c) = \frac{1}{2}[\widehat{V^2}c^2 + \widehat{\sigma^2}]\varphi''(c). \quad [8.175]$$

The limit operator [8.171] has a representation [8.142]–[8.144], which completes the proof of Lemma 8.11. □

The completion of the proof of Theorem 8.4 is realized by applying the model limit theorem (Korolyuk and Limnios 2005, Theorem 6.3), which ensures the convergence of finite-dimensional distributions [8.141]. □

CONCLUSION 8.2.– The stationary nature of the limit process $\zeta(t)$, $t \geq 0$, given by a solution of the stochastic differential equation [8.148], provides the inequality $\widehat{V_0^2} < \widehat{V^2}$ (see [8.144]).

8.5. Adapted random evolutionary systems

Here, we consider adapted statistical experiments with a random change of time, which transforms a discrete stochastic basis into a continuous one. The adapted statistical experiments are studied in a continuous stochastic basis in a series scheme (Korolyouk 2017). The passage to limit, by a parameter of series, generates an approximation by diffusion processes with evolution (Jacod and Shiryaev 1987, Chapter I).

As before, the discrete Markov diffusion is given by the averaged sums of random variables that take a finite number of possible values.

The statistical experiments are considered in a discrete stochastic basis

$$\mathfrak{B}_{\mathbb{N}} = (\Omega,\, \mathfrak{F},\, (\mathfrak{F}_k,\, k \in \mathbb{N}),\, \mathcal{P})$$

with filtration $(\mathfrak{F}_k,\, k \in \mathbb{N} = \{0, 1, ...\})$ on a probability space $(\Omega,\, \mathfrak{F},\, \mathcal{P})$.

Next, we investigate the discrete Markov diffusion determined by a random change of time (Nielsen and Shiryaev 2010, Chapter 1), which transforms a discrete stochastic basis $\mathfrak{B}_{\mathbb{N}}$ into a continuous one:

$$\mathfrak{B}_T = (\Omega,\, \mathfrak{G},\, (\mathfrak{G}_t,\, t \in \mathbb{R}_+),\, \mathcal{P})$$

8.5.1. *Bernoulli approximation of the discrete Markov diffusion stochastic component*

Consider discrete Markov diffusion in a discrete stochastic basis $\mathfrak{B}_{\mathbb{N}}$. The evolutionary (predictable) part and the stochastic part are described.

The stochastic component is the martingale differences:

$$\Delta\mu_N(k+1) := \Delta S_N(k+1) - E[\Delta S_N(k+1) \mid S_N(k)],\; k \geq 0. \qquad [8.176]$$

Given the difference evolution equation [1.30], the martingale differences [8.176] have a representation

$$\Delta\mu_N(k+1) = \Delta S_N(k+1) + V_0(S_N(k)),\; k \geq 0. \qquad [8.177]$$

CONCLUSION 8.3.– The increments of discrete Markov diffusion are determined by the sum of two components

$$\Delta S_N(k+1) = -V_0(S_N(k)) + \Delta\mu_N(k+1),\; k \geq 0. \qquad [8.178]$$

The predictable component $V_0(S_N(k))$, $k \geq 0$ is given by the regression function of increments.

The martingale differences [8.176] are characterized by the first two moments

$$E\Delta\mu_N(k+1) = 0,$$
$$E[(\Delta\mu_N(k+1))^2 \mid S_N(k)] = \sigma^2(S_N(k))/N, \; k \geq 0. \quad [8.179]$$

The dispersion of the stochastic component has the form (Korolyouk 2015b)

$$\sigma^2(c) = 1 - V^2(c), \; V(c) = c - V_0(c), \; |c| \leq 1. \quad [8.180]$$

Stochastic dynamics of the discrete Markov diffusion $S_N(k)$, $k \geq 0$ is given by the difference stochastic equations [8.178]–[8.180].

Properties of the stochastic component are allowed specification.

LEMMA 8.12.– *The stochastic component determined by martingale differences* [8.177] *has the following representation:*

$$\Delta\mu_N(k+1) = \frac{1}{N}\sum_{n=1}^{N} \beta_n(k+1), \; k \geq 0. \quad [8.181]$$

The sample variables $\beta_n(k+1)$, $0 \leq n \leq N$, $k \geq 0$ *take two values:*

$$\beta_n(k+1) = \Big\{\pm 1 - V(C(k)) \;\; \textit{with probability} \;\; P_\pm(k+1)\Big\}, \quad [8.182]$$

where

$$P_\pm(k+1) = \frac{1}{2}\big[1 \pm C(k+1)\big] = \frac{1}{2}\big[1 \pm V(C(k))\big],$$
$$C(k) := P_+(k) - P_-(k), \; k \geq 0.$$

PROOF OF LEMMA 8.12.– Since

$$\Delta S_N(k+1) = S_N(k+1) - S_N(k) = \frac{1}{N}\sum_{n=1}^{N}[\delta_n(k+1) - \delta_n(k)],$$

then

$$\Delta\mu_N(k+1) = \frac{1}{N}\sum_{n=1}^{N}\beta_n(k+1) = \frac{1}{N}\sum_{n=1}^{N}[\delta_n(k+1) - \delta_n(k)] + V_0(\Delta S_N(k)).$$

Taking into account that

$$\beta_n(k+1) = \delta_n(k+1) - \delta_n(k) + V_0(\delta_n^0(k)) = \delta_n(k+1) - \delta_n^0(k);$$
$$\delta_n^0(k) := \delta_n(k) - V_0(\delta_n^0(k)), \; E[\delta_n^0(k) \mid S_N(k) = c] = V(c),$$

and also, for all n:

$$E[\delta_n(k+1) \mid S_N(k)=c] = P_+(k+1) - P_-(k+1) = C(k+1) = V(C(k)),$$

we have the values of the first two moments $\forall c$: $|c| \leq 1$

$$\begin{aligned} E[\beta_n(k+1) \mid S_N(k) = c] &= E[\delta_n(k+1) \mid S_N(k) = c] - V(c) \equiv 0, \\ E[\beta_n^2(k+1) \mid S_N(k) = c] & \\ = [1 - V(c)]^2 P_+(k+1) &+ [1 + V(c)]^2 P_-(k+1) = 1 - V^2(c), \end{aligned}$$

which completes the proof of Lemma 8.12. □

CONCLUSION 8.4.– The stochastic component [8.181] is given by the Bernoulli distribution:

$$\begin{aligned} B_N(\nu; V(C(k))) &= P\{\Delta\mu_N(k+1) = \nu - V(C(k)) \mid S_N(k) = C(k)\} \\ &= \frac{N!}{N_+!N_-!} P_+^{N_+}(k+1) P_-^{N_-}(k+1), \end{aligned} \qquad [8.183]$$

where

$$N_\pm/N = \frac{1}{2}[1 \pm \nu], \ \nu = \nu_+ - \nu_-, \ \nu_\pm = N_\pm/N.$$

with the first two moments:

$$\begin{aligned} &E[\beta_n(k+1) \mid S_N(k)] = 0, \ \forall k \geq 0, \\ &E[\beta_n^2(k+1) \mid S_N(k)] = \sigma^2(S_N(k)) = 1 - V^2(S_N(k)). \end{aligned}$$

Now, the dynamics of discrete Markov diffusion has two interpretations:

– The increments of discrete Markov diffusion $\Delta S_N(k)$ are determined by the difference equation [8.178], in which the stochastic component is given by the Bernoulli distribution [8.183].

– The probabilities [8.182] determine the Bernoulli distribution [8.183] of the stochastic component, at a given kth stage.

8.5.2. *Adapted random evolutionary systems*

The passage from a discrete stochastic basis $\mathfrak{B}_{\mathbb{N}} = (\Omega, \mathfrak{F}, (\mathfrak{F}_k, k \in \mathbb{N}), \mathcal{P})$ to a continuous stochastic basis $\mathfrak{B}_T = (\Omega, \mathfrak{G}, (\mathfrak{G}_t, t \in \mathbb{R}_+), \mathcal{P})$ is realized by a *random change of time*

$$\nu(t), \, t \geq 0, \ \nu(0) = 0.$$

The increasing process $\nu(t)$, $t \geq 0$, right continuous with left limits, is given by Markov moments of jumps:

$$\tau_k := \inf\{t \,:\, \nu(t) \geq k\}, \; k \in \mathbb{N}.$$

The regularity of the process $\nu(t)$, $t \geq 0$, is ensured by the condition:

$$P\{\tau_k < +\infty\}, \; \forall k > 0. \qquad [8.184]$$

The counting renewal process $\nu(t)$, $t \geq 0$ is considered (for simplicity) as a stationary Poisson process with exponentially distributed renewal intervals $\theta_{k+1} := \tau_{k+1} - \tau_k$, $k \geq 0$:

$$P\{\theta_{k+1} \geq t\} = \exp(-qt), \; 0 < q < +\infty, \; t \geq 0.$$

It is known (Korolyuk 1999) that the compensator of the Poisson process $\nu(t)$, $t \geq 0$ is given by the formula:

$$E\nu(t) = qt, \; t \geq 0.$$

DEFINITION 8.2.– *The random change of time in a discrete basis* $\mathfrak{B}_{\mathbb{N}}$ *is given by the filtration*

$$\mathfrak{G}_t = \mathfrak{F}_{\nu(t)}, \; t \geq 0. \qquad [8.185]$$

According to Lemma 3.8 (Jacod and Shiryaev 1987, Chapter II, section 3b), there are the properties: $\tau_0 = 0$, $\mathfrak{G}_0 = \mathfrak{F}_0$ and $\mathfrak{G}_{\tau_k} = \mathfrak{F}_k$ on the set $\tau_k < \infty$ and $\mathfrak{F}_{k-1} = \mathfrak{G}_{\tau_k -}$, $k > 0$.

In particular, if $\nu(t) = [t]$ is the integer part of a positive number $t > 0$, then the basis $\mathfrak{B}_T$ coincides with the basis $\mathfrak{B}_{\mathbb{N}}$.

DEFINITION 8.3.– *The adapted discrete Markov diffusion with the random change of time* [8.185] *is given by the relation:*

$$\alpha_N(t) := S_N(\nu(t)), \; t \geq 0,$$

or, otherwise:

$$\alpha_N(t) := S_N(k), \; \tau_k \leq t < \tau_{k+1}, \; t \geq 0.$$

CONCLUSION 8.5.– The adapted discrete Markov diffusion is a *special semimartingale*, for which:

– the predictable component, given by the increment regression function $V_0(c)$, $|c| \leq 1$;

– the stochastic component is given by the Bernoulli distribution [8.183] of increments $\Delta\mu_N(k+1)$, $k \geq 0$.

Namely, there is the following representation (see [8.178]):

$$\alpha_N(t) = \alpha_N(0) + V_N(t) + M_N(t), \ t \geq 0,$$

$$V_N(t) := -\sum_{k=0}^{\nu(t)} V_0(\alpha_N(\tau_k)), \ M_N(t) := \sum_{k=0}^{\nu(t)} \Delta\mu_N(k+1).$$

8.5.3. *Adapted random evolutionary systems in a series scheme*

The adapted discrete Markov diffusion in a series scheme with a small parameter $\varepsilon \to 0$, ($\varepsilon > 0$), is determined by the random change of time

$$\nu_\varepsilon(t) := \nu(t/\varepsilon^2), \ t \geq 0, \qquad [8.186]$$

as well as by the following normalization of increments [8.178] by a series parameter ε^2:

DEFINITION 8.4.– *The adapted discrete Markov diffusion with the random change of time* [8.186] *in a series scheme*

$$\alpha^\varepsilon(t) := S_N(\nu_\varepsilon(t)), \ \varepsilon^2 = 1/N, \ t \geq 0,$$

is determined by the following three predictable characteristics (Jacod and Shiryaev 1987, Chapter 2):

– *evolutionary component*

$$V_t^\varepsilon := -\varepsilon^2 \sum_{k=0}^{\nu_\varepsilon(t)} V_0(\alpha^\varepsilon(\tau_k^\varepsilon)), \ t \geq 0; \qquad [8.187]$$

– *dispersion of the stochastic component*

$$\sigma_t^\varepsilon := \varepsilon^2 \sum_{k=0}^{\nu_\varepsilon(t)} \sigma^2(\alpha^\varepsilon(\tau_k^\varepsilon)), \ \sigma^2(c) := 1 - V^2(c), \ t \geq 0, \qquad [8.188]$$

– *compensating jump measures*

$$\Gamma_t^\varepsilon(g) := \varepsilon^2 \sum_{k=0}^{\nu_\varepsilon(t)} E\big[g(\alpha^\varepsilon(\tau_{k+1}^\varepsilon)) \,|\, \mathfrak{G}_{\tau_k^\varepsilon}^\varepsilon\big], \ g \in C_2(\mathbb{R}), \ t \geq 0. \qquad [8.189]$$

Here, the class of functions $C_2(R)$ is defined in Jacod and Shiryaev (1987); the filtration in the condition [8.189] *is $\mathfrak{G}_t^\varepsilon := \mathfrak{F}_{\nu_\varepsilon(t)}^\varepsilon$, $t \geq 0$ (see* [8.185]*).*

The predictable characteristics of the adapted discrete Markov diffusion [8.187]–[8.189] depend on the current value of discrete Markov diffusion $\alpha^\varepsilon(\tau_k^\varepsilon)$, $k \geq 0$ at renewal moments $\tau_k^\varepsilon = \varepsilon^2 \tau_k$, $k \geq 0$, of the random change of time $\nu_\varepsilon(t)$, $t \geq 0$. Therefore, the study of the convergence problem for the adapted discrete Markov diffusion in a series scheme, by $\varepsilon \to 0$, is realized in the following two stages (see Liptser (1994)):

Stage 1. Determination of the compactness conditions of the adapted discrete Markov diffusion $\alpha^\varepsilon(t)$, $0 \leq t \leq T$, $\varepsilon \geq 0$.

Stage 2. Under additional conditions on the predictable characteristics, i.e. functions $V_0(c)$, $\sigma^2(c)$, $|c| \leq 1$, identify the limit process given by the corresponding predictable characteristics.

At the first stage, the approach proposed in Liptser (1994) is used (see also Limnios and Samoilenko (2013)). Namely, the condition of *compact embedding* takes place:

$$\lim_{c \to \infty} \sup_{\varepsilon > 0} P\{\sup_{0 \leq t \leq T} |\alpha^\varepsilon(t)| > c\} = 0. \qquad [8.190]$$

LEMMA 8.13.– *By the condition on the initial values $E|\alpha^\varepsilon(0)| \leq c_0$ with a constant that does not depend on ε, there is the following estimate:*

$$E \sup_{0 \leq t \leq T} |\alpha^\varepsilon(t)|^2 \leq C. \qquad [8.191]$$

PROOF OF LEMMA 8.13.– The semimartingale representation of the adapted discrete Markov diffusion and the properties of the predictable characteristics [8.187]–[8.189] are used.

The evolutionary component V_t^ε, $t \geq 0$ has an estimate:

$$\sup_{0 \leq t \leq T} |V_t^\varepsilon| \leq C_V \int_0^T [1 + |V_0(\alpha_*^\varepsilon(t))|] dt,$$

$$\alpha_*^\varepsilon(t) := \sup_{0 \leq s \leq t} |\alpha^\varepsilon(s)|.$$

The dispersion of the stochastic component M_t^ε, $t \geq 0$ is uniformly bounded in mean square:

$$E \sup_{0 \leq t \leq T} |M_t^\varepsilon|^2 \leq C_\sigma \int_0^T [1 + \sigma^2(\alpha_*^\varepsilon(t))] dt.$$

Therefore, for the adapted discrete Markov diffusion, there is the following estimate:

$$E \sup_{0 \le t \le T} |\alpha^\varepsilon(t)|^2 \le 3E\left[[\alpha^\varepsilon(0)]^2 + \sup_{0\le t\le T} |V_t^\varepsilon| + \sup_{0 \le t \le T} |M_t^\varepsilon|^2\right].$$

The last three inequalities provide the following estimate:

$$E[\alpha_*^\varepsilon(T)]^2 \le C_1 + C_2 \int_0^T E|\alpha_*^\varepsilon(t)|^2 dt.$$

Finally, the last inequality – Gronwall–Bellman inequality – establishes the inequality [8.191]. Lemma 8.13 is proved. □

Now, the Kolmogorov inequality for the adapted discrete Markov diffusion $\alpha^\varepsilon(t)$, $0 \le t \le T, \varepsilon \ge 0$ establishes the condition of compact embedding [8.190].

REMARK 8.6.– *Another approach to establishing the condition of compact embedding* [8.190] *is shown in Ethier and Kurtz (1986, section 4.5).*

CONCLUSION 8.6.– By Lemma 8.13, there is the following estimate:

$$E|\alpha^\varepsilon(t) - \alpha^\varepsilon(t')|^2 \le C_T|t - t'|,\ 0 \le t,\ t' \le T. \qquad [8.192]$$

Under the conditions [8.190] and [8.192], there is a compactness of processes $\alpha^\varepsilon(t), 0 \le t \le T, \varepsilon > 0$.

At the second stage, under the condition of compactness of the adapted discrete Markov diffusion in a series scheme $\alpha^\varepsilon(t)$, $0 \le t \le T$, $\varepsilon > 0$, the verification of the limit process is realized by the study of convergence (by $\varepsilon \to 0$) of predictable characteristics [8.187]–[8.189]. First, the convergence of the compensating jump measures [8.189] is established:

$$\sup_{0 \le t \le T} \Gamma_t^\varepsilon(g) \xrightarrow{D} 0,\ \varepsilon \to 0,$$

which is provided by the Lindeberg condition:

$$\sum_{n=1}^N E[\beta_n^\varepsilon(k+1)]^2 \cdot I(|\beta_n(k+1)|) \ge h/\varepsilon \,|\, S_N(k) = c] \to 0,\ \varepsilon \to 0,$$

for the samples sum

$$\Delta\mu^\varepsilon(k+1) := \varepsilon \sum_{n=1}^N \beta_n(k+1),$$

with dispersion

$$D_N^2 := E[(\Delta\mu^\varepsilon(k+1))^2 \,|\, S_N(k) = c] = \sigma^2(c).$$

Next, we need to prove the convergence of the evolutionary component and the dispersion of the stochastic component of semimartingale $\alpha^\varepsilon(t), t \geq 0$.

LEMMA 8.14.– *Under the conditions of Lemma 8.13, there are the convergences, in distribution, as* $\varepsilon \to 0$*:*

$$V_t^\varepsilon \xrightarrow{D} V_t^0 = -\int_0^{qt} V_0(\alpha^0(u))du,\ 0 \leq t \leq T,$$

$$\sigma_t^\varepsilon \xrightarrow{D} \sigma_t^0 = \int_0^{qt} \sigma^2(\alpha^0(u))du,\ 0 \leq t \leq T,\ \sigma^2(c) = 1 - V^2(c).$$

Here, the limit process $\alpha^0(t), t \geq 0$ is determined by the condition of compactness (see Lemma 8.13):

$$\alpha^{\varepsilon_r}(t) \xrightarrow{D} \alpha^0(t),\ \varepsilon_r \to 0,\ r \to \infty.$$

PROOF OF LEMMA 8.14.– Since both predictable characteristics [8.187] and [8.188] have the same structure of the integral functional on the process $\alpha^\varepsilon(t),\ t \geq 0$, it is sufficient to investigate the convergence of one of them, for example, the evolutionary component [8.187].

We use the martingale characterization:

$$\mu_t^\varepsilon = \varphi(V_t^\varepsilon) - \int_0^t L_V^\varepsilon \varphi(V_u^\varepsilon)du,\ t \geq 0.$$

The integral functional generator of [8.187] is

$$L_V^\varepsilon \varphi(c) = \varepsilon^{-2} q[\varphi(c - \varepsilon^2 V_0(c)) - \varphi(c)],\ \varphi(c) \in C^2(\mathbb{R}),$$

which has an asymptotic representation on a class of test functions:

$$L_V^\varepsilon \varphi(c) = L_V^0 \varphi(c) + R_\varepsilon \varphi(c),\ \varphi(c) \in C^2(\mathbb{R})$$

with a negligible term:

$$R_\varepsilon \varphi(c) \to 0,\ \varepsilon \to 0,\ \varphi(c) \in C^2(\mathbb{R}).$$

The limit operator L_V^0 determines the evolution

$$L_V^0 \varphi(c) = -qV_0(c)\varphi'(c),\ \varphi(c) \in C^2(\mathbb{R}).$$

The limit evolution is given by the relation:

$$V_t^0 = -\int_0^{qt} V_0(\alpha^0(u))du, \ t \geq 0.$$

Consequently, the compactness of the adapted discrete Markov diffusion [8.191], established by Lemma 8.13, ensures the convergence of martingales:

$$\mu_t^\varepsilon \Rightarrow \mu_t^0 = \varphi(V_t^0) - \int_0^t L_V^0 \varphi(V_u^0)du, \ 0 \leq t \leq T, \ \varepsilon \to 0.$$

Similarly, the convergence of the quadratic characteristic [8.188] is established using the martingale characterization:

$$\mu_t^\varepsilon = \varphi(\sigma_t^\varepsilon) - \int_0^t L_\sigma^\varepsilon \varphi(\sigma_u^\varepsilon)du, \ t \geq 0,$$

with the generator

$$L_\sigma^\varepsilon \varphi(c) = \varepsilon^{-2} q[\varphi(c + \varepsilon^2 \sigma_0^2(c)) - \varphi(c)], \ \varphi(c) \in C^2(\mathbb{R}),$$

which has an asymptotic representation on the class of test functions $\varphi(c) \in C^2(\mathbb{R})$:

$$L_\sigma^\varepsilon \varphi(c) = q\sigma_0^2(c)\varphi'(c) + R_\varepsilon \varphi(c), \ \varphi(c) \in C^2(\mathbb{R}).$$

with a negligible term:

$$R_\varepsilon \varphi(c) \to 0, \ \varepsilon \to 0, \ \varphi(c) \in C^2(\mathbb{R}).$$

Consequently, the limit quadratic characteristic has a representation:

$$\sigma_t^0 = \int_0^{qt} \sigma^2(\alpha_0(u))du, \ t \geq 0, \ \sigma^2(c) = 1 - V^2(c).$$

At the final stage, the condition of uniqueness of semimartingale characterization of the diffusion Markov process with evolution $\alpha^0(t), \ t \geq 0$ is used, given by the generator (Jacod and Shiryaev 1987, Chapter IX):

$$L_V^0 \varphi(c) = -V_0(c)\varphi'(c) + \frac{1}{2}\sigma^2(c)\varphi''(c), \ \varphi(c) \in C^2(\mathbb{R}).$$

□

THEOREM 8.6.– *The adapted discrete Markov diffusion* $\alpha^\varepsilon(t), \ t \geq 0$, *in a series scheme with a small parameter* $\varepsilon \to 0, \ (\varepsilon > 0)$, *given by predictable characteristics* [8.187]–[8.189], *under the additional condition of convergence of initial values:*

$$\alpha^\varepsilon(0) \xrightarrow{D} \alpha_0, \ E\alpha^\varepsilon(0) \to E\alpha_0, \ \varepsilon \to 0.$$

converges, in distribution, to a diffusion process with evolution with the random change of time

$$\alpha^\varepsilon(t) \xrightarrow{D} \alpha^0(t),\ 0 \le t \le T,\ \varepsilon \to 0.$$

In this case, the predictable characteristics of the limit process $\alpha^0(t)$*,* $t \ge 0$*, have the representation:*

$$V_t^0 = \int_0^{qt} V_0(\alpha^0(u))du,\ \sigma_t^0 = \int_0^{qt} \sigma^2(\alpha^0(u))du,\ 0 \le t \le T,$$

and there is no compensating jump measure:

$$B_t^\varepsilon(g) \to 0,\ \varepsilon \to 0,\ g(c) \in C_1(\mathbb{R}).$$

CONCLUSION 8.7.– The limit diffusion process with evolution $\alpha^0(t)$, $t \ge 0$ is given by the following stochastic differential equation:

$$d\alpha(t) = -V_0(\alpha(t))dt + \sigma(\alpha(t))dW_t,\ t \ge 0,$$

with the scaling change of time:

$$\alpha^0(t) = \alpha(qt),\ t \ge 0.$$

The intensity q is determined by the equality $E\nu(t) = qt$.

References

de Acosta, A. (1997). Exponential tightness and projective systems in large deviation theory. *Festschrift for Lucien Le Cam*. Springer, New York.

Anisimov, V.V. (2008). *Switching Processes in Queuing Models*. ISTE Ltd, London and John Wiley & Sons, New York.

Barndorff-Nielsen, O.E. and Shiryaev, A.N. (2010). *Change of Time and Change of Measure*. World Scientific, New York/London.

Bertoin, J. (1996). *Cambridge Tracts in Mathematics 121: Lévy Processes*. Cambridge University Press, Cambridge.

Boichuk, A.A. and Samoilenko, A.M. (2004). *Generalized Inverse Operators and Fredholm Boundary-Value Problems*. Walter de Gruyter, Utrecht/Boston, MA.

Borovkov, A.A. and Mogulskii, A.A. (1992). Large deviations and testing statistical hypotheses I. Large deviations of sums of random vectors. *Siberian Adv. Math.*, 2(3), 52–120.

Borovkov A.A. and Mogulskii, A.A. (2012). Large deviation principles for random walk trajectories I. *Theory Probab. Appl.*, 56(4), 538–561.

Borovskikh, Y.V. and Korolyuk, V.S. (1997). *Martingale Approximation*. VSP, Utrecht.

Bryc, W. (1990). Large deviations by the asymptotic value method. *Diffusion Processes and Related Problems in Analysis*. Birkhauser, Basel.

Chabaniuk Y. (2007). Stability of a dynamical system with semi-Markov switchings under conditions of diffusion approximation. *Ukr. Math. J.*, 59, 1441–1452.

Chabaniuk Y., Korolyuk, V.S., Limnios, N. (2007). Fluctuation of stochastic systems with average equilibrium point. *C. R. Math.*, 345(7), 405–410.

Chernoff, H. (1952). Measure of asymptotic efficiency for tests of a hypothesis based on the sum of observations. *Ann. Math. Stat.*, 23(4), 493–655.

Çinlar, E., Jacod, J., Protter, P., Sharpe, M.J. (1980). Semimartingale and Markov processes. *Z. Wahrschein. verw. Gebiete*, 54, 161–219.

Cole, J.D. (1951). On a quasi-linear parabolic equation occuring in aerodynamics. *Quart. Appl. Math.*, 9, 225–236.

Cramér, H. (1938). Sur un nouveau théoreme-limite de la théorie des probabilités. *Acta Sci. Ind.*, 736, 5–23.

Crandall, M.G. and Lions, P.-L. (1983). Viscosity solutions of Hamilton–Jacobi equations. *Trans. Am. Math. Soc.*, 277, 1–42.

Davis, M.H.A. (1993). *Markov Models and Optimization.* Chapman&Hall/CRC Press, Boca Raton, FL.

Dellacherie, C. and Meyer, P.-A. (1978). *Probabilities and Potential.* North-Holland Publishing, Amsterdam/New York.

Dembo, A. and Zeitouni, O. (2010). *Large Deviations Techniques and Applications.* Springer-Verlag, Berlin Heidelberg.

Deuschel, J.-D. and Stroock, D.W. (1989). *Large Deviations.* Academic Press, Boston, MA.

Donsker M.D. and Varadhan S.R.S. (1975a). Asymptotic evaluation of certain Markov process expectations for large time I. *Comm. Pure Appl. Math.*, 27, 1–47

Donsker M.D. and Varadhan S.R.S. (1975b). Asymptotic evaluation of certain Markov process expectations for large time II. *Comm. Pure Appl. Math.*, 28, 279–301.

Donsker M.D. and Varadhan S.R.S. (1976). Asymptotic evaluation of certain Markov process expectations for large time III. *Comm. Pure Appl. Math.*, 29, 389–461.

Doob, J.L. (1990). *Stochastic Processes.* John Wiley & Sons, New York.

Dupuis, P. and Ellis, R.S. (1997). *A Weak Convergence Approach to the Theory of Large Deviations.* Wiley, New York.

Ethier, S.N. and Kurtz, T.G. (1986). *Markov Processes: Characterization and Convergence.* Wiley, New Jersey.

Evans, L.C. (1989). The perturbed test function method for viscosity solutions of nonlinear PDE. *Proc. Roy. Soc. Edinburgh Sect. A.*, 111(3–4), 359–375.

Feng, J. (1999). Martingale problems for large deviations of Markov processes. *Stochastic Processes and Applications*, 81(2), 165–216.

Feng, J. and Kurtz, T.G. (2006). *Large Deviations for Stochastic Processes.* AMS New York/Boston, MA.

Fleming W.H. and Souganidis P.E. (1986). A PDE approach to some large deviations problems. Nonlinear systems of partial differential equations in applied mathematics, Part 1 (Santa Fe, N.M., 1984). *Lectures Appl. Math.*, 23, AMS Providence, RI.

Freidlin, M.I. (1978). The averaging principle and theorems on large deviations. *Russian Math. Surveys*, 33, 117–176.

Freidlin, M.I. and Ventzell, A.D. (2012). *Random Perturbation of Dynamical Systems.* Springer, Berlin/Heidelberg.

Fukushima, M. and Stroock, D.W. (1986). Reversibility of solutions to martingale problems. *Adv. Math. (Suppl. Stud.)*, 9, 107–123.

Gikhman, I.I. and Skorohod, A.V. (1979). *Controlled Stochastic Processes.* Springer-Verlag, New York.

Gorostiza, L. (1972). A central limit theorem for a class of d-dimensional random motions with constant speed. *Bull. Amer. Math. Soc.*, 78, 575–577.

Gorostiza, L. (1973a). The central limit theorem for random motions of d-dimensional Euclidean space. *Ann. Probab.*, 1, 603–612.

Gorostiza, L. (1973b). An invariance principle for a class of d-dimensional polygonal random functions. *Trans. Am. Math. Soc.*, 177, 413–445.

Griego, R. and Hersh, R. (1969). Random evolutions, Markov chains, and systems of partial differential equations. *Proc. Natl. Acad. Sci.*, 62, 305–308.

Griego, R. and Hersh, R. (1971). Theory of random evolutions with applications to partial differential equations. *Trans. Am. Math. Soc.*, 156, 405–418.

Hersh, R. (1974). Random evolutions: Survey of results and problems. *Rocky Mount. J. Math.*, 4, 443–477.

Hersh, R. (2003). The birth of random evolutions. *Math. Intell.*, 25(1), 53–60.

Hersh, R. and Papanicolaou, R. (1972). Non-commuting random evolutions, and an operator-valued Feynman–Kac formula. *Comm. Pure and Appl. Math.*, XXX, 337–367.

Hersh, R. and Pinsky, M. (1972). Random evolutions are asymptotically Gaussian. *Comm. Pure and Appl. Math.*, XXV, 33–34.

Hopf, E. (1950). The partial differential equation. *Comm. Pure Appl. Math.*, 3, 201–230.

Jacod, J. and Shiryaev, A.N. (1987). *Limit Theorems for Stochastic Processes.* Springer, Berlin/Heidelberg.

Jacod, J. and Shiryaev, A.N. (2003). *Limit Theorems for Stochastic Processes.* Springer-Verlag, Berlin.

Khasminskii, R.Z. (1966a). On stochastic processes defined by differential equations with a small parameter. *Theory Probab. Appl.*, 11(2), 211–228.

Khasminskii, R.Z. (1966b). A limit theorem for the solutions of differential equations with random right-hand sides. *Theory Probab. Appl.*, 11(3), 390–406.

Khasminskii, R.Z. (2012). *Stochastic Stability of Differential Equations.* Springer-Verlag, Berlin/Heidelberg.

Koroliouk, D. (2015a). Two component binary statistical experiments with persistent linear regression. *Theor. Prob. and Math. Stat.*, 90, 103–114.

Koroliouk, D. (2015b). Binary statistical experiments with persistent nonlinear regression. *Theor. Probab. Math. Statist.*, 91, 71–80.

Koroliouk, D. (2015c). Binary statistical experiments with persistent linear regression in the Markov random medium. *Rep. Nat. Acad. Sci. of Ukraine*, 4, 12–17.

Koroliouk, D. (2015d). Classification of binary deterministic statistical experiments with persistent regression. *Cyber. Syst. Anal.*, 51(4), 644–649.

Korolіouk, D. (2016a). Stationary statistical experiments and the optimal estimator for a predictable component. *J. Math. Sci.*, 214(2), 220–228.

Koroliouk, D. (2016b). Multivariate statistical experiments with persistent linear regression and equilibrium. *Theor. Probab. Math. Statist.*, 92, 71–79.

Koroliouk, D. (2016c). The problem of discrete Markov diffusion leaving an interval. *Cyber. Syst. Anal.*, 52(4), 571–576.

Koroliouk, D. (2017). Adapted statistical experiments. *J. Math. Sci.*, 220(5), 615–623.

Koroliouk, D. (2020). *Dynamics of Statistical Experiments*. ISTE Ltd, London and John Wiley & Sons, New York.

Koroliouk, D. and Korolyuk, V.S. (2019). Equilibrium in Wright–Fisher models of population genetics. *Cyber. Syst. Anal.*, 55(2), 253–258.

Koroliouk, D., Korolyuk, V.S., Rosato, N. (2014). Equilibrium process in biomedical data analysis: The Wright–Fisher model. *Cyber. and Syst. Anal.*, 50(6), 890–897.

Koroliouk, D., Bertotti, M.L., Korolyuk, V.S. (2016a). Stochastic behavioral models. Classification. *Cyber. Syst. Anal.*, 52(6), 884–895.

Koroliouk, D., Korolyuk, V.S., Nicolai, E., Bisegna, P., Stella, L., Rosato, N. (2016b). A statistical model of macromolecules dynamics for Fluorescence Correlation Spectroscopy data analysis. *Stat. Optim. Inf. Comput. (SOIC)*, 4, 233–242.

Korolyuk, V.S. (1969). On asymptotical estimate for time of a semi-Markov process being in the set of states. *Ukr. Mat. Zh.*, 21, 842–845.

Korolyuk, V.S. (1991). Stability of an autonomous dynamic system with fast Markov switching. *Ukr. Math. J.*, 43, 1101–1105.

Korolyuk, V.S. (1998). Stability of stochastic systems in the diffusion-approximation scheme. *Ukr. Math. J.*, 50, 40–54.

Korolyuk, V.S. (2010a). Markov random evolutions with independent increments in the asymptotically small diffusion scheme. *Dopovidi NAN Ukrainy*, 6, 22–26 (in Russian).

Korolyuk, V.S. (2010b). Problem of large deviations for Markov random evolutions with independent increments in the scheme of asymptotically small diffusion. *Ukr. Math. J.*, 62(5), 739–747.

Korolyuk, V.S. (2011a). Large deviation problems for Markov random evolution with independent increments in the scheme of asymptotically small diffusion. *Commun. Stat. Theory Methods*, 40(19–20), 3385–3395.

Korolyuk, V.S. (2011b). Large deviations principle for random evolutions. *Bulletin of KNU im. Shevchenka*, 25, 4–6 (in Russian).

Korolyuk, V.S. (2011c). Random evolutions with locally independent increments on increasing time intervals. *J. Math. Sci.*, 179(2), 273–289.

Korolyuk, V.S. (2012). Dynamic random evolutions on increasing time intervals. *Theor. Probab. Math. Statist.*, 85, 83–91.

Korolyuk V.S. and Koroliouk D. (1995). Diffusion approximation of stochastic Markov models with persistent regression. *Ukr. Math. J.*, 47(7), 1065–1073.

Korolyuk, V.S. and Korolyuk, V.V. (1999). *Stochastic Models of Systems*. Kluwer Academic Publishers, Dodrecht/Boston, MA.

Korolyuk, V.S. and Limnios, N. (2005). *Stochastic Systems in Merging Phase Space*. World Scientific, New Jersey/London.

Korolyuk, V.S. and Samoilenko, I.V. (2014a). Large deviations for impulsive accumulation processes in the phase merging scheme. *Dopovidi NAN Ukrainy*, 7, 28–35 (in Ukrainian).

Korolyuk, V.S. and Samoilenko, I.V. (2014b). Large deviations for random evolutions in the scheme of asymptotically small diffusion. In *Modern Stochastics and Applications*, Korolyuk, V., Limnios, N., Mishura, Y., Sakhno, L., Shevchenko, G. (eds). Springer Optimization and Its Applications, Cham.

Korolyuk, D.V. and Silvestrov D.S. (1983). Entry times into asymptotically receding domains for ergodic Markov chains. *Theory Probab. Appl.*, 28, 432–442.

Korolyuk, D.V. and Silvestrov D.S. (1984). Entry times into asymptotically receding regions for processes with semi-Markov switchings. *Theory Probab. Appl.*, 29, 558–563.

Korolyuk, V.S. and Swishchuk A.V. (1995a). *Semi-Markov Random Evolutions*. Kluwer Academic Publishers, Dordrecht.

Korolyuk, V.S. and Swishchuk A.V. (1995b). *Evolution of Systems in Random Media*. CRC Press, Boca Raton, FL.

Korolyuk, V.S. and Turbin, A.F. (1993). *Mathematical Foundations of the State Lumping of Large Systems*. Kluwer Academic Publishers, Dordrecht.

Korolyuk, V.S., Brodi, S.M., Turbin, A.F. (1975). Semi-Markov processes and their applications. *J. Sov. Math.*, 4, 244–280.

Korolyuk, V.S., Manca, R., D'Amico, G. (2014a). Storage impulsive processes in the merging phase space. *J. Math. Sci.*, 196(5), 644–651.

Korolyuk, V.S., Manca, R., D'Amico, G. (2014b). Storage impulsive processes on increasing time intervals. *Theor. Probab. Math. Statist.*, 89, 71–81.

Kurtz, T.G. (1973). A limit theorem for perturbed operator semigroups with applications to random evolutions. *J. Funct. Anal.*, 12, 55–67.

Kurtz, T.G. (1987). *Martingale Problems for Controlled Processes*. Lecture Notes in Control and Information Sciences, 91, Springer, Berlin, 75–90.

Kushner, H.J. (1990). *Weak Convergence Methods and Singular Perturbed Stochastic Control and Filtering Problems*. Birkhäuser, Boston, MA.

Liggett, T.M. (2010). Continuous-time Markov processes: An introduction, AMS. *Graduate Studies in Mathematics Volume 113*, Providence, Rhode Island.

Limnios, N. and Samoilenko, I. (2013). Poisson approximation of processes with locally independent increments with Markov switching. *Teor. Imovir. ta Matem. Statyst.*, 89, 104–114.

Liptser, R. (1994). The Bogoliubov averaging principle for semimartingales. *Proc. Steklov Math. Institute*, 202, 143–153.

Liptser, R. and Shiryayev, A.N. (1989). *Theory of Martingales*. Springer, Amsterdam.

Liptser, R. and Shiryaev, A.N. (2001). *Statistics of Random Processes II. Applications*. Springer, Berlin/Heidelberg.

Maslov, V.P. (2003). Nonlinear averaging axioms in financial mathematics and stock price dynamics. *Teor. Veroyatnost. i Primenen.*, 48(4), 800–810.

Maslov, V.P. (2004). Nonlinear averaging axioms in financial mathematics and stock price dynamics. *Theory Probab. Appl.*, 48(4), 723–733.

Maslov, V.P. (2005a). Nonlinear averages in economics. *Mat. Zametki*, 78(3), 377–395.

Maslov, V.P. (2005b). Nonlinear averages in economics. *Math. Notes*, 78(3), 347–363.

Mogulskii, A.A. (1993). Large deviation for processes with independent increments. *Ann. Prob.*, 21, 202–215.

Moklyachuk, M.P. (1994). Stochastic autoregressive sequences and minimax interpolation. *Theory Probab. Math. Stat.*, 48, 95–104.

Nashed, M.Z. (1976). *Generalized Inverses and Applications*. Academic Press, New York.

Nevelson, M.B. and Hasminskii, R.Z. (1976). Stochastic approximation and recursive estimation. *Am. Math. Soc., Transl. Math. Monogr.*, 244.

Nisio, M. (1976a). On a non-linear semi-group attached to stochastic optimal control. *Publ. RIMS, Kyoto Univ.*, 13, 513–537.

Nisio, M. (1976b). On stochastic optimal controls and envelope of Markovian semi-groups. *Proc. Intern. Symp. SDE*, Kinokuniya, Kyoto, 297–325.

O'Brien, G.L. and Vervaat, W. (1995). Compactness in the theory of large deviations. *Stochastic Processes and their Applications*, 57(1), 1–10.

Papanicolaou, G.C. (1971a). Motion of a particle in a random field. *J. Math. Phys.*, 12, 1494–1496.

Papanicolaou, G.C. (1971b). Wave propagation in a one-dimensional random medium. *SIAM J. Appl. Math.*, 21, 13–18.

Papanicolaou, G.C. (1973). Stochastic equations and their applications. *Am. Math. Monthly*, 80, 526–544.

Papanicolaou, G.C. (1975). Asymptotic analysis of transport processes. *Bull. Am. Math. Soc.*, 81, 330–392.

Papanicolaou, G.C. (1976). Probabilistic problems and methods in singular perturbations. *Rocky Mountain J. Math.*, 6, 653–673.

Papanicolaou, G.C. and Hersh, R. (1972). Some limit theorems for stochastic equations and applications. *Indiana U. Math. J.*, 21, 815–840.

Papanicolaou, G.C. and Kellek, J. (1971). Stochastic differential equations with applications to random harmonic oscillations and wave propagation in random media. *SIAM J. Appl. Math.*, 21, 287–305.

Papanicolaou, G.C., Stroock, D., Varadhan, S.R.S. (1977). Martingale approach to some limit theorems. *Duke Univ. Math. Ser.*, 21, 202–215.

Pavliotis, G.A. and Stuart, A.M. (2008). *Multiscale Methods: Averaging and Homogenization*. Springer-Verlag, New York.

Pinsky, M. (1968). Differential equations with a small parameter and the central limit theorem for functions defined on a finite Markov chain. *Z. Wahrschein. verw. Gebiete*, 2, 101–111.

Pinsky, M. (1991). *Lectures on Random Evolutions*. World Scientific, Singapore.

Protter, P. (1990). Stochastic integration and differential equations. A new approach. *Applications of Mathematics*, 21, Springer-Verlag, Berlin.

Puhalskii A. (1991). On functional principle of large deviations. *New Trends in Probability and Statistics*, 1, VSP, Utrecht.

Robbins H. and Monro S. (1951). A stochastic approximation method. *Ann. Math. Stat.*, 22(1), 400–407.

Samoilenko, I.V. (2012). Large deviation for random evolutions with independent increments in the scheme of Poisson approximation. *Theor. Probab. Math. Statist.*, 85, 107–114.

Samoilenko, I.V. (2013). Large deviations for impulsive processes in the scheme of Poisson approximation. *Ukr. Math. J.*, 64(11), 1727–1738.

Samoilenko, I.V. (2014). Large deviations for impulsive processes in the scheme of the Lévy approximation. *Theor. Probability Math. Statist.*, 88, 151–160.

Samoilenko, I.V. (2015a). Large deviations for random evolutions with independent increments in the scheme of Lévy approximation with split and double merging. *Random Operators and Stochastic Equations*, 23(2), 137–149.

Samoilenko, I.V. (2015b). Large deviations for random evolutions with independent increments in a scheme of Lévy approximation. *J. Math. Sci.*, 210, 52–66.

Schied, A. (1995). Criteria for exponential tightness in path spaces. *Probability Theory and Related Fields*, 104, 319–347.

Shiryaev, A.N. (1999). *Essentials of Stochastic Finance: Facts, Models, Theory*. World Scientific, Singapore.

Shiryaev, A.N. (2011). *Probability 1*. Springer, Berlin.

Silvestrov, D.S. (2004). Limit theorems for randomly stopped stochastic processes. *Springer Optimization and Its Applications Series*, Berlin.

Skorokhod A.V. (2008). Asymptotic methods in the theory of stochastic differential equations. *Translations of Mathematical Monographs*, 78. AMS, Providence, RI.

Skorokhod, A.V., Hoppensteadt, F.C., Salehi, H. (2002). *Random Perturbation Methods with Applications in Science and Engineering*. Springer-Verlag, New York.

Stratonovich, R.L. (1967). *Topics in the Theory of Random Noise 2*. CRC Press, Boca Raton, FL.

Stroock, D.W. and Varadhan, S.R.S. (1979). *Multidimensional Diffusion Processes*. Springer-Verlag, Berlin.

Sviridenko, M.N. (1998). Martingale characterization of limit distributions in the space of functions without discontinuities of second kind. *Math. Notes*, 43(5), 398–402.

Tataru, D. (1992). Viscosity solutions of Hamilton–Jacobi equations with unbounded nonlinear terms. *J. Math. Anal. Appl.*, 163, 345–392.

Ventsel, A.D. (1976a). Rough limit theorems on large deviations for Markov stochastic processes I. *Theory Probab. Appl.*, 21(2), 227–242.

Ventsel, A.D. (1976b). Rough limit theorems on large deviations for Markov stochastic processes II. *Theory Probab. Appl.*, 21(3), 499–512.

Ventsel, A.D. (1979). Rough limit theorems on large deviations for Markov stochastic processes III. *Theory Probab. Appl.*, 24(4), 675–672.

Wentzell, A.D. (1990). *Limit Theorems on Large Deviations for Markov Stochastic Processes*. Springer, The Netherlands.

Yin, G.G. and Zhang, Q. (1998). *Continuous-time Markov Chains and Applications. A Singular Perturbation Approach*. Springer, New York.

Yin, G.G. and Zhang, Q. (2005). *Discrete-time Markov Chains. Two-time-scale Methods and Applications*. Springer, New York.

Index

N, O

P

R

S

V, W, Z

Other titles from

iSTE

in

Mathematics and Statistics

2021

MOKLYACHUK Mikhail
Convex Optimization: Introductory Course

POGORUI Anatoliy, SWISHCHUK Anatoliy, RODRÍGUEZ-DAGNINO Ramón M.
Random Motions in Markov and Semi-Markov Random Environments 1: Homogeneous Random Motions and their Applications
Random Motions in Markov and Semi-Markov Random Environments 2: High-dimensional Random Motions and Financial Applications

2020

BARBU Vlad Stefan, VERGNE Nicolas
Statistical Topics and Stochastic Models for Dependent Data with Applications

CHABANYUK Yaroslav, NIKITIN Anatolii, KHIMKA Uliana
Asymptotic Analyses for Complex Evolutionary Systems with Markov and Semi-Markov Switching Using Approximation Schemes

KOROLIOUK Dmitri
Dynamics of Statistical Experiments

MANOU-ABI Solym Mawaki, DABO-NIANG Sophie, SALONE Jean-Jacques
Mathematical Modeling of Random and Deterministic Phenomena

2019

BANNA Oksana, MISHURA Yuliya, RALCHENKO Kostiantyn, SHKLYAR Sergiy
Fractional Brownian Motion: Approximations and Projections

GANA Kamel, BROC Guillaume
Structural Equation Modeling with lavaan

KUKUSH Alexander
Gaussian Measures in Hilbert Space: Construction and Properties

LUZ Maksym, MOKLYACHUK Mikhail
Estimation of Stochastic Processes with Stationary Increments and Cointegrated Sequences

MICHELITSCH Thomas, PÉREZ RIASCOS Alejandro, COLLET Bernard, NOWAKOWSKI Andrzej, NICOLLEAU Franck
Fractional Dynamics on Networks and Lattices

VOTSI Irene, LIMNIOS Nikolaos, PAPADIMITRIOU Eleftheria, TSAKLIDIS George
Earthquake Statistical Analysis through Multi-state Modeling (Statistical Methods for Earthquakes Set – Volume 2)

2018

AZAÏS Romain, BOUGUET Florian
Statistical Inference for Piecewise-deterministic Markov Processes

IBRAHIMI Mohammed
Mergers & Acquisitions: Theory, Strategy, Finance

PARROCHIA Daniel
Mathematics and Philosophy

2017

CARONI Chysseis
First Hitting Time Regression Models: Lifetime Data Analysis Based on Underlying Stochastic Processes
(Mathematical Models and Methods in Reliability Set – Volume 4)

CELANT Giorgio, BRONIATOWSKI Michel
Interpolation and Extrapolation Optimal Designs 2: Finite Dimensional General Models

CONSOLE Rodolfo, MURRU Maura, FALCONE Giuseppe
Earthquake Occurrence: Short- and Long-term Models and their Validation
(Statistical Methods for Earthquakes Set – Volume 1)

D'AMICO Guglielmo, DI BIASE Giuseppe, JANSSEN Jacques, MANCA Raimondo
Semi-Markov Migration Models for Credit Risk
(Stochastic Models for Insurance Set – Volume 1)

GONZÁLEZ VELASCO Miguel, del PUERTO GARCÍA Inés, YANEV George P.
Controlled Branching Processes
(Branching Processes, Branching Random Walks and Branching Particle Fields Set – Volume 2)

HARLAMOV Boris
Stochastic Analysis of Risk and Management
(Stochastic Models in Survival Analysis and Reliability Set – Volume 2)

KERSTING Götz, VATUTIN Vladimir
Discrete Time Branching Processes in Random Environment
(Branching Processes, Branching Random Walks and Branching Particle Fields Set – Volume 1)

MISHURA YULIYA, SHEVCHENKO Georgiy
Theory and Statistical Applications of Stochastic Processes

NIKULIN Mikhail, CHIMITOVA Ekaterina
Chi-squared Goodness-of-fit Tests for Censored Data
(Stochastic Models in Survival Analysis and Reliability Set – Volume 3)

SIMON Jacques
Banach, Fréchet, Hilbert and Neumann Spaces
(Analysis for PDEs Set – Volume 1)

2016

CELANT Giorgio, BRONIATOWSKI Michel
Interpolation and Extrapolation Optimal Designs 1: Polynomial Regression and Approximation Theory

CHIASSERINI Carla Fabiana, GRIBAUDO Marco, MANINI Daniele
Analytical Modeling of Wireless Communication Systems
(Stochastic Models in Computer Science and Telecommunication Networks Set – Volume 1)

GOUDON Thierry
Mathematics for Modeling and Scientific Computing

KAHLE Waltraud, MERCIER Sophie, PAROISSIN Christian
Degradation Processes in Reliability
(Mathematial Models and Methods in Reliability Set – Volume 3)

KERN Michel
Numerical Methods for Inverse Problems

RYKOV Vladimir
Reliability of Engineering Systems and Technological Risks
(Stochastic Models in Survival Analysis and Reliability Set – Volume 1)

2015

DE SAPORTA Benoîte, DUFOUR François, ZHANG Huilong
Numerical Methods for Simulation and Optimization of Piecewise Deterministic Markov Processes

DEVOLDER Pierre, JANSSEN Jacques, MANCA Raimondo
Basic Stochastic Processes

LE GAT Yves
Recurrent Event Modeling Based on the Yule Process
(Mathematical Models and Methods in Reliability Set – Volume 2)

2014

COOKE Roger M., NIEBOER Daan, MISIEWICZ Jolanta
Fat-tailed Distributions: Data, Diagnostics and Dependence (Mathematical Models and Methods in Reliability Set – Volume 1)

MACKEVIČIUS Vigirdas
Integral and Measure: From Rather Simple to Rather Complex

PASCHOS Vangelis Th
Combinatorial Optimization – 3-volume series – 2nd edition
Concepts of Combinatorial Optimization / Concepts and Fundamentals – volume 1
Paradigms of Combinatorial Optimization – volume 2
Applications of Combinatorial Optimization – volume 3

2013

COUALLIER Vincent, GERVILLE-RÉACHE Léo, HUBER Catherine, LIMNIOS Nikolaos, MESBAH Mounir
Statistical Models and Methods for Reliability and Survival Analysis

JANSSEN Jacques, MANCA Oronzio, MANCA Raimondo
Applied Diffusion Processes from Engineering to Finance

SERICOLA Bruno
Markov Chains: Theory, Algorithms and Applications

2012

BOSQ Denis
Mathematical Statistics and Stochastic Processes

CHRISTENSEN Karl Bang, KREINER Svend, MESBAH Mounir
Rasch Models in Health

DEVOLDER Pierre, JANSSEN Jacques, MANCA Raimondo
Stochastic Methods for Pension Funds

2011

MACKEVIČIUS Vigirdas
Introduction to Stochastic Analysis: Integrals and Differential Equations

MAHJOUB Ridha

Recent Progress in Combinatorial Optimization – ISCO2010

RAYNAUD Hervé, ARROW Kenneth
Managerial Logic

2010

BAGDONAVIČIUS Vilijandas, KRUOPIS Julius, NIKULIN Mikhail
Nonparametric Tests for Censored Data

BAGDONAVIČIUS Vilijandas, KRUOPIS Julius, NIKULIN Mikhail
Nonparametric Tests for Complete Data

IOSIFESCU Marius *et al.*
Introduction to Stochastic Models

VASSILIOU PCG
Discrete-time Asset Pricing Models in Applied Stochastic Finance

2008

ANISIMOV Vladimir
Switching Processes in Queuing Models

FICHE Georges, HÉBUTERNE Gérard
Mathematics for Engineers

HUBER Catherine, LIMNIOS Nikolaos *et al.*
Mathematical Methods in Survival Analysis, Reliability and Quality of Life

JANSSEN Jacques, MANCA Raimondo, VOLPE Ernesto
Mathematical Finance

2007

HARLAMOV Boris
Continuous Semi-Markov Processes

2006

CLERC Maurice
Particle Swarm Optimization